I0044702

Advances in Seed Science

Advances in Seed Science

Edited by Lisa Swinton

SYRAWOOD
PUBLISHING HOUSE
New York

Published by Syrawood Publishing House,
750 Third Avenue, 9th Floor,
New York, NY 10017, USA
www.syrawoodpublishinghouse.com

Advances in Seed Science
Edited by Lisa Swinton

© 2019 Syrawood Publishing House

International Standard Book Number: 978-1-68286-746-4 (Hardback)

This book contains information obtained from authentic and highly regarded sources. Copyright for all individual chapters remain with the respective authors as indicated. All chapters are published with permission under the Creative Commons Attribution License or equivalent. A wide variety of references are listed. Permission and sources are indicated; for detailed attributions, please refer to the permissions page and list of contributors. Reasonable efforts have been made to publish reliable data and information, but the authors, editors and publisher cannot assume any responsibility for the validity of all materials or the consequences of their use.

Trademark Notice: Registered trademark of products or corporate names are used only for explanation and identification without intent to infringe.

Cataloging-in-Publication Data

Advances in seed science / edited by Lisa Swinton.
    p. cm.
Includes bibliographical references and index.
ISBN 978-1-68286-746-4
1. Seeds. 2. Seed technology. I. Swinton, Lisa.
SB117 .A38 2019
631.521--dc23

TABLE OF CONTENTS

PREFACE

It is often said that books are a boon to mankind. They document every progress and pass on the knowledge from one generation to the other. They play a crucial role in our lives. Thus I was both excited and nervous while editing this book. I was pleased by the thought of being able to make a mark but I was also nervous to do it right because the future of students depends upon it. Hence, I took a few months to research further into the discipline, revise my knowledge and also explore some more aspects. Post this process, I begun with the editing of this book.

The scientific study of seeds, its structure and processes falls under the domain of seed science. A seed is an essential part of the plant. It is formed during the reproduction process in spermatophytes, including gymnosperm and angiosperm plants. Angiosperm seeds consist of three chief constituents- the embryo, the endosperm and the seed coat. In angiosperms, seed development begins with double fertilization. They are essential for the nourishment of the embryo. Seed production in plants is affected by weather conditions, diseases, insects, etc. They are found in various shapes such as bean shaped, triangular, square, spherical and lenticular among many others. Many crops, forest trees, legumes, turf grasses, etc. are propagated through seeds. This book includes some of the vital pieces of work being conducted across the world, on various topics related to seed science. From theories to research to practical applications, case studies related to all contemporary topics of relevance to this field have been included in this book. It will provide comprehensive knowledge to the readers.

I thank my publisher with all my heart for considering me worthy of this unparalleled opportunity and for showing unwavering faith in my skills. I would also like to thank the editorial team who worked closely with me at every step and contributed immensely towards the successful completion of this book. Last but not the least, I wish to thank my friends and colleagues for their support.

Editor

Alternating temperature and accelerated aging in mobilization of reserves during germination of *Carica papaya* L. seeds

Liana Hilda Golin Mengarda[1], José Carlos Lopes[1*], Rodrigo Sobreira Alexandre[2],
Rafael Fonsêca Zanotti[1], Pedro Ramon Manhone[1]

ABSTRACT – The aim of this study was to identify the phases of water absorption during germination of *Carica papaya* seeds and evaluate the influence of alternating temperature and accelerated aging on mobilization of reserves during germination. Weight gain was evaluated, obtaining the imbibition curve. Phase I of germination comprises the period from zero to five hours; phase II, from five to 120 hours; and phase III begins after 144 hours. Seeds were subjected to the germination test under temperatures of 25 °C and of 20-30 °C (16/8 h), and before and after accelerated aging (43 °C / 72 h). During the germination test, at initial time, after 4, 10, 120, and 240 hours of soaking, we determined the levels of soluble sugars, starch, lipids, and total proteins. Greater germination was observed under alternating temperature. Under this condition, there is reduction of carbohydrates, lipids, and proteins in phase I, and fluctuations in lipid levels, and an increase in protein levels during phases II and III. Mobilization of lipids in papaya seeds is not influenced by accelerated aging, but the seeds subjected to aging have lower protein content in phase III.

Index terms: soaking, papaya, soluble sugars, starch, lipids, proteins.

Alternância de temperatura e envelhecimento acelerado na mobilização de reservas durante a germinação de sementes de *Carica papaya* L.

RESUMO – Objetivou-se, neste trabalho, identificar as fases da absorção de água durante a germinação de sementes de *Carica papaya* e avaliar a influência da alternância de temperatura e do envelhecimento acelerado na mobilização de reservas durante a germinação. Foi avaliado o ganho de peso, obtendo-se a curva de embebição. A fase I da germinação compreende o período entre zero e cinco horas; a fase II, entre cinco e 120 horas; e a fase III se inicia após 144 horas. Sementes foram submetidas ao teste de germinação sob temperaturas de 25 °C e de 20-30 °C (16/8 h), e antes e após o envelhecimento acelerado (43 °C / 72 h). Durante o teste de germinação, no tempo inicial, após 4, 10, 120 e 240 horas de embebição, foram determinados os teores de açúcares solúveis, amido, lipídeos e proteínas totais. Observou-se maior germinação sob alternância de temperatura. Nesta condição há redução dos carboidratos, lipídeos e proteínas na fase I, flutuações nos teores de lipídeos e aumento nos teores de proteínas durante as fases II e III. A mobilização dos lipídeos em sementes de mamão não sofre influência do envelhecimento acelerado, mas as sementes submetidas ao envelhecimento apresentam menor teor de proteínas na fase III.

Temos para indexação: embebição, mamão, açúcares solúveis, amido, lipídeos, proteínas.

Introduction

Papaya (*Carica papaya* L.) seeds, just as other orthodox seeds, complete their maturation under a condition of desiccation in which metabolic activity is greatly reduced, requiring reabsorption of water (imbibition) for germination to occur. Germination depends on an ordered sequence of metabolic events that undergo the influence of external stimuli, among which are temperature and aging (Bewley and Black, 1994; Ferreira and Borghetti, 2004; Marcos-Filho, 2005).

Bewley and Black (1994) proposed a triphasic pattern of water absorption during germination, relating the speed of water absorption to digestion of reserves, enzymatic activity, and the formation of macromolecules. The duration of the phases varies a great deal among species. The delimitation of the different phases and understanding of the metabolic events related to each phase carry important information since they may identify more critical stages in the plant life cycle,

[1]Departamento de Produção Vegetal, Universidade Federal do Espírito Santo, 29500-000, Alegre, ES, Brasil.

[2]Departamento de Ciências Florestais, Universidade Federal do Espírito Santo, 29550-00, Jerônimo Monteiro, ES, Brasil.
*Corresponding author <jcufes@bol.com.br>

assisting in crop management (Marcos-Filho, 2005). Moreover, identification of the duration of each phase is also relevant for studies on osmoconditioning and for characterization of the mobilization patterns of reserves during germination.

Adequate mobilization of seed reserve compounds is responsible for spurring seedling growth until the seedling acquires photosynthetic efficiency. Mobilization consists of digestion of the reserves and of synthesis of new compounds used for the formation of new protoplasms and cell walls (Nonogakia et al., 2010). Seeds with greater vigor have more reserves of proteins, lipids, and carbohydrates. This provides the seeds with greater storage potential, greater tolerance to stresses, and greater capacity for mobilization of reserves in germination, resulting in seedlings with greater initial performance (Henning et al., 2010; Kim et al., 2011; Zhu and Chen, 2007).

Ideal germination conditions, which involve adequate supply of water, temperature, oxygen, substrate, and light, allow a continual process of mobilization of reserves during germination. Nevertheless, mobilization undergoes the effect of osmotic components (Reis et al., 2012), of salinity (Silva et al., 2008; Dantas et al., 2014), of light and growth regulators (Ren et al., 2007; Zanotti et al., 2014), of accelerated aging (Zhu and Chen, 2007; Menezes et al., 2014), and of storage (Abbade and Takaki, 2014), factors which may accelerate or slow the process, as well as hurt the development of normal seedlings.

Conditions unfavorable to water supply (Reis et al., 2012) and temperature, as well as deterioration processes (Marcos-Filho, 2005), may alter or interrupt mobilization of reserves, hurting germination. Thus, the aim of this study was to identify the phases of water absorption during germination of papaya (*Carica papaya* L.) seeds and evaluate the effect of alternating temperature and accelerated aging on mobilization of reserves during germination.

Material and Methods

The study was developed at the Seed Analysis Laboratory of the Agrarian Sciences Center of the Universidade Federal do Espírito Santo, Vitoria, ES, Brazil. Hybrid papaya (*Carica papaya* L.) seeds, JS12 x Waimanalo, were used, and the following evaluations were made:

Water absorption during germination: seeds were used with and without sclerotesta. The removal of the sclerotesta was performed manually, with the aid of a tweezers. The seeds were weighed and distributed in rolls of paper moistened with distilled water and kept in a germination chamber at a temperature of 20-30 °C (16/8 h). Weight gain was evaluated every 15 minutes in the first hour, every hour for 12 hours, and every 24 hours for 10 days, calculating the moisture content in

the seeds. The imbibition curve was determined and, associated with the observation of growth of the primary root, the water absorption phases were delimited (Bewley and Black, 1994).

Constant temperature and alternating temperature: intact seeds were germinated under temperatures of 25 °C and from 20-30 °C, and samples were collected for biochemical evaluations.

Accelerated aging: intact seeds were distributed in a single layer over a screen in a gerbox type box with 40 mL of distilled water in a germination chamber regulated to the temperature of 43 °C for 72 h. Aged (AS) and non-aged (NAS) seeds were germinated under a temperature of 20-30 °C, with samples being collected for biochemical evaluations.

The physiological quality of the seeds was determined based on the following evaluations:

Percentage of seedlings with radicle protrusion (RP): considering the seeds with radicle protrusion ≥ 0.2 cm up to 28 days after seeding;

First and second count of seedlings with radicle protrusion (FC and SC): performed at seven and 14 days, respectively, considering the seedlings with radicle protrusion and expressed in percentage. Since they did not exhibit normality and homogeneity, making statistical analysis impossible, the data in regard to FC were not presented;

Germination speed index (GSI): considering the seedlings with radicle protrusion, calculated according to the formula proposed by Maguire (1962);

Percentage of normal seedlings (NS): calculated in relation to the total of seedlings with radicle protrusion (SP), 28 days after sowing. Under normal seedling conditions, the following evaluations were made:

Length of the shoots (SL) and of the roots (RL): expressed in cm.seedling^{-1};

Fresh matter (FM) and dry matter (DM): expressed in mg.seedling^{-1}. Dry matter was obtained in an air circulation laboratory oven at 70 °C for 72 h.

For biochemical evaluations, seed samples of 0.1 g of dry matter were collected at times zero (initial), 4 (phase I), 10 (beginning of phase II), 120 (end of phase II), and 240 hours (phase III) after seeding, with quantification of:

Soluble sugars (A): extraction was carried out by the methanol, chloroform, and water (1:1:1) method, obtaining two liquid phases and one solid phase. The sugars soluble in the solution of the upper portion of the liquid phase were quantified by the anthrone method (Yemn and Willis, 1954);

Lipids (L): lipids were quantified as of the lower portion of the liquid phase of the extraction (Bligh and Dyer, 1959, modified);

Starch (St): the solid phase of the extraction underwent digestion in 3% HCl in a water bath and centrifugation, collecting the supernatant, with which the anthrone method was carried out;

Total proteins (P): this was carried out by inference based on the nitrogen level (Galvani and Gaertner, 2006). The results were expressed in percentage (% m/m) of dry matter;

Experimental design and statistical analysis: a completely randomized design was adopted, with four replications of 25 seeds per treatment. Treatments consisted of temperatures, constant (25 ºC) and alternating (20-30 ºC), in the first experiment; and of non-aged seeds (NAS) and aged seeds (AS), in the second experiment. The experiments were analyzed separately. Analysis of variance was carried out, and the mean values between the treatments were compared by the Tukey test (p ≤ 5%). For analysis of soaking times, a 2 x 5 split-plot design was adopted, with two treatments [two temperatures (25 ºC and 20-30 ºC) in the first experiment; and two aging conditions (NAS and AS), in the second experiment], and five evaluation times (0, 4, 10, 120, and 240 hours), carrying out analysis of regression and adopting models which were significant (p ≤ 5%) and of greater order (R^2) (The R Foundation for Statistical Computing Platform, 2014).

Results and Discussion

Water absorption during germination:

The initial moisture content of the seeds with sclerotesta and without sclerotesta was 9.11 and 4.95%, respectively. In the first hour of imbibition, the seeds with sclerotesta reached a moisture content of 62.98%, while the seeds without sclerotesta reached 18.44% moisture (Figure 1A).

The presence of the sclerotesta affected the imbibition rate of papaya seeds, which was, on average, 43% greater than in the seeds without sclerotesta. The initial imbibition rate may vary in accordance with the characteristics of the seed coat, the testa, and the pericarp that surrounds the embryo (Ferreira and Borghetti, 2004). In *Crotalaria juncea*, different moisture absorption rates during imbibition were attributed to the type of coating and to the thickness of the endosperm of different morphotypes of seeds (Pascualides and Planchuelo, 2007).

According to Bewley and Black (1994), imbibition begins with rapid water gain (phase I), a physical process that occurs whether or not the seed is viable or inviable, as long as there is no impermeability of the seed coat. Then, stabilization is observed (phase II) and, after that, the seed returns to gaining moisture with greater speed (phase III). In this study, the absorption pattern was similar among the seeds with and without sclerotesta, with stabilization being observed after five hours of soaking. That way, it was possible to identify the period from zero to five hours as phase I of water absorption during papaya seed germination (Figure 1A).

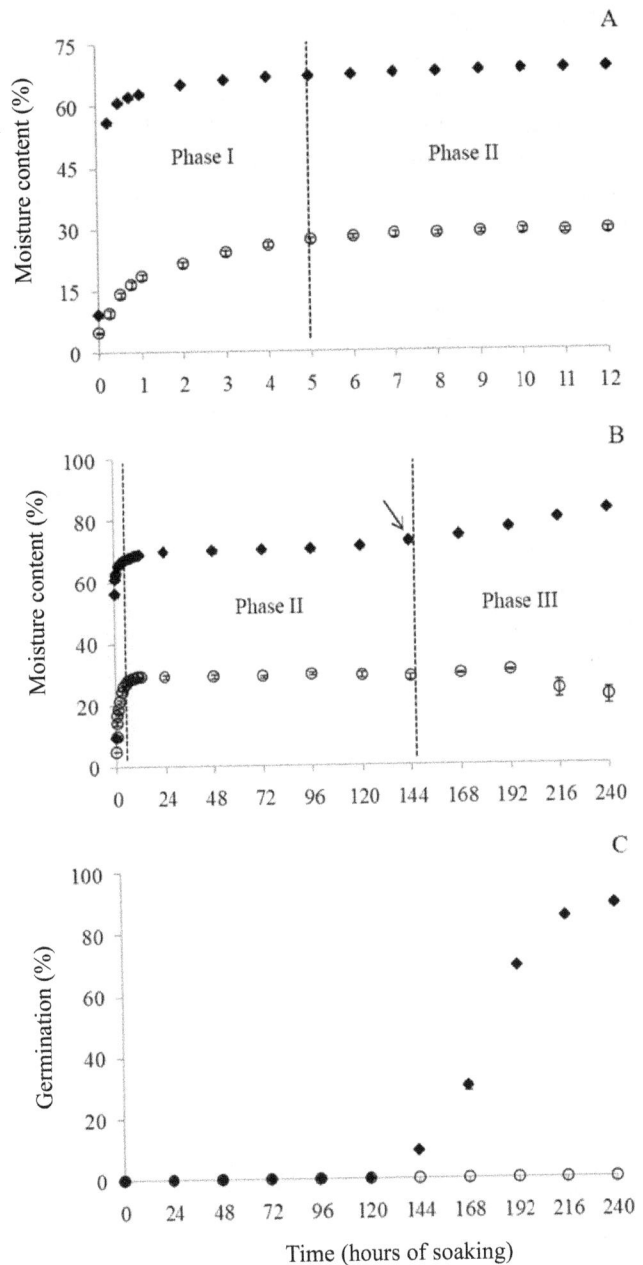

Figure 1. A – moisture content (%) of papaya seeds with sclerotesta (♦) and without sclerotesta (○) during the first 12 hours of soaking; B – moisture content (%) of seeds with sclerotesta (♦) and without sclerotesta (○) during 240 hours of imbibition. The arrow indicates the beginning of primary root protrusion; C – percentage of seedlings with radicle protrusion from seeds with sclerotesta (♦) and without sclerotesta (○) during the soaking period. Bars represent the standard error of the mean (n= 4).

In the beginning of imbibition of viable seeds, important physiological changes occur. When the seeds reach around

20% moisture, respiration and production of ATP begins; with approximately 40% moisture, the synthesis of mRNA and repair of DNA, the synthesis of polysomes and of proteins, begins (Bewley and Black, 1994). When the cells achieve full hydration, the moisture content remains constant or increases very slowly, this being a period of preparation for reactivation of metabolism, called phase II (Ferreira and Borghetti, 2004). In this study, the interval from five to 120 hours was observed as phase II (Figures 1A and B), in which the moisture content of seeds with sclerotesta ranged from 67.12 to 71.21%. In the seeds with sclerotesta, this variation was from 27.10 to 29.28% moisture.

Phase II culminates with protrusion of the primary root, which, for many species, occurs with moisture content near 60%. At this point, phase III begins, with a new increase in imbibition, associated with embryo growth (cell division). The passage from phase II to phase III is a point of no return in the germination process; the seed may not undergo stress through water deficiency, which may definitively interrupt germination (Ferreira and Borghetti, 2004; Marcos-Filho, 2005).

For seeds with sclerotesta, the beginning of phase III occurred after 144 hours of soaking, when the seeds reached 73% moisture, with the beginning of primary root protrusion and a new increase in water absorption being observed. After 240 hours of imbibition, the seeds with sclerotesta exhibited 89% of seedlings with radicle protrusion (Figures 1B and 1C). Protrusion of the primary root stabilized after 13 days, reaching 93%. Thus, it was possible to delimit the water absorption phases during germination: phase I occurred in the interval from zero to five hours of soaking; phase II, from five to 120 hours; the transition from phase II to phase II, from 120 to 144 hours; and phase III, as of 144 hours.

The seeds without sclerotesta did not exhibit an increase in moisture content nor root protrusion, but rather deterioration. After 216 hours of imbibition, they exhibited a decrease in weight gain (moisture content) (Figures 1B and C).

Temperature in mobilization of reserves during germination:

Papaya seeds exhibited a greater percentage of seedlings with radicle protrusion under alternating temperature (91%) as compared to constant temperature (69%). The temperature of 20-30 °C also led to a greater speed of germination and greater percentage of normal seedlings (Table 1).

In *Annona emarginata* seeds, a greater percentage of germination was also observed under alternating temperature (20-30 °C) (Costa et al., 2011). Temperature has an effect on water absorption speed and on the speed and uniformity of germination. In addition, temperature is also a determining factor for the biochemical reactions that determine the entire process (Ferreira and Borghetti, 2004; Marcos-Filho, 2005).

Table 1. Physiological quality of papaya seeds subjected to constant temperature (25 °C) and alternating temperature (20-30 °C): percentage of seedlings with radicle protrusion (RP), second count of seedlings with radicle protrusion (SC), germination speed index (GSI), percentage of normal seedlings (NS), shoot length (SL), root length (RL), fresh matter (FM), and dry matter (DM) of the normal seedlings.

Temperature	RP	SC	GSI	NS
25 °C	69 b*	67 b	2.13 b	66 b
20-30 °C	91 a	84 a	2.63 a	88 a
F_{cal}	38.21 **	20.16 **	12.44 *	27.92 **
CV (%)	6.29	7.09	8.36	7.65
	SL	RL	FM	DM
25 °C	6.75 a	1.73 b	75.10 b	3.16 b
20-30 °C	7.26 a	5.26 a	96.25 a	4.70 a
F_{cal}	1.33 ns	40.19 **	24.38 **	21.67 **
CV (%)	9.01	22.58	7.07	11.87

Mean values followed by the same letters (columns) are statistically equal by the Tukey test., ** Significant at the level of 5% and 1%, respectively. ns = not significant. F_{cal} = F value calculated obtained by analysis of variance. CV (%) = coefficient of variation (n = 4).

Concentration of sugars was greater in seeds subjected to constant temperature (Table 2). Under the temperature 20-30 °C it was observed that after four hours of soaking (phase I), there was a sharp decrease (from 5.93 to 3.63%) in the concentration of soluble sugars. That suggests that under alternating temperature, there was mobilization of soluble sugars (Ferreira and Borghetti, 2004).

With alternating temperature (20 - 30 °C), a lower concentration of starch was observed after 240 hours of soaking (phase III). The lipid concentration exhibited lower mean values after four and 240 hours of soaking (phase I and phase III, respectively). The concentration of total proteins exhibited a lower mean value after four hours of soaking (phase I), and greater mean values in the following evaluations (Table 2).

The mean content of total soluble sugars in the different phases was 5.47% (Figure 2A). There was no fit to the regression models and, thus, it was not possible to identify the mobilization pattern of these reserves. In *Apuleia leiocarpa*, it was observed that the mean contents of soluble sugars did not differ significantly during the soaking of the seeds, leading to the conclusion that germination was not dependent on carbohydrates as a source of energy or to create physical structures (Pontes et al., 2002). Nevertheless, interconversion among the different sugars is common, with simultaneous increases in one and decreases in another

(Ferreira and Borghetti, 2004; Magalhães et al., 2009). Therefore, isolated analysis of the concentrations of total soluble sugars does not allow conclusions to be drawn in regard to their mobilization.

Table 2. Concentrations of soluble sugars, starch, lipids, and total proteins, expressed in percentage (m/m) during the germination phases of papaya seeds subjected to constant temperature (25 °C) and alternating temperature (20-30 °C).

Time (h)	0	4	10	120	240	
Temperature (°C)			Soluble sugars (%)			Mean
25	6.03	5.30	6.71	6.48	5.59	6.02 a
20-30	5.93	3.63	5.66	4.98	4.34	4.90 b
Mean	5.98	4.47	6.19	5.73	4.97	
F_{cal} (temperature)		26.85**				
F_{cal} (time)		9.15**				
CV (%)		12.43				
			Starch (%)			Mean
25	0.48 a*	0.55 a	0.57 a	0.55 a	0.76 a	0.58
20-30	0.59 a	0.45 a	0.53 a	0.51 a	0.52 b	0.52
Mean	0.54	0.50	0.55	0.53	0.64	
F_{cal}		4.07**				
CV (%)		15.66				
			Lipids (%)			Mean
25	4.15 a	3.77 a	4.29 a	3.88 a	4.11 a	4.04
20-30	4.32 a	2.01 b	2.52 b	2.43 b	2.91 b	2.84
Mean	4.24	2.89	3.40	3.16	3.51	
F_{cal}		5.29**				
CV (%)		14.37				
			Total proteins (%)			Mean
25	28.14 a	27.96 a	21.53 b	20.41 b	21.45 b	23.90
20-30	29.12 a	22.53 b	28.07 a	27.17 a	30.52 a	27.48
Mean	28.63	25.25	24.80	23.79	25.99	
F_{cal}		121.48**				
CV (%)		2.92				

*Mean values followed by the same letters (columns) are statistically equal by the Tukey test. ** Significant at the level of 1%. Fcal = F value calculated obtained by analysis of variance. CV (%) = coefficient of variation (n = 4).

There was a linear increase in the starch contents under constant temperature (25 °C). As for alternating temperature (20-30 °C), the starch content remained constant (0.52%) (Figure 2B). Maintenance of a constant starch content may have occurred due to the small quantity present in papaya seeds, which suggests that starch may not play a relevant role as an energy or structural reserve. Furthermore, degradation of starch may be inhibited by the accumulation of its final product (maltose and glucose) since they inhibit beta-amylase activity (Ferreira and Borghetti, 2004).

The concentration of lipids remained constant (4.04%) at the temperature of 25 °C. As for 20-30 °C, the condition in which a greater percentage of seedlings with radicle protrusion were observed, the initial content showed express reduction, from 4.32% to 2.01% after four hours of imbibition (phase I) (Figure 2C). At the beginning of germination, the initial substrates used as

an energy source are the soluble sugars. Nevertheless, for many seeds, the energy for germination comes from the breakdown of starch and lipids in sucrose (Ferreira and Borghetti, 2004).

After initial reduction in the concentration of lipids, an increase was observed at the beginning of phase II, reduction at the end of phase II, and an increase in phase III, indicating mobilization of these compounds throughout the germination process. As of quantification and evaluation of the mobilization of lipids during germination of *Apuleia leiocarpa*, it was seen that the energetic and structural needs of the germination are supplied by lipids (Pontes et al., 2002). Corte et al. (2008) suggest the direct involvement of the lipid reserves as energy supply for germination and growth of *Caesalpinia peltophoroides* seedlings. In papaya seeds, the use of substances like KNO_3, effective in increasing germination and in overcoming dormancy, promote greater mobilization of lipids (Zanotti et al., 2014). Therefore, it is suggested that lipids exercise an important effect on germination of papaya seeds.

At 25 °C, a fit to the cubic model was observed, with a tendency to reduction in the total protein contents in phase III. Under alternating temperature (20-30 °C), the total protein content fit the quadratic model, with a tendency to increase in the final phases of germination (Figure 2D). While in phase I there is energy consumption and degradation of macromolecules, during phase II, the synthesis of numerous compounds begins. At the end of phase II and in phase

III, amino acids, sucrose, and ATP are translocated to the embryonic axis, which shows an increase in protein contents (Marcos-Filho, 2005; Kim et al., 2011), as observed for the papaya seeds in 20-30 °C.

Thus, under alternating temperature (20-30 °C), it was observed that the initial contents of reserves (sugars, starch, lipids, and proteins) showed reduction at four hours (phase I). In addition, an increase was observed in the protein contents during phases II and III, which may have favored the papaya germination process.

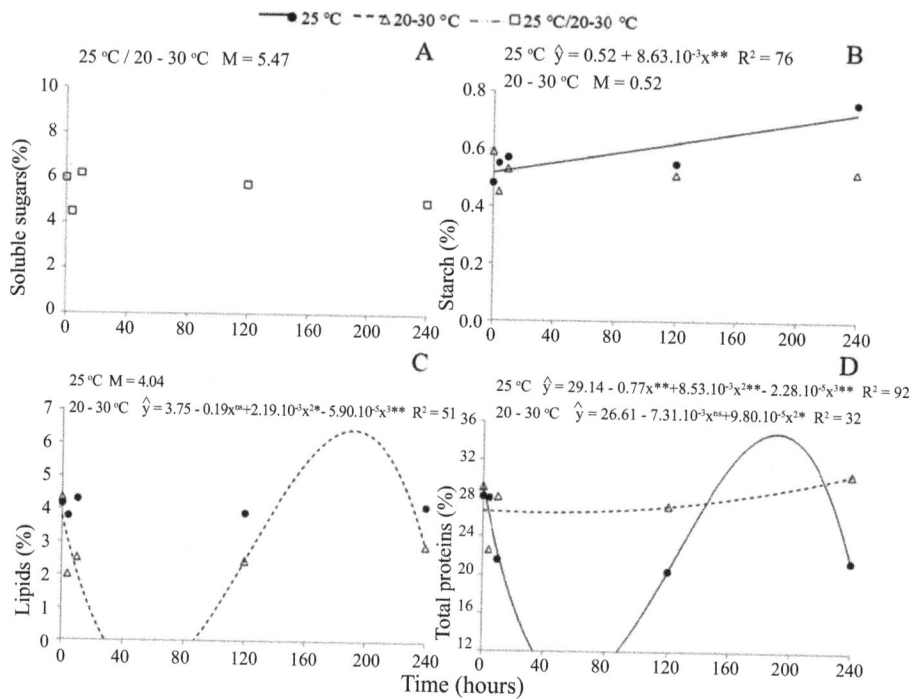

Figure 2. Mobilization of reserves during germination of papaya seeds subjected to constant temperature (25 °C), to alternating temperature (20-30 °C), or to the mean of the two treatments (25 °C/20-30 °C) (interaction not significant): A – concentration of soluble sugars; B – concentration of starch; C – concentration of lipids; D – concentration of total proteins. Values expressed in percentage (m/m). $\hat{y}$ – estimated y; M – mean of the treatment(s). *, ** Significant at the level of 5% and 1%, respectively; ns - not significant.

Accelerated aging in mobilization of reserves during germination:

Non-aged seeds (NAS) exhibited greater mean values for second count (SC), germination speed index (GSI), root length (RL), fresh matter (FM), and dry matter (DM) of the seedlings, compared to the seeds subjected to accelerated aging (AS) (Table 3).

The non-aged seeds (NAS) exhibited a lower concentration of soluble sugars after 240 hours of imbibition, and greater concentration of total proteins after 10 and after 240 hours of imbibition. The starch and lipid concentrations,

for their part, were not affected by aging (Table 4).

For the non-aged seeds (NAS), a linear reduction was observed in the concentration of sugars, with express reduction after four hours of imbibition, from 5.93 to 3.63%. In the aged seeds (AS), fluctuations were observed in the concentrations of soluble sugars, fitting with the cubic model and with a tendency to increase at the end of germination (Figure 3A). In general, the soluble sugars represent the initial substrates used as an energy source in germination, and a reduction in the concentrations of these compounds is expected (Ferreira and Borghetti, 2004), as observed in the non-aged seeds (NAS).

Table 3. Physiological quality of non-aged papaya seeds (NAS) subjected to accelerated aging (AS): percentage of seedlings with radicle protrusion (RP), second count of seedlings with radicle protrusion (SC), germination speed index (GSI), percentage of normal seedlings (NS), shoot length (SL), root length (RL), fresh matter (FM), and dry matter (DM) of normal seedlings.

Treatments	RP	SC	GSI	NS
NAS	91 a*	84 a	2.13 a	88 a
AS	82 a	67 b	1.61 b	80 a
F_{cal}	4.41^{ns}	6.24*	37.64**	3.00^{ns}
CV (%)	7.00	12.75	6.38	7.78
	SL	RL	FM	DM
NAS	7.26 a	5.26 a	96.25 a	4.70 a
AS	6.25 a	3.56 b	76.15 b	4.14 b
F_{cal}	5.52^{ns}	7.51*	29.72**	13.48*
CV (%)	9.00	19.89	6.05	4.90

*Mean values followed by the same letters (columns) are statistically equal by the Tukey test. *, ** Significant at the level of 5% and 1%, respectively. ^{ns} = not significant. F_{cal} = F value calculated obtained by analysis of variance. CV (%) = coefficient of variation (n = 4).

Table 4. Concentrations of soluble sugars, starch, lipids, and total proteins, expressed in percentage (m/m) during the germination phases of non-aged papaya seeds (NAS) subjected to accelerated aging (AS).

Time (h)	0	4	10	120	240	Mean
Treatments			Soluble sugars (%)			
NAS	5.93 a*	3.63 a	5.66 a	4.98 a	4.34 b	4.90
AS	4.80 a	4.80 a	5.81 a	4.20 a	5.36 a	4.99
Mean	5.37	4.22	5.74	4.59	4.85	
F_{cal}	7.37 **					
CV (%)	10.91					
			Starch (%)			Mean
NAS	0.59	0.45	0.53	0.51	0.52	0.52 a
AS	0.60	0.56	0.55	0.46	0.52	0.54 a
Mean	0.60	0.51	0.54	0.49	0.52	
F_{cal} (temperature)	0.59^{ns}					
F_{cal} (time)	2.27^{ns}					
CV (%)	14.61					
			Lipids (%)			Mean
NAS	4.32	2.01	2.52	2.43	2.91	2.84 a
AS	3.67	2.93	2.51	2.48	3.65	3.04 a
Mean	4.00	2.47	2.52	2.46	3.28	
F_{cal} (temperature)	1.43^{ns}					
F_{cal} (time)	12.44 **					
CV (%)	18.65					
Treatments			Total proteins (%)			
NAS	29.12 a	22.53 b	28.07 a	27.17 b	30.52 a	27.48
AS	28.28 a	28.68 a	27.13 b	29.79 a	29.31 b	28.64
Mean	28.70	25.61	27.60	28.48	29.92	
F_{cal}	55.49 **					
CV (%)	2.17					

*Mean values followed by the same letters (columns) are statistically equal by the Tukey test. *, ** Significant at the level of 5% and 1%, respectively. F_{cal} = F value calculated obtained by analysis of variance. CV (%) = coefficient of variation (n = 4).

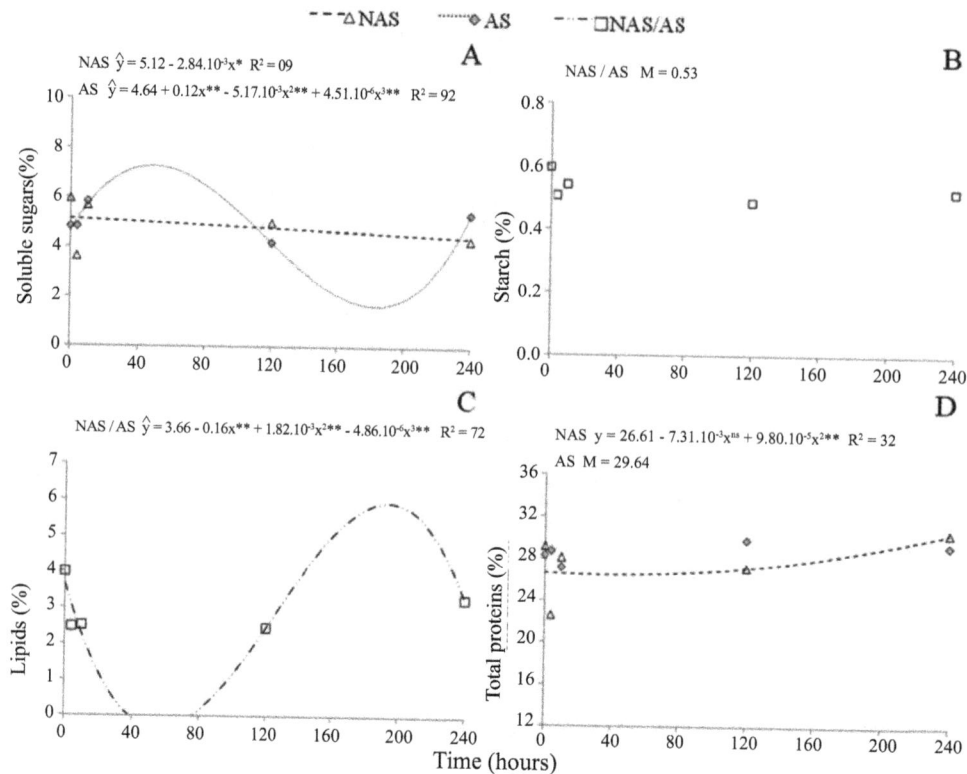

Figure 3. Mobilization of reserves during germination of non-aged papaya seeds (NAS), seeds subjected to accelerated aging (AS), or to the mean of the two treatments (NAS/AS) (interaction not significant): A – concentration of soluble sugars; B – concentration of starch; C – concentration of lipids; D – concentration of total proteins. Values expressed in percentage (m/m). $\hat{y}$ – estimated y; M – mean of the treatment(s). *, ** Significant at the level of 5% and 1%, respectively; ns - not significant.

The starch was not affected by the soaking times, remaining constant throughout the germination phases (0.53%) (Figure 3B).

The concentrations of lipids showed reduction after four and 10 hours (phase I and beginning of phase II), an increase after 120 hours (end of phase II), and a tendency to reduction after 240 hours of soaking (phase III), fitting with the cubic model of regression (Figure 3C). Thus, apart from AS, the mobilization of lipids showed behavior similar to that observed for seeds subjected to alternating temperature (Figure 2C), which showed high germination (Table 2). Therefore, the mobilization of lipids was not affected by accelerated aging and, although the seeds subjected to AS showed reduction in vigor (lower germination speed and reduction in seedling development), they did not show reduction in the percentages of seedlings with radicle protrusion and of normal seedlings.

The concentration of total proteins in the non-aged seeds (NAS) showed a tendency to increase during phases II and III, fitting the quadratic model. The aged seeds (AS) did not fit the models of regression (Figure 3D). In the seed deterioration process, there may be a decrease in the concentration and synthesis of proteins, as well as denaturation brought about by high temperatures (Bewley and Black, 1994; Marcos-Filho, 2005; Kim et al., 2011). As a consequence, the number of cells in division and the mitoctic index may be affected, as observed in *Triticum aestivum* seeds (Menezes et al., 2014).

In *Tabebuia roseoalba* seeds, a reduction in physiological quality of the seeds was observed, brought about by storage, with reduction in protein levels (Abbade and Takaki, 2014). Accelerated aging provides a favorable condition to seed deterioration, similar to storage. Thus, the low concentration of proteins observed in the aged seeds in phase III suggests that the reduction in the vigor of seeds subjected to aging may be related to the compromising of mobilization of proteins.

Conclusions

Phase I of water absorption during the germination of papaya (*Carica papaya* L.) seeds includes the period from zero to five hours of soaking; phase II, from five to 120 hours; and phase III begins after 144 hours of soaking.

Papaya seeds show greater germination under alternating temperature, with reduction in the concentrations of sugars, starch, lipids, and proteins during phase I of germination, mobilization of lipids, and increase in the concentrations of proteins during phases II and III.

The mobilization of lipids in papaya seeds is not affected by accelerated aging. Nevertheless, the seeds subjected to aging have a lower concentration of proteins in phase III, which may be related to reduction in vigor.

Acknowledgments

Thanks to Caliman Agrícola S/A for supplying plant material. To the CAPES and to the CNPq for granting scholarships for graduate studies and for research productivity, respectively.

References

ABBADE, L.C.; TAKAKI, M. Biochemical and physiological changes of *Tabebuia roseoalba* (Ridl.) Sandwith (Bignoniaceae) seeds under storage. *Journal of Seed Science*, v.36, n.1, p.100-107, 2014. http://www.scielo.br/pdf/jss/v36n1/a13v36n1.pdf

BEWLEY, J.D.; BLACK, M. *Seeds:* physiology of development and germination.New York: Plenum Press, 1994. 445p.

BLIGH, E.G.; DYER, W.J. A rapid method of total lipid extraction and purification. *Canadian Journal Biochemistry and Physiology*, v.37, n.8, p. 911-917, 1959.

CORTE, V.B.; BORGES, E.E.L.; VENTRELLA, M.C.; LEITE, I.T.A.; BRAGA, A.J.T. Histochemical aspects of reserves mobilization of *Caesalpinia peltophoroides* (Leguminosae) seeds during germination and seedlings early growth. *Revista Árvore*, v.32, n.4, p.641-650, 2008. http://www.scielo.br/pdf/rarv/v32n4/a05v32n4.pdf

COSTA, P.N.; BUENO, S.S.C.; FERREIRA, G. Fases da germinação de sementes de *Annona emarginata* (Schltdl.) H. Rainer em diferentes temperaturas. *Revista Brasileira de Fruticultura*, v.33, n.1, p.253-260, 2011. http://www.scielo.br/pdf/rbf/v33n1/aop01011.pdf

DANTAS, B.F.; RIBEIRO, R.C.; MATIAS, J.R.; ARAÚJO, G.G.L. Germinative metabolism of Caatinga forest species in biosaline agriculture. *Journal of Seed Science*, v.36, n.2, p.194-203, 2014. www.scielo.br/pdf/jss/v36n2/v36n2a08.pdf

FERREIRA, A.G.; BORGHETTI, F. *Germinação*: do básico ao aplicado. Porto Alegre: Artmed, 2004, 323p.

GALVANI, F.; GAERTNER, E. *Adequação da metodologia Kjeldahl para determinação de nitrogênio total e proteína bruta* (Circular Técnica 63). Corumbá: Embrapa Pantanal, 2006. 9p.

HENNING, F.A.; MERTZ, L.M.; JACOB JUNIOR, E.A.; MACHADO, R.D.; FISS, G.; ZIMMER, P.D. Composição química e mobilização de reservas em sementes de soja de alto e baixo vigor. *Bragantia*, v.69, n.3, p.727-734, 2010. http://www.scielo.br/pdf/brag/v69n3/26.pdf

KIM, H.T.; CHOI, U.; RYU, H.S.; LEE, S.J.; KWON, O. Mobilization of storage proteins in soybean seed (*Glycine max* L.) during germination and seedling growth. *Biochimica et Biophysica Acta*, v.1814, p.1178–1187, 2011. http://www.sciencedirect.com/science/article/pii/S1570963911001282

MAGALHÃES, S.R.; BORGES, E.E.L.; BERGER, A.P.A. Alterações nas atividades das enzimas alfa-galactosidase e poligalacturonase e nas reservas de carboidratos de sementes de *Schizolobium parahyba* (Vell.) Blake (Guapuruvú) durante a germinação. *Revista Brasileira de Sementes*, v.31, n.2, p.253-261, 2009. http://www.scielo.br/pdf/rbs/v31n2/v31n2a30.pdf

MAGUIRE, J.D. Speeds of germination-aid selection and evaluation for seedling emergence and vigor. *Crop Science*, v.2, p.176-177, 1962.

MARCOS-FILHO, J. *Fisiologia de sementes de plantas cultivadas*. Piracicaba: FEALQ, 2005. 495p.

MENEZES, V.O.; LOPES, S.J.; TEDESCO, S.B.; HENNING, F.A.; ZEN, H.D.; MERTZ, L.M. Cytogenetic analysis of wheat seeds submitted to artificial aging stress. *Journal of Seed Science*, v.36, n.1, p.71-78, 2014. http://www.scielo.br/pdf/jss/v36n1/a09v36n1.pdf

NONOGAKIA, H.; BASSELB, G.W.; BEWLEY, J.D. Germination - Still a mystery. *Plant Science*, v.179, p.574-581, 2010. http://www.sciencedirect.com/science/article/pii/S016894521000040

PASCUALIDES, A.L.; PLANCHUELO, A.M. Seed morphology and imbibition pattern of *Crotalaria juncea* L. (Fabaceae). *Seed Science & Technology*, v.35, p.760-764, 2007. http://www.ingentaconnect.com/content/ista/sst/2007/00000035/00000003/art00024

PONTES, C.A.; BORGES, E.E.L.; BORGES, R.C.G.; SOARES, C.P.B. Mobilização de reservas em sementes de *Apuleia leiocarpa* (Vogel) J.F. Macbr. (Garapa) durante a embebição. *Revista Árvore*, v.26, n.5, p. 593-601, 2002. http://www.scielo.br/pdf/rarv/v26n5/a10v26n5.pdf

REIS, R.C.R.; DANTAS, B.F.; PELACANI, C.R. Mobilization of reserves and germination of seeds of *Erythrina velutina* Willd. (Leguminosae-Papilionoideae) under different osmotic potentials. *Revista Brasileira de Sementes*, v.34, n.4, p.580-588, 2012. http://www.scielo.br/pdf/rbs/v34n4/08.pdf

REN, C.; BILYEU, K.D.; ROBERTS, C.A.; BEUSELINCK, P.R. Factors regulating the mobilization of storage reserves in soybean cotyledons during post-germinative growth. *Seed Science & Technology*, v.35, p.303-317, 2007. http://www.ingentaconnect.com/content/ista/sst/2007/00000035/00000002/art00006

SILVA, R.N.; DUARTE, G.L.; LOPES, N.F.; MORAES, D.M.; PEREIRA, A.L.A. Composição química de sementes de trigo (*Triticum aestivum* L.) submetidas a estresse salino na germinação. *Revista Brasileira de Sementes*, v.30, n.1, p.215-220, 2008. http://www.scielo.br/pdf/rbs/v30n1/a27v30n1.pdf

THE R FOUNDATION FOR STATISTICAL COMPUTING PLATAFORM. *R*. A language and environment for statistical computing. Version 3.1.1 Vienna: R Foundation for Statistical Computing, 2014. http://www.r-project.org/

YEMN, E.W.; WILLIS, A.J. The estimation of carbohydrate in plant extracts by anthrone. *The Biochemical Journal*, v.57, p.508-514, 1954. http://www.ncbi.nlm.nih.gov/pmc/articles/PMC1269789/pdf/biochemj01083-0159.pdf

ZANOTTI, R.F.; DIAS, D.C.F.S.; BARROS, R.S.; DAMATTA, F.M.; OLIVEIRA, G.L. Germination and biochemical changes in 'Formosa' papaya seeds treated with plant hormones. *Acta Scientiarum Agronomy*, v.36, n.4, p.435-442, 2014. http://periodicos.uem.br/ojs/index.php/ActaSciAgron/article/view/18057/pdf_38

ZHU, C.; CHEN, J. Changes in soluble sugar and antioxidant enzymes in peanut seeds during ultra dry storage and after accelerated aging. *Seed Science & Technology*, v.35, n.2, p. 387-401, 2007. http://www.ingentaconnect.com/content/ista/sst/2007/00000035/00000002/art00014

2

Environmental factors on seed germination, seedling survival and initial growth of sacha inchi (*Plukenetia volubilis* L.)

Amanda Ávila Cardoso[1*], Amana de Magalhães Matos Obolari[2],
Eduardo Euclydes de Lima e Borges[2], Cristiane Jovelina da
Silva[3], Haroldo Silva Rodrigues[5]

ABSTRACT – Sacha inchi (*Plukenetia volubilis* L.) is an Amazon species of elevated agro-industrial potential due the high content of omega-3 and omega-6 in its seeds. Despite of it, little information about its propagation by seeds is currently available. Thus, the aim of this study was to assess seed germination, seedling survival and growth of this species under different conditions of substrate (on paper, between papers and paper roll), light (continuous darkness, 12-h photoperiod and continuous light) and temperature (continuous temperature at 20, 25, 30, 35 and 40 °C). Germination is stimulated by substrates with increased surface contact with the seeds, presence of light and temperatures between 25 and 35 °C. Survival and initial growth of seedlings are favored by vermiculite, continuous light and 30 °C temperature. These conditions allow rapid and uniform germination of seeds and better establishment and development of seedlings. We encourage the propagation of sacha inchi by seeds, since we consider it a feasible technique.

Index terms: *Plukenetia volubilis*, substrate, light, temperature, seedling production.

Fatores ambientais na germinação de sementes e na sobrevivência e crescimento inicial de plântulas de sacha inchi (*Plukenetia volubilis* L.)

RESUMO – A sacha inchi (*Plukenetia volubilis* L.) é uma espécie amazônica de elevado potencial agroindustrial devido ao alto teor de ômega-3 e ômega-6 em suas sementes. Apesar disto, pouca informação acerca de sua propagação seminífera é encontrada. Dessa forma, o objetivo do trabalho foi avaliar a germinação das sementes, sobrevivência e crescimento das plântulas da espécie em diferentes condições de substrato (sobre papel, entre papel e em rolos de papel), luz (escuro contínuo, fotoperíodo de 12 horas e luz contínua) e temperatura (constante nas temperaturas de 20, 25, 30, 35 e 40 °C). A germinação é estimulada por substratos que possuam maior superfície de contato com a semente, pela presença de luz e por temperaturas entre 25 e 35 °C. A sobrevivência e o crescimento inicial das plântulas são favorecidos pela vermiculita, luz contínua e temperatura de 30 °C. Estas condições permitem uma germinação rápida e uniforme e um bom estabelecimento e desenvolvimento das plântulas. Nós encorajamos a propagação seminífera de sacha inchi, uma vez que ela se mostra uma técnica viável.

Termos para indexação: *Plukenetia volubilis*, substrato, luz, temperatura, produção de mudas.

Introduction

Information about the influence of abiotic factors on germination is essential for the study of seeds. Temperature, light and substrate affect seed germination in a species-dependent manner and may, in many cases, inhibit the germination process (Carvalho and Nakagawa, 2000). The same environmental factors also influence the seedlings survival and growth and the better understanding of these elements is essential in order to produce high-quality seedlings (Nogueira et al., 2003). Studies indicate a wide variability in the requirements of these components for the best growth and development of seedlings. This fact justifies this type of analysis for lesser known species and especially the native ones (Zamith and Scarano, 2004).

Sacha inchi (*Plukenetia volubilis* L. – Euphorbiaceae)

[1]Departamento de Biologia Vegetal, Universidade Federal de Viçosa, 36570-000 - Viçosa, MG, Brasil.
[2]Departamento de Engenharia Florestal, Universidade Federal de Viçosa, 36570-000 - Viçosa, MG, Brasil.
[3]Departamento de Botânica, Universidade Federal de Minas Gerais, 31270-901 - Belo Horizonte, MG, Brasil.
[4]Departamento de Fitotecnia, Universidade Federal de Viçosa, 36570-000 - Viçosa, MG, Brasil.
*Corresponding author <amandaavilacardoso@gmail.com>

is native to the Amazon region, occurring in Brazil, Peru, Colombia, and Venezuela (Céspedes, 2006). Its seeds have high agro-industrial potential due their elevated content of fatty acids, such as α-linolenic acid (omega-3) and linoleic acid (omega-6) (Follegatti-Romero et al., 2009). The species also has desirable features for reforestation and slope protection and it is appointed as an alternative to the recovery of degraded areas and to family farming programs (Bordignon et al., 2012).

Although studies for *in vitro* propagation of sacha inchi can be found (Bordignon et al., 2012; Rodrigues et al., 2014), information about seed germination and seedling growth is incipient (Rosa and Quijada, 2013). Therefore, the objective of this study was to evaluate the substrate, light and temperature requirements for seed germination and for seedlings survival and initial growth of sacha inchi.

Material and Methods

Sacha inchi seeds were acquired from Germplasm Bank of Sacha inchi of Embrapa Amazônia Ocidental, Manaus, AM, Brazil (they were collected in November 2012) and stored at 20 °C until the beginning of the experiments. Before implementing the tests, seeds were treated with Captan 0.5%.

Seeds were eliminated when they were deformed, shriveled and visibly attacked by insects. Since little is known about sacha inchi seeds, prior characterization was performed on water content (%), fresh weight (g), length (mm), width (mm) and thickness (mm). Water content was obtained from four replications of 10 seeds by the standard method in an oven at 105 °C for 24 hours (Brasil, 2009). Mass, length, width and thickness data were obtained with the aid of analytical balance and digital caliper. For this purpose, 50 randomly selected seeds were used from the total seeds.

Seed germination

To assess the effect of the substrate on germination, seeds were sown on paper, between papers and in paper rolls as recommended by the Regras para Análise de Sementes – Brazilian Rules for Seed Analysis (Brasil, 2009). Germitest papers were placed in petri dishes and the seeds distributed on or between the sheets. The substrates were moistened with an equivalent volume of water at 2.5 times its dry weight. The dishes and rolls were sealed to prevent excessive water losses and they were maintained in B.O.D. chambers under continuous light and at 30 °C.

The effect of different light conditions (continuous darkness, 12-h photoperiod and continuous light) on seed germination was also evaluated. Seeds were distributed in germitest paper rolls and maintained in B.O.D. chambers at

30 °C. The assessment of the germination rate in the absence of light was carried out in a dark room with the aid of green light.

In order to evaluate the effect of temperature on seed germination, seeds were distributed in germitest paper rolls and maintained in B.O.D. chambers under continuous light, and at the continuous temperatures of 20, 25, 30, 35 and 40 °C.

In each germination test, four replicates of 25 seeds per treatment were used. Germination tests lasted seven days and at the end, germination percentage and germination speed index (GSI) were assessed (Maguire, 1962). Seeds with primary root longer than 1 mm were considered germinated. The light of the germination tests was provided by four 20-W lamps.

Seedling survival and initial growth

Survival and growth of seedlings were assessed in different substrate, light and temperature conditions at the end of 15 days, since this species has a short vegetative phase. Sacha inchi seeds were sown in germitest paper rolls and maintained in B.O.D. chambers under continuous light and at 30 °C for five days. The seedlings obtained were transferred to plastic trays and placed in B.O.D. in the conditions described below.

To check the influence of the substrate, 200 seedlings were transplanted to plastic trays containing similar amounts of sand or fine vermiculite, wet to field capacity. The sand used was prior washed and sifted in a square mesh sieve number 4. The experiment was conducted under continuous light and at 30 °C for 15 days.

To evaluate the influence of light, another sample of 200 seedlings was transplanted to plastic trays containing vermiculite wet to field capacity. The trays were stored at 30 °C, under a 12-h photoperiod or continuous light for 15 days.

The effect of temperatures was also assessed. For this, a sample of 300 seedlings was transplanted to plastic trays containing vermiculite wet to field capacity. The trays remained in continuous light and at 25, 30 or 35 °C for 15 days.

In the survival tests 100 seedlings distributed in four replicates of 25 per treatment were assessed and in the growth tests 40 seedlings distributed in four replicates of 10 per treatment were used. At the end of 15 days, survival rates and seedlings growth parameters were evaluated. The number of leaves (N_L), seedling dry weight (W_S), leaves dry weight (W_L), roots dry weight (W_R), the leaf mass ratio (LMR=W_L/W_S), and root mass ratio (RMR=W_R/W_S) were estimated. Measurements were performed using an analytical balance. To determine the dry weight, plant material was dried in an oven at 70 °C for 72 hours. The light of the survival and growth tests was provided by four 20-W lamps.

Experimental design and statistical analysis

The statistical design was completely randomized and the data were submitted to analysis of variance (ANOVA). For variables with significant F (P < 0.05) and treatment's degree of freedom bigger than one, a grouping of averages was performed by the Scott-Knott test (P < 0.05). For variables with significant F and treatment's degree of freedom equal to one, the F test was conclusive. Both germination and survival percentage were transformed to arcsin $(x/100)^{1/2}$ prior to analysis and all data were checked for normality and homoscedasticity.

Results and Discussion

Seed characteristics and germination

Sacha inchi seeds had a 7.09 ± 0.25% water content and the following biometric characteristics: 0.97 ± 0.2 g weight, 1.8 ± 0.1 cm length, 1.5 ± 0.1 cm width, and 0.8 ± 0.1 cm thickness. Unlike many Amazonian species seeds, which have high humidity (O'Neill et al., 2001), sacha inchi seeds exhibited low water content in this study. The low moisture content of seeds and its large size indicate that sacha inchi seeds may be orthodox (Roberts, 1973). In order to confirm this classification desiccation and cold storage tests should be performed.

Seed germination started on the 3rd day after sowing and it varied depending on the substrate, light and temperature (Figure 1). About the substrates, both between paper and paper roll promoted significantly higher germination percentages (80 and 88%) relative to the substrate on paper (21%). In addition, the paper roll promoted higher GSI (4.5) compared to other substrates (Figure 1A). The substrates between paper and paper roll have greater contact surface with the seeds and therefore may favor their hydration and germination.

Assessing sacha inchi germination under different substrates, Rosa and Quijada (2013) found 70% and 3.0 as maximum germination rate and GSI, when moss was used as substrate. These values are low compared with values obtained in our study, when paper roll (88% and 4.5) and between papers (80% and 3.6) were used. Thus, these substrates show up as more suitable for sacha inchi seed germination.

The presence of light has not shown to be an essential factor for sacha inchi germination and its seeds may be considered as neutral photoblastic (Vázquez-Yanes and Orozco-Segovia, 1990). However, sacha inchi seed germination was improved in the presence of light (88% in continuous light and 83% in 12-h photoperiod) regarding its absence (69%) (Figure 1B). Similarly, higher GSIs were found in the presence of light (4.5 in continuous light and 4.4 in 12-h photoperiod) compared

to the continuous darkness (3.8). In resume, although the presence of light is not a necessary factor for sacha inchi germination to occur, it allows greater speed and percentage of seed germination.

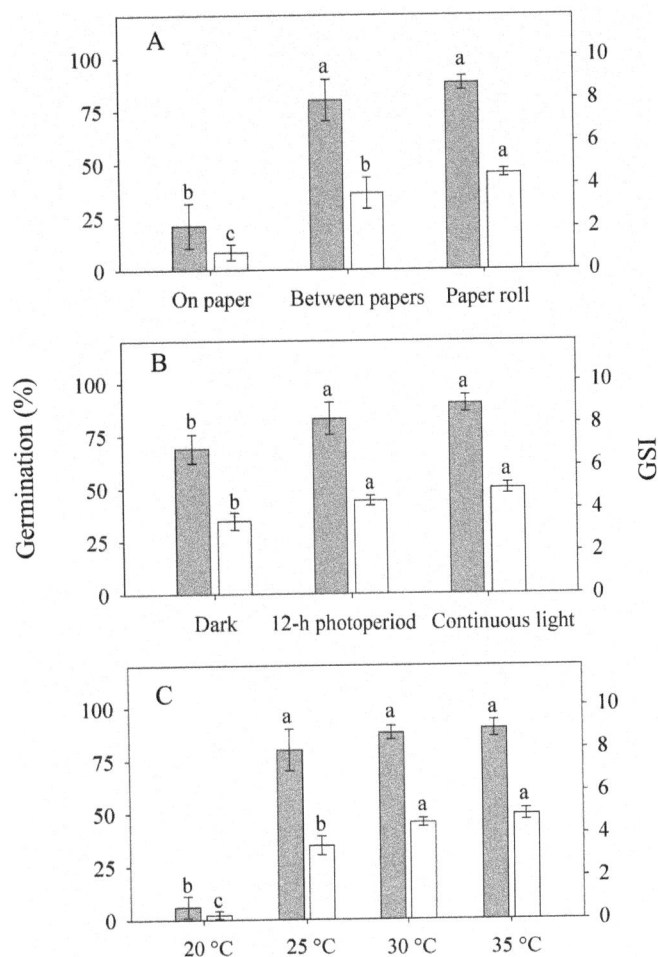

Figure 1. Germination (%) (gray bars) and GSI (white bars) of sacha inchi seeds under different conditions of substrate (A), light (B) and temperature (C). Bars with the same letter do not differ by the Scott-Knott test (P < 0.05). Mean ± SD, n = 4.

The highest germination percentages were found at 25, 30 and 35 °C (80, 88 and 90%, respectively), and they did not differ statistically from each other. However, GSIs at 30 and 35 °C (4.5 and 4.9) were superior to GSI at 25 °C (3.5). The 20 °C temperature negatively affected the germination rate and GSI (6% and 0.2); and the 40 °C temperature completely inhibited germination during the evaluation period (Figure 1C). It can be inferred that the minimum temperature for the sacha inchi seed germination is lower than 20 °C and the maximum temperature is between 35 and 40 °C. Although sacha inchi is native and occurs in Amazon biome, typical of elevated temperatures, its germination was inhibited at 40 °C. This result is in line with the information from Brancalion et al. (2010) that suggest temperatures around 30 °C for germination of species from that biome.

Survival and growth of seedlings

Sacha inchi seedlings survival was affected by the substrate and temperature, but not by the light. Seedling survival was higher in vermiculite (98%) than in the sand (79%). The temperature of 35 °C significantly reduced seedling survival (52%) compared to temperatures of 25 and 30 °C (94 and 97%), which did not differ significantly from each other (Figure 2).

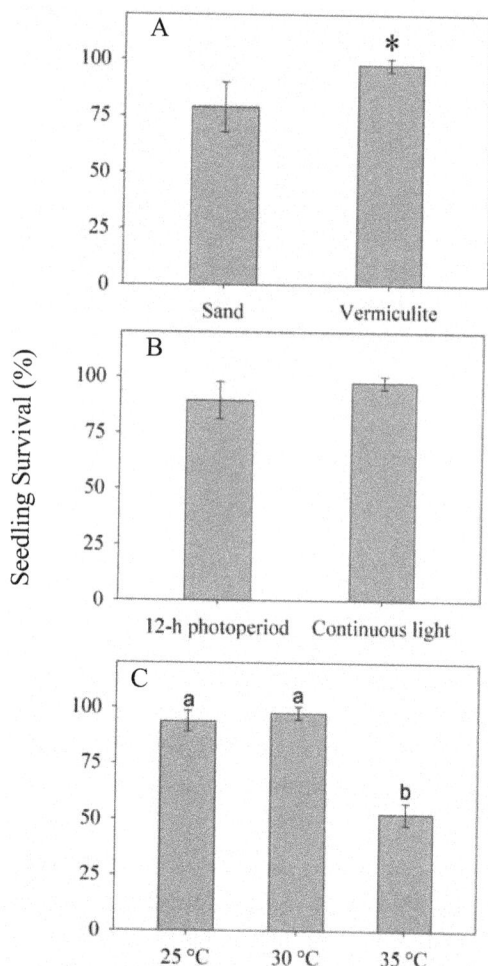

Figure 2. Seedling survival (%) of sacha inchi under different substrate (A), light (B) and temperature (C) conditions. Bars with an asterisk indicate that the F test was significant (P < 0.05) and bars with the same letter do not differ by the Scott-Knott test (P < 0.05). Mean ± SD, n = 4.

Although all evaluated factors have promoted changes in sacha inchi seedling growth, temperature was the component that most affected the assessed parameters (Table 1 and Figure 3). Substrate did not show significant difference between sand and vermiculite for N_L and W_S (Table 1). However, significant differences between both substrates were observed on the other parameters. The development of roots from seedlings grown in sand was higher (larger W_R and RMR) relative to the seedlings grown in vermiculite. On the other hand, seedlings grown in vermiculite exhibited greater development of the leaves compared to seedlings maintained in sand (higher W_L and LMR).

Sand has higher aeration in relation to vermiculite (Sodré et al., 2007), which may have allowed the roots to develop better in that substrate. Furthermore, sand has lower water retention, which may have stimulated greater root growth, aiming to favor the process of water uptake.

Fine vermiculite has a low particle size (0.3 to 0.5 mm) and, therefore, higher water retention and low aeration (Martins et al., 2012). Thus, root growth was disfavored and the leaves became the main sink of the plant.

The light condition affected only the leaves development (Table 1). The continuous light allowed to obtain higher W_L and LMR compared to the 12-h photoperiod. This fact indicates that seedlings under continuous light invest more on leaves, the photosynthetic organ of higher interest. It is possible that the seedlings maintained in continuous light are the most promising during the initial establishment.

Temperature was the factor with higher interference on seedlings development and it affected all evaluated parameters (Table 1 and Figure 3). At 30 °C seedlings developed well and showed the high N_L. Seedlings maintained at 25 and 30 °C exhibited similar W_S, but at 30 °C they showed higher W_L and W_R (due to increased LMR and RMR) than at 25 °C. Although the total biomass of seedlings maintained at 25 and 30 °C were similar, the seedlings maintained at 30 °C had more developed leaves (photosynthetic organ) and roots (organ responsible for the water and nutrients uptake), which characterizes them as better established.

The temperature at 35 °C resulted in high mortality of seedlings and dramatically affected the growth of the surviving seedlings, reducing the parameters: N_F, W_T, W_F and W_R (Table 1). High temperatures negatively affect important physiological mechanisms such as photosynthesis, photorespiration and transpiration, which are directly and indirectly related to the carbon balance of plants (Machado et al., 2002). It is likely that the constant temperature at 35 °C occasions an intense stress, enhancing the transpiration and photorespiration events, and reducing photosynthesis, which disfavors seedling establishment and development.

The propagation of sacha inchi by seeds is presented as a feasible alternative, since under suitable conditions of substrate, light and temperature the species shows rapid and uniform germination. Furthermore, both seedling survival and growth are satisfactory when appropriate environmental conditions are given.

Table 1. Number of leaves (N_L), seedling dry weight (W_S) (g), leaves dry weight (W_L) (g), roots dry weight (W_R) (g), leaf mass ratio (LMR) and root mass ratio (RMR) of sacha inchi seedlings maintained under different substrate, light and temperature conditions.

	Substrate		Light		Temperature		
	Sand	Vermiculite	Photoperiod	Continuous light	25 °C	30 °C	35 °C
N_L	4.15	3.75	3.75	3.75	1.85 b	3.75 a	0.20 c
W_S	0.51	0.46	0.40	0.46	0.49 a	0.46 a	0.22 b
W_L	0.11*	0.15	0.10*	0.15	0.09 b	0.15 a	0.01 b
W_R	0.23*	0.15	0.15	0.15	0.12 b	0.15 a	0.07 c
LMR	0.24*	0.33	0.27*	0.33	0.18 b	0.33 a	0.03 c
RMR	0.46*	0.33	0.37	0.34	0.24 b	0.34 a	0.33 a

Means followed by an asterisk indicate that the F test was significant ($P < 0.05$) and means followed by the same letter do not differ by the Scott-Knott test ($P < 0.05$). Mean, n = 4.

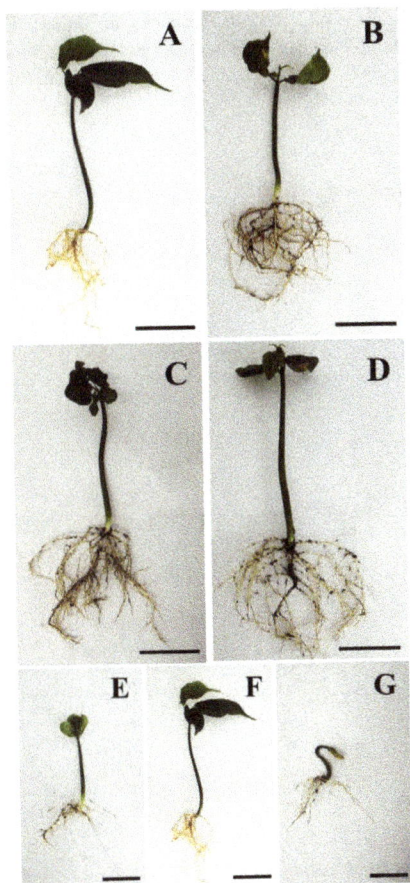

Figure 3. Seedlings of sacha inchi maintained in different substrate (sand – A and vermiculite – B), light (continuous light – C and 12-h photoperiod – D) and temperature (25 – E, 30 – F and 35 °C – G) conditions. Bars = 5 cm.

Conclusions

The sacha inchi seed germination is improved by substrates with higher surface contact, presence of light and by temperatures between 25 and 35 °C. Both survival and growth of sacha inchi seedlings are favored by vermiculite, continuous light and temperature at 30 °C. We consider the propagation of sacha inchi by seeds a feasible technique.

Acknowledgments

Thanks are due to CNPq (Brazilian Research Council), FAPEMIG (Foundation for Research Support of Minas Gerais State) and CAPES (Coordination for Scientific Support for Post-Graduate Level) for the scholarships and financial support during the conduct of this research. We also thank the researcher at Embrapa Amazônia Ocidental, Francisco Célio Maia Chaves, for sending the seeds.

References

BORDIGNON, S.R.; AMBROSANO, G.M.B.; RODRIGUES, P.H.V. Propagação in vitro de Sacha inchi. Ciência Rural, v.42, n.7, p.1168-1172, 2012. http://www.scielo.br/scielo.php?script=sci_arttext&pid=S0103-84782012000700005

BRANCALION, P.H.S.; NOVEMBRE, A.D.L.C.; RODRIGUES, R.R. Temperatura ótima de germinação de sementes de espécies arbóreas brasileiras. Revista Brasileira de Sementes, v.32, n.4, p.15-21, 2010. http://www.scielo.br/scielo.php?pid=S0101-31222010000400002&script=sci_arttext

BRASIL. Ministério da Agricultura, Pecuária e Abastecimento. Regras para análise de sementes. Ministério da Agricultura, Pecuária e Abastecimento. Secretaria de Defesa Agropecuária. Brasília: MAPA/ACS, 2009. 395p. http://www.agricultura.gov.br/arq_editor/file/2946_regras_analise__sementes.pdf

CARVALHO, N.M.; NAKAGAWA, J. *Sementes*: ciência, tecnologia e produção. 4 ed. Jaboticabal: FUNEP, 2000. 424p.

CÉSPEDES, E.I.M. Cultivo de Sacha Inchi. Tarapoto, San Martin, Peru: INIIA, Subdirección De Recursos Geneticos Y Biotecnología, 2006. 11p.

FOLLEGATTI-ROMERO, L.A.; PIANTINO, C.R.; GRIMALDI, R.; CABRAL, F.A. Supercritical CO_2 extraction of omega-3 rich oil from Sacha inchi (*Plukenetia volubilis* L.) seeds. *Journal of Supercritical Fluids*, v.49, n.3, p.323-329, 2009. http://www.sciencedirect.com/science/article/pii/S0896844609001119

MACHADO, E.C.; MEDINA, C.L.; GOMES, M.M.A.; HABERMANN, G. Variação sazonal da fotossíntese, condutância estomática e potencial da água na folha de laranjeira 'valência'. *Scientia Agricola*, v.59, n.1, p.53-58, 2002. http://www.scielo.br/pdf/sa/v59n1/8073.pdf

MAGUIRE, J.D. Speed of germination-aid in selection and evaluation for seedling emergence and vigor. *Crop Science*, v.2, n.1, p.176-177, 1962.

MARTINS, C.C.; MACHADO, C.G.; SANTANA, D.G.; ZUCARELI, C. Vermiculita como substrato para o teste de germinação de sementes de ipê-amarelo. *Semina: Ciências Agrárias*, v.33, n.2, p.533-540, 2012. http://www.uel.br/revistas/uel/index.php/semagrarias/article/view/6370

NOGUEIRA, R.J.M.C.; ALBUQUERQUE, M.B.; SILVA JUNIOR, JF. Efeito do substrato na emergência, crescimento e comportamento estomático em plântulas de mangabeira. *Revista Brasileira de Fruticultura*, v.25, n.1, p.15-18, 2003. http://www.scielo.br/pdf/rbf/v25n1/a06v25n1

O'NEILL, G.A; DAWSON, I.; SOTELO-MONTES, C.; GUARINO, L.; GUARIGUATA, M.; CURRENT, D.; WEBER, J.C. Strategies for genetic conservation of trees in the Peruvian Amazon. *Biodiversity & Conservation*, v.10, n.6, p.837-850, 2001. http://link.springer.com/article/10.1023%2FA%3A1016644706237

ROBERTS, E.H. Predicting the storage life of seeds. *Seed Science and Technology*, v.1, n.4, p.499-514, 1973. http://www.scielo.br/scielo.php?script=sci_nlinks&ref=000079&pid=S0101-31222007000300024000018&lng=en

RODRIGUES, P.H.V.; BORDIGNON, S.V.; AMBROSANO, G.M.B. Desempenho horticultural de plantas propagadas *in vitro* de Sacha inchi. *Ciência Rural*, v.44, n.6, p.1050-1053, 2014. http://www.scielo.br/scielo.php?pid=S0103-84782014000600016&script=sci_arttext

ROSA, R.L.; QUIJADA, J. Germinación del sacha inchi, *Plukenetia volubilis* L. (McBride, 1951) (Malpighiales, Euphorbiaceae) bajo cuatro diferentes condiciones. *The Biologist*, v.11, n.1, p.9-14, 2013. http://sisbib.unmsm.edu.pe/bvrevistas/biologist/v11_n1/pdf/a2v11n1.pdf

SODRÉ, G.A.; CORÁ, J.E.; SOUZA-JÚNIOR, J.O. Caracterização física de substratos à base de serragem e recipientes para crescimento de mudas de cacaueiro. *Revista Brasileira de Fruticultura*, v.29, n.2, p.339-344, 2007. http://www.scielo.br/pdf/rbf/v29n2/27.pdf

VÁZQUEZ-YANES, C.; OROZCO-SEGOVIA, A. Ecological significance of light controlled seed germination in two contrasting tropical habitats. *Oecologia*, v.83, n.2, p.171-175, 1990. http://link.springer.com/article/10.1007%2FBF00317748

ZAMITH, L.R.; SCARANO, F.R. Produção de mudas de espécies das restingas do município do Rio de Janeiro, RJ, Brasil. *Acta Botanica Brasilica*, v.18, p.161-176, 2004. http://www.scielo.br/pdf/abb/v18n1/v18n1a14.pdf

3

Aryl removal methods and passion fruit seed positions: Germination and emergence

Sérgio Macedo Silva[1], Roberta Camargos de Oliveira[1],
Risely Ferraz de Almeida[2], Adílio de Sá Júnior[1]*, Carlos Machado dos Santos[1]

ABSTRACT – Yellow passion fruit has conquered a significant position in the agribusiness of tropical fruits; therefore, farmers have been interested in expanding their groves, and technical information that guarantees the high yield of the farmings is needed. Aiming to observe factors related to the propagative material targeted at quality and maximization of the genetic potential, the objective of the study consisted in assessing germination and emergence of passion fruit seeds collected in three positions in the fruits and submitted to different methods of mucilage removal (aryl). The passion fruits were split in three parts: distal, medial and proximal and the contents of each part was divided in two. The content of each part of the fruit was submitted to mechanical and by fermentation methods for removal of aryl. It is concluded that the seed position in the fruit does not interferes with the germination. The aryl removal method by mechanical extraction affects the germination and seedling emergence.

Index terms: *Passiflora edulis*, mechanical extraction, fermentation, propagation, mucilage.

Métodos de remoção do arilo e posições das sementes no fruto de maracujá: Germinação e emergência

RESUMO – O maracujá-amarelo conquistou significativa posição no agronegócio de frutas tropicais, com isso, produtores despertaram interesse para expandir seus pomares, sendo necessárias informações técnicas que garantam a alta produtividade das lavouras. No intuito de observar fatores relacionados ao material propagativo visando a qualidade e a maximização do potencial genético objetivou-se avaliar a germinação e a emergência de sementes de maracujá coletadas em três posições nos frutos e submetidas a métodos de retirada da mucilagem (arilo). Os frutos de maracujá foram seccionados em três partes: distal, mediana e proximal. Cada um destes conteúdos foi submetido aos métodos mecânico e de fermentação para remoção do arilo. Podendo concluir que a posição da semente no fruto do maracujazeiro não interfere na germinação e o método de remoção do arilo por extração mecânica prejudica a germinação e a emergência de plântulas.

Termos para indexação: *Passiflora edulis*, extração mecânica, fermentação, propagação, mucilagem.

Introduction

Passion fruit, *Passiflora edulis* Sims f. *flavicarpa* Degener, is a species widely marketed both for processing and for consumption "in natura" (Zeraik et al., 2010). The high price of yellow passion fruit juice in the international market has made this species achieve a higher position in the agribusiness of tropical fruits and, therefore, it has raised interest for farmers in expanding their groves (Meletti et al., 2002; Almeida, 2012a).

Investment in seedlings and/or selected seeds is an important component of the production process as it is a fundamental prerequisite to the success of the activity, especially in projects aimed at achieving the noblest portions of the consumer market, such as exports (David et al., 1999).

Passion fruit seedlings may be obtained by means of seeds or vegetative state through root cuttings. Sowing takes precedence over the asexual method due to the ease of the process and the shorter period for forming the seedlings. However, care should be taken since germination problems are very common in the *Passiflora genus* (Lima et al., 2006).

Usually the spread is carried out without regard to factors related to the steps between harvesting and sowing. As a consequence, low percentage of germination, little vigorous plants, low crop yield and great variability in the physical characteristics of the fruits can occur (Carlesso et al., 2008).

In this regard, in order to obtain seedlings in greater

[1]Submitted on 02/24/2015. Accepted for publication on 05/25/2015. [1]
[1]Universidade Federal de Uberlândia, Instituto de Ciências Agrárias, Caixa Postal 593, 38400-902 – Uberlândia, MG, Brasil.

[2]Universidade Estadual Paulista Júlio de Mesquita Filho, Caixa Postal 237, 18610-307 – Jaboticabal, SP, Brasil.
*Corresponding author <adilio.junior@yahoo.com.br>

quantity and better quality, studies with passion fruit seeds are needed, in all sectors of seed technology. This knowledge would ensure high crop yields (Meletti et al., 2002).

Removal of aryl is directly related to methods for obtaining high quality seeds (Aguiar et al., 2014; Araújo et al., 2015). Besides other techniques conducted in the development of plants, such as pruning (Almeida, 2012a), soil fertilization (Almeida, 2012b), seed storage (Araújo et al., 2009) and the use of a suitable substrate in seedling formation (Wagner Junior et al., 2007; Aguiar et al., 2014).

The germination behavior and the emergence of passion fruit seeds as well as the position of the seeds in the fruits and the methods for removal of aryl were the focus of this study. In Brazil, research on seed technology in fruit species are uncommon (Osipi et al., 2011). In this sense, the objective was to assess the germination and emergence of passion fruit seeds collected in three positions in fruits and subjected to mucilage removal methods (aryl).

Material and Methods

Characterization of the experiment

The experiment was carried out in the Seeds Laboratory of Federal University of Uberlândia, city of Uberlândia, Minas Gerais, MG, Brazil. Passion fruits were collected in a clonal grove in the city of Araguari, MG, obtained in the same position in plants, maturation stage and uniform format, weighing from 20 to 22 g.fruit^{-1} (Figure 1A). The experiment was established in a a randomized block design (RBD) with four replications; treatments were distributed in a (3 x 2 + 2) factorial arrangement, whose first factor was the seed position in the fruit (distal, medial and proximal), Figure 1B, the second factor was the aryl removal mode (mechanical and fermentation) and additional treatments consisting of seeds of all the fruit with aryl removal by the fermentation method (additional 1) and mechanical (additional 2).

Figure 1. (A) Passion fruit (*Passiflora edulis*) collected and standardized for the treatments. (B) Details of the seed positions in passion fruit: distal, medial and proximal.

Experiment design

The fruits, after sectioned and removed the contents of each part, were divided into two equal parts, which were subjected to mechanical and fermentation methods for aryl removal.

In the aryl mechanical extraction method the seeds were first centrifuged at 200 rpm per minute. Subsequently, mucilage was separated from the seeds and washed to remove traces and placed to dry in the shade at room temperature for 24 hours.

The removal by the seeds fermentation method was accomplished using the 10% solution of slaked lime (calcium hydroxide), as recommended by Osipi et al. (2011). After 48 hours, seeds were washed in water to remove lime. Subsequently they were placed to dry under the same

conditions of the seeds prepared by the mechanical method.

The seeds germination test was carried out using germitest paper substrate, moistened with a water volume (mL) 2.5 times the weight of the dry paper in grams. 50 seeds were sown per plot, using two sheets of paper in rolls that were placed in a Mangelsdorf germinator, set at an alternating temperature between 25 and 30 °C and 8 hours of light, as done by Bornhofen et al. (2015).

The emergence test was conducted in tubes used in commercial seedling formation. Trays were used with 96 tubes with a volumetric capacity of 110.0 cm^3, filled with Plantimax® substrate, and the plot was made up of 16 tubes. Sowing was done manually by placing three seeds at a depth of 0.5 cm. Then the trays were transferred to the greenhouse.

Variables analyzed

Germination assessments were on the 7th and 28th days after sowing. The seedlings were classified into normal and non-germinated seeds (hard, numb and dead).

The percentage of emergence count was performed at intervals of eight hours from the emergence of the first seedling until the stabilization that occurred thirty days after sowing. This procedure was carried out according to the Rules for Seed Testing (Brasil, 2009), considering as emerged seedling the one that had expanded cotyledons, as described by Osipi et al. (2011).

The emergence speed index (ESI) seeds was performed according to the formula proposed by Maguire (1962), Equation 1: where *Ngi* refers to the number of seeds germinated on day i.; *Ti*: the time in days after sowing to germination. This equation was also tested for other species such as *Archontophoenix cunninghamii*, popularly known in Brazil as seafórtia (Pivetta et al., 2008) and cotton seeds, *Gossypium hirsutum* L. (Faria et al., 2003).

$$IVE = \frac{\sum Ngi}{Ti}$$

The mean emergence time (MET) was calculated using the formula proposed by Bianchetti and Amaral (1978), by emergence percentage ratio by the emergence speed index (ESI). Whereas to calculate the average speed of germination was used the formula proposed by Throneberry and Smith (1956).

Statistical analyses

The results of the variables were submitted to the waste normality tests (Shapiro-Wilk test, SPSS Inc., USA) and homogeneity of variances (Bartlett test, SPSS Inc., USA) and additivity of the blocks to check the quality of the database.

Then the variables were subjected to analysis of variance (ANOVA) and when H0 was rejected averages were compared by Tukey test at 0.05 probability (Sisvar Inc., Brasil), according to the recommendations from Ferreira (2011).

Results and Discussion

There was a distinction between the methods for seed extraction. Fermentation allowed better germination percentage, unlike the mechanically aryl removal. Values of 82.0% of germination obtained by treatment with fermentation (Figure 2) were similar to those obtained by Martins et al. (2006), who have also assessed the influence of fermentation in seed viability.

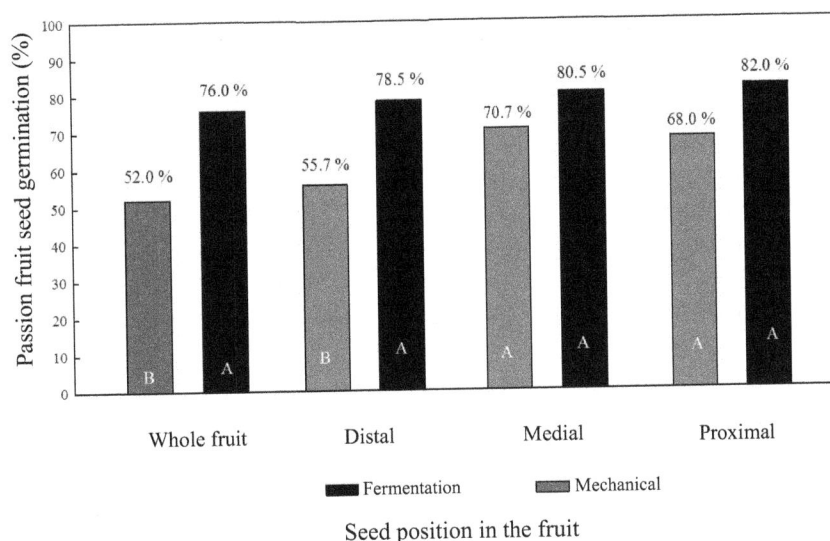

Figure 2. Seed germination (%) of *Passiflora edulis* in relation to the seed position in the fruit (distal, medial and proximal), aryl removal (by fermentation and mechanical) and control treatment (whole fruit). Figure: Means followed by the same letter are not statistically different from each other by Tukey test at 5% probability. (Coefficient of variation (%): 11.84; Significant average difference: 16.78).

The use of lime in the fermentation allowed for the maintenance of the germination potential when compared to mechanical extraction, since in this method damage to the seed coat occurs (Martins et al., 2006; Cardoso et al., 2001). According to Cardoso et al. (2001) lime helps remove the mucilage (aryl) without harming emergence.

Aryl adhering to the seeds attracts microorganisms due to its chemical composition. These may promote the destruction of the barrier in immediate contact with the seed, which promotes germination (Osipi et al., 2011).

The presence of aryl is related to the exposure of the seeds to very negative water potential, which hinders the absorption of water by the seeds and hence prolongs the early stages of germination. Furthermore, the presence of unsaturated triglycerides in the aryl form a barrier to the water inlet and the presence of steroids, which may act as hormones, compete for the active site and/or inhibit the action of hormones necessary for the germination and contained in the seeds (Martins et al., 2010).

Thus, seeds without aryl contribute to a higher percentage of normal seedlings, ensuring higher germination rates and an inverse relationship of aryl presence with the harshness of seeds (Lopes et al., 2007).

Mechanical aryl extraction damages the integument of passion fruit seeds; then the final yield of seeds is reduced. In the intense damage, the seeds are discarded or do not germinate; as for light damage, there is a decrease in the genetic potential to produce high vigor seedlings. This is because lesions or damage to the seed coat, endosperm and embryo disrupt the water absorption and, therefore, all the initial germination stages are changed. In the long term, this is reflected in the field, in lower yield and/or delay in fruiting.

For non-germinated seeds (hard, numb and dead) statistical assessment was not performed due to low occurrence; in this case, as the incidence was timely, the application of mathematical tests could express unrealistic results, proven by the high coefficient of variation.

Regarding the emergence test, treatment with the extraction of seeds by the fermentation method had higher means. The significant difference was observed in the distal and proximal areas for the emergence (E) and emergence speed index (ESI), respectively 85.93% and 0.04053 for distal, and 90.62% and 0.03861 for proximal. The mean emergence time (MET) for the mechanical treatment showed a small increase of 7.87% relative to fermenting (for additionals). Between additional treatments had significant difference just for the variable ESI (Table 1). This result is due to characteristics of Passifloraceae that are considered among the families whose seeds present dormancy due to water inlet control mechanisms into it. One can attribute this effect to a possible interference of fermentation in this process (Cardoso et al., 2001).

Table 1. Emergence (E), emergence speed index (ESI), average emergence speed (AES), mean emergence time (MET) of *Passiflora edulis* seedlings in relation to seed position in the fruit (distal, medial and proximal), aryl (mechanical and by fermentation) removal, and additional with mechanical removal (A1) and fermentation (A2).

Position	Aryl removal method	E (%)	ESI	AES	MET (h)
Distal	Mechanical	40.62 b	0.01704 b	0.00243 a	420.99 a
	Fermentation	85.93 a	0.04053 a	0.00279 a	360.42 a
Medial	Mechanical	65.62 a	0.02745 a	0.00244 a	411.05 a
	Fermentation	95.31 a	0.04170 a	0.00257 a	389.75 a
Proximal	Mechanical	43.75 b	0.01754 b	0.00169 a	339.06 a
	Fermentation	90.62 a	0.03861 a	0.00244 a	415.95 a
A1	Mechanical	57.81 a	0.02323 b	0.00237 a	422.79 a
A2	Fermentation	90.62 a	0.03989 a	0.00258 a	391.91 a
SAD[1]		36.39	0.01584	0.00092	186.46
CV[2] (%)		25.37	25.61	18.99	23.52

[1]SAD: significant average difference; [2]CV: Coefficient of variation. Means followed by similar letters are not statistically different from each other by Tukey test at 5 % probability for the aryl removal method.

ESI of the seeds is important because it determines the seed lot vigor both in field conditions as in a greenhouse (Oliveira et al., 2009). The more vigorous the seed lot, the faster the seedling emergence will be (Aguiar et al., 2014) and the greater the speed in the metabolic processes of the plant; consequently, there will be a faster and more uniform emission of the primary root in the germination process (Munizzi et al., 2010).

Among the positions of seeds in the fruit there was no significant difference for variables AES and MET, obtaining better results in the medial position, with averages of 0.002 and 400.4, respectively. Similar results were found by Freitas et al. (2013), when working with *Mimosa caesalpiniifolia*.

However, in other species such as: *Moringa oleifera* Lam (Oliveira et al., 2013) and *Caesalpinia ferrea* Mart ex Tul (Nogueira et al., 2010) difference occurs between seed positions in the fruit (Marcos-Filho, 2005).

Conclusions

Passion fruit seed position does not interfere with germination. The aryl removal method by mechanical extraction affects the germination and seedling emergence.

Acknowledgments

The authors thank Conselho Nacional de Desenvolvimento Científico e Tecnológico (CNPq; National Counsel of Technological and Scientific Development) and Coordenação de Aperfeiçoamento de Pessoal de Nível Superior (CAPES; Coordination of Improvement of Higher Education) for supporting the research.

References

AGUIAR, R.S.; YAMAMOTO, L.Y.; PRETI, E.A.; SOUZA, G.R.B.; SBRUSSI, C.A.G.; OLIVEIRA, E.A.P.; ASSIS, A.M.; ROBERTO, S.R.; NEVES, C.S.V.J. Extração de mucilagem e substratos no desenvolvimento de plântulas de maracujazeiro-amarelo. *Semina: Ciências Agrárias*, v.35, n.2, p.605-612, 2014. http://www.uel.br/revistas/uel/index.php/semagrarias/article/view/10418/pdf_261

ALMEIDA, R.F. Características da poda em maracujazeiro. *Revista Verde de Agroecologia e Desenvolvimento Sustentável*, v.7, n.5, p.53-58, 2012a. http://gvaa.com.br/revista/index.php/RVADS/article/viewFile/1156/pdf_619

ALMEIDA, R.F. Nutrição de maracujazeiro. *Revista Verde de Agroecologia e Desenvolvimento Sustentável*, v.7, n.3, p.12-17, 2012b. http://www.gvaa.com.br/revista/index.php/RVADS/article/view/1155/pdf_545

ARAÚJO, L.R.; ALVES, E.U.; RODRIGUES, C.M.; RODRIGUES, A.A.M. Emergência e crescimento inicial de plântulas de *Eugenia jambolana* Lam. após remoção da polpa. *Revista Ciência Rural*, v.45, n.1, p.14-18, 2015. http://www.scielo.br/pdf/cr/v45n1/0103-8478-cr-45-01-00014.pdf

ARAÚJO, E.C.; SILVA, R.F.; BARROSO, D.G.; CARVALHO, A.J.C. Efeito do armazenamento e do progenitor masculino sobre a qualidade e micromorfologia de sementes de maracujá. *Revista Brasileira de Sementes*, v.31, n.4, p.110-119, 2009. http://www.scielo.br/pdf/rbs/v31n4/13.pdf

BIANCHETTI, A.; AMARAL, E. Dia médio e velocidade de germinação de sementes de cebola (*Allium cepa* L.). *Pesquisa Agropecuária Brasileira*, v.13, p.33-44, 1978. http://seer.sct.embrapa.br/index.php/pab/article/view/16748

BORNHOFEN, E.; BENIN, G; GALVAN, D; FLORES, M.F. Épocas de semeadura e desempenho qualitativo de sementes de soja. *Revista Pesquisa Agropecuária Tropical*, v. 45, n.1, p. 46-55, 2015. http://www.scielo.br/pdf/pat/v45n1/1983-4063-pat-45-01-0046.pdf

BRASIL. Ministério da Agricultura, Pecuária e Abastecimento. *Regras para análises de sementes*. Ministério da Agricultura, Pecuária e Abastecimento. Secretaria de Defesa Agropecuária. Brasília: MAPA/ACS, 2009. 395p. http://www.agricultura.gov.br/arq_editor/file/2946_regras_analise__sementes.pdf

CARDOSO, G.D.; TAVARES, J.C.; FERREIRA, R.L.F.; CÂMARA, F.A.A.; CARMO, G.A. Desenvolvimento de mudas de maracujazeiro-amarelo obtidas de sementes extraídas por fermentação. *Revista Brasileira de Fruticultura*, v.23, n.3, p.639-642, 2001. http://www.scielo.br/pdf/rbf/v23n3/8042.pdf

CARLESSO, V.O.; BERBERT, P.A.; SILVA, R.F.; DETMANN, E. Secagem e armazenamento de sementes de maracujá amarelo (*Passiflora edulis* Sims f. *flavicarpa* Degener). *Revista Brasileira de Sementes*, v.30, n.2, p.065-074, 2008. http://www.scielo.br/pdf/rbs/v30n2/a09v30n2.pdf.

DAVID, D.V.; SILVA, J.M.A.; SILVA, P.M. (Coord.). *Diagnóstico de produção e comercialização de mudas e semente de espécies frutíferas na região Nordeste do Brasil.* Viçosa, MG: UFV; 1999. 215p. http://www.scielo.br/pdf/rbs/v25n1/19640.pdf

FARIA, A.Y.K.; ALBUQUERQUE, M.C.F.; CASSETARI NETO, D. Qualidade fisiológica de sementes de algodoeiro submetidas a tratamentos químico e biológico. *Revista Brasileira de Sementes*, v.25, n.1, p.121-127, 2003. www.abrates.org.br/revista/artigos/2003/v25n1/artigo19.pdf

FERREIRA, D. F. Sisvar: a computer statistical analysis system. *Ciência e Agrotecnologia*, v.35, n.6, p.1039-1042, 2011. http://www.scielo.br/pdf/cagro/v35n6/a01v35n6.pdf

FREITAS, T.P.; FREITAS, T.A.S.; CAMPOS, B.M.; FONSECA, M.D.S.; MENDONÇA, A.V.R. Morfologia e caracterização da germinação em função da posição das sementes no fruto de sabiá. *Scientia Plena*, v.9, n.3, p.1-9, 2013. http://scientiaplena.org.br/sp/article/view/790/670

LIMA, A.A.; CALDAS, R.C.; SANTOS, V.S. Germinação e crescimento de espécies de maracujá. *Revista Brasileira de Fruticultura*, v.28, n.1, p.125-127, 2006. http://www.scielo.br/pdf/rbf/v28n1/29708.pdf

LOPES, J.C.; BONO, G.M.; ALEXANDRE, R.S.; MAIA, V.M. Germinação e vigor de plantas de maracujazeiro-amarelo em diferentes extratos de maturação do fruto, arilo e substrato. *Ciência e Agrotecnologia*, v.31, n.5, p.1340-1346, 2007. http://www.scielo.br/pdf/cagro/v31n5/10.pdf

MAGUIRE, J.D. Speed of germination-aid in selection and evaluation for seedlingemergence and vigor. *Crop Science*, v.2, n.1, p.176-177, 1962.

MARCOS-FILHO, J. *Fisiologia de sementes de plantas cultivadas.* Piracicaba: FEALQ, 2005. 495p.

MARTINS, M.R.; REIS, M.C.; NETO, J.A.M.; GUSMÃO, L.L.; GOMES, J.J.A. Influência de diferentes métodos de remoção do arilo na germinação de sementes de maracujazeiro-amarelo (*Passiflora edulis* Sims F. *flavicarpa* Deg). *Revista da FZVA*, v.13, n.2, p. 28-38, 2006. http://revistaseletronicas.pucrs.br/ojs/index.php/fzva/article/view/2362/1849

MARTINS, C.M.; VASCONCELLOS, M.A.S; ROSSETTO, C.A.V.; CARVALHO, M.G. Prospecção fitoquímica do arilo de sementes de maracujá-amarelo e influência em germinação de sementes. *Ciência Rural*, v.40, n.9, p.1934-1940, 2010. http://www.scielo.br/pdf/cr/v40n9/a720cr2740.pdf

MELETTI, L.M.M.; FURLANI, P.R.; ÁLVARES, V.; SOARES-SCOTT, M.D.; BERNACCI, L.C.; AZEVEDO FILHO, J.A. Novas tecnologias melhoram a produção de mudas de maracujá. *O Agronômico*, v.54, n.1, p.30-33, 2002. http://www.iac.sp.gov.br/publicacoes/agronomico/pdf/541_08t72.pdf

MUNIZZI, A.; BRACCINI, A.L.; RANGEL, M.A.S.; SCAPIM, C.A.; BARBOSA, M.C.; ALBRECHT, L.P. Qualidade de sementes de quatro cultivares de soja, colhidas em dois locais no estado de Mato Grosso do Sul. *Revista Brasileira de Sementes*, v.32, n.1, p.176-185, 2010. http://www.scielo.br/pdf/rbs/v32n1/v32n1a20.pdf

NOGUEIRA, N.W.; MARTINS, H.V.G.; BATISTA, D.S.; RIBEIRO, M.C.C.; BENEDITO, C.P. Grau de dormência das sementes de jucá em função da posição na vagem. *Revista Verde de Agroecologia e Desenvolvimento Sustentável*, v.5, n.1, p.39-42, 2010. http://www.gvaa.com.br/revista/index.php/RVADS/article/viewFile/242/242

OLIVEIRA, A.C.S.; MARTINS, G.N.; SILVA, R.F.; VIEIRA, H.D. Testes de vigor em sementes baseados no desempenho de plântulas. *Inter Science Place*, v.2, n.4, 2009. Disponível em: < http://www.interscienceplace.org/interscienceplace/article/view/37/43 >. Accessed on: Jan. 12th, 2015.

OLIVEIRA, F.A.; OLIVEIRA, M.K.T.; SILVA, R.C.P.; SILVA, O.M.P.; MAIA, P.M.E.; CÂNDIDO, W.S. Crescimento de mudas de moringa em função da salinidade da água e da posição das sementes nos frutos. *Revista Árvore*, v. 37, p.79-87, 2013. http://www.scielo.br/pdf/rarv/v37n1/v37n1a09.pdf

OSIPI, E.A.F; LIMA, C.B.; COSSA, C.A. Influência de métodos de remoção do arilo na qualidade fisiológica de sementes de *Passiflora alata* Curtis. *Revista Brasileira de Fruticultura*, v.33, n.1, p.680-685, 2011. http://www.scielo.br/pdf/rbf/v33nspe1/a95v33nspe1.pdf

PIVETTA, K.F.L.; SARZI, I.; ESTELLITA, M.; BECKMANN-CAVALCANTE, M.Z. Tamanho do diásporo, substrato e temperatura na germinação de sementes de *Archontophoenix cunninghamii* (Arecaceae). *Revista de Biologia e Ciências da Terra*, v. 8, n.1, p.126-134, 2008. http://joaootavio.com.br/bioterra/workspace/uploads/artigos/pivetta-518173973deb6.pdf

THRONEBERRY, G.0.; SMITH, F.G. Relation of respiratory and enzimatic activity to corri seed viability. *Plant Physiology*, v.30, p.337-43, 1956. http://www.ncbi.nlm.nih.gov/pmc/articles/PMC540658/

WAGNER JUNIOR, A.; SANTOS, C.E.M.; ALEXANDRE, R.S.; SILVA, J.O.C.; NEGREIROS, J.R.S.; PIMENTEL, L.D.; ÁLVARES, V.S.; BRUCKNER, C.H. Efeito da pré-embebição das sementes e do substrato na germinação e no desenvolvimento inicial do maracujazeiro-doce. *Revista Ceres*, v.54, n.311, p.1-6, 2007. http://www.redalyc.org/pdf/3052/305226663011.pdf

ZERAIK, M.L.; PEREIRA, C.A.M.; ZUIN, V.G.; YARIWAKE, J.H. Maracujá: um alimento funcional? *Revista Brasileira de Farmagnosia*, v.20, n.3, p.459-471, 2010. http://www.scielo.br/pdf/rbfar/v20n3/a26v20n3.pdf

4

Maturation of seeds of *Poincianella pluviosa* (Caesalpinoideae)

João Paulo Naldi Silva[1*], Danilo da Cruz Centeno[2],
Rita de Cássia Leone Figueiredo-Ribeiro[3], Claudio José Barbedo[1]

ABSTRACT - The persistence of viable seeds in the soil is an important way to assure plant propagation, especially for species which produce seeds with short lifespan. *Poincianella pluviosa* is a tree species which seeds have short to medium storability at room temperature. The comprehension of the maturation process is crucial to understand its strategy for propagation and it could provide tools to improve seed viability in *ex situ* conditions. Flowers were tagged in two consecutive cycles of maturation and pods were periodically harvested until dispersion. Seeds were classified based on their morpho-physiological features, capability to germinate and develop seedlings. The complete maturation process was attained 315-330 days after anthesis, a period longer than reported for seed viability at room temperature. The maximum dry mass and seed vigor were reached at the end of maturation, although elevated seedling production was obtained before physiological maturity. We suggest that the precocious ability to produce seedlings in a long maturation could be a strategy to overcome environmental constraints, as the species is distributed in a wide range of phytogeographic domains in Brazil.

Index terms: *ex situ* conservation, physiological maturity, seed shedding, seed maturation, sibipiruna.

Maturação de sementes de *Poincianella pluviosa* (Caesalpinoideae)

RESUMO – A manutenção de sementes viáveis no solo é uma forma importante para garantir a propagação, especialmente para espécies que produzem sementes de vida útil curta. *Poincianella pluviosa* é uma espécie arbórea cujas sementes apresentam curto a médio período de armazenamento em condições de laboratório. Assim, a compreensão do processo de maturação é fundamental para entender as estratégias de propagação da espécie, representando uma ferramenta para aprimorar a viabilidade das sementes em condições *ex situ*. As flores foram marcadas em dois ciclos de maturação consecutivos e os frutos foram colhidos até dispersão. As sementes foram classificadas de acordo com suas características morfofisiológicas e a capacidade de germinar e desenvolver plântulas. Observou-se que o processo completo de maturação (315-330 dias após antese) é maior do que o tempo descrito na literatura para manutenção da viabilidade destas sementes em temperatura ambiente. Valores máximos de massa seca e vigor foram alcançados no final da maturação, embora elevada produção de plântulas tenha sido obtida antes da maturidade fisiológica. Sugere-se que a capacidade precoce de produzir plântulas em um longo período de maturação poderia ser uma estratégia para superar limitações ambientais, já que a espécie está distribuída em diversos domínios fitogeográficos no Brasil.

Termos para indexação: conservação *ex situ*, dispersão de sementes, maturação de sementes, maturidade fisiológica, sibipiruna.

Introduction

Some seeds can maintain viability for thousands of years, representing an important feature for plant survival (Sallon et al., 2008). However, the longevity of seeds can be regulated by how advanced is the maturation process at the moment they are dispersed. An incomplete process can result in both poor seed germination and low persistence in the seed bank and produce seed lots with low vigor (Kermode, 1990; Barbedo et al., 2002). On the other hand, late harvesting can accelerate seed deterioration due to adverse environmental conditions, which are frequently unsuitable for seed storage (Butler et al., 2009). Therefore, characterization of the maturation cycle is important to infer how long the soil seed bank will be available for propagation, as well as for defining the harvest timing to maintain longer seed viability during storage.

[1]Instituto de Botânica, Núcleo de Pesquisa em Sementes, Caixa Postal, 68041, 04301012 – São Paulo, SP, Brasil.
[2]Universidade Federal do ABC, Centro de Ciências Naturais e Humanas, 09606-070 - São Bernardo do Campo, SP, Brasil.
[3]Instituto de Botânica, Centro de Pesquisa em Ecologia e Fisiologia, Caixa-Postal, 68041, 04301902 - São Paulo, SP, Brasil.
*Corresponding author <silvajpn7@gmail.com>

Seed development starts with the fertilization of the ovule followed by a period of extensive cell division and differentiation. After that, dry matter is transferred from the mother plant to the seed which has its water content progressively reduced (Bewley et al., 2013) until abscission indicating, for most species, the end of reserve accumulation and the achievement of mass maturity (Ellis and Pieta Filho, 1992).

Poincianella pluviosa (DC.) L.P.Queiroz (= *Caesalpinia pluviosa* DC.; = *Caesalpinia peltophoroides* Benth.) is a member of Leguminosae, one of the most numerous and important neo-tropical families. Its taxonomic classification was reviewed (Lewis, 1998; Queiroz, 2009) but is still under discussion (Souza et al., 2013). This species shows a wide distribution in Brazil, covering various phytogeographic domains (Amazon, Caatinga, Cerrado, Atlantic forest and Pantanal – Queiroz, 2009; Lewis, 2015). Except for an ontogenetic study (Souza et al., 2013), limited information about seeds of this species is available. They are neither photosensitive nor show physical dormancy (Ferraz-Grande and Takaki, 2006) and are viable at room temperature for 240 days (Figliolia et al., 2001; Pontes et al., 2006). These seeds accumulate predominantly lipids, approximately 50% of total weight (Corte et al., 2006).

In this study we characterized the physiological maturity of seeds of *P. pluviosa* and described morphological and physiological changes occurring during the whole maturation period aiming to understand this process and to contribute to the conservation strategies of an underexploited neo-tropical species.

Material and Methods

Plant material

Trees (*ca.* 35) planted in the Rubião Jr *campus* of the Universidade Estadual Paulista (UNESP) in Botucatu, SP, Brazil (22°52'20''S 48°26'37''W) had inflorescences tagged at the beginning of their anthesis in two consecutive maturation periods: September/2009 to August/2010 (referred as 2009/10) and September/2010 to July/2011 (referred as 2010/11). Pods were harvested directly from the branches at 80, 161, 203, 217, 245, 259, 285, 301 and 315 days after anthesis (DAA) in 2009/10 and at 97, 174, 202, 216, 230, 245, 258, 286, 301, 317 and 323 DAA, in 2010/11. Fruits and seeds obtained for each stage were characterized as described below. The estimated time to reach each stage was calculated and considered as the age of fruits/seeds. Additionally, seeds were obtained directly from the ground, not exceeding 48 h after shedding, which were named recently-dispersed seeds (RDS). A voucher specimen from this species is deposited in the SP Herbarium of the Universidade Estadual Paulista

(UNESP) in Botucatu, SP, Brazil, number BOTU 1745 (Col. Garde Filho, D. 05/10/1971).

Fruit and seed analysis

Fruits and seeds were manually removed from the pods and characterized biometrically (three replicates of five fruits or fifteen seeds) with respect to color, texture and size (length, width and thickness, in mm). The water content (g of water per g of dry weight, $g. g^{-1}$) and dry mass (g. seed^{-1}) were determined gravimetrically in an oven at 103 °C ± 3 °C for 17 h, according to ISTA (2015) with four replicates of five seeds. The water potential was measured with a Decagon WP4 potentiometer (Pullman, USA) based on the dew point temperature of the air after equilibrium with the sample, according to Bonjovani and Barbedo (2008). The assessment of the water potential was performed by sorption isotherms in solutions of polyethylene glycol (PEG 6000) at different osmotic potentials, based on Michel and Kaufmann (1973).

Germination tests were carried out in four replications of 15 seeds each, using two sheets for the base and one for the covering of Germitest paper previously moistened with tap water (ISTA, 2015). The rolls were maintained in a germination chamber (25 ± 1 °C) and were evaluated every 2 days until 40 days, registering the percentage of germinated seeds (protrusion of at least 5 mm of primary root), normal seedling development (seedlings with at least 3 cm and no visual malformations) and, at the end of the analysis the speed of germination-aid was calculated according to Maguire (1962).

Meteorological data

The harvest site is classified as a semideciduous forest and the climate is Cwa, *i.e.*, humid temperate with dry winter and hot summer, according to Köppen classification, and it is located at 827 m above sea level (CEPAGRI, 2015). Values of minimum and maximum temperatures and rainfall from flowering to shedding for both periods were obtained in an automatic meteorological station located in the Rubião Jr *campus* in Botucatu (SP).

Statistical analysis

Data are means ± standard deviation of four replicates of 20 seeds per treatment for the physical and physiological analyses. The experiments were carried out with a completely randomized design and data were statistically analyzed by ANOVA, using *t*-test at 5% probability for the comparison of the two maturation periods (Santana and Ranal, 2004).

Results and Discussion

The maturation process of seeds of *P. pluviosa* is characterized by 13 distinct stages distributed in a complete cycle of 330 DAA or 11 months (Figure 1). Long periods of seed maturation were also described for other species, as exampled by *Cedrela fissilis* (8-9 months), *Coffea sp.* (7-12 months), *Hypodiscus aristatus* (18 months), and some palms (from 180 to 400 days - Chapin, 1999; Corvello et al., 1999; Dussert et al., 2000; Newton et al., 2002; Eira et al., 2006; Pérez et al., 2012).

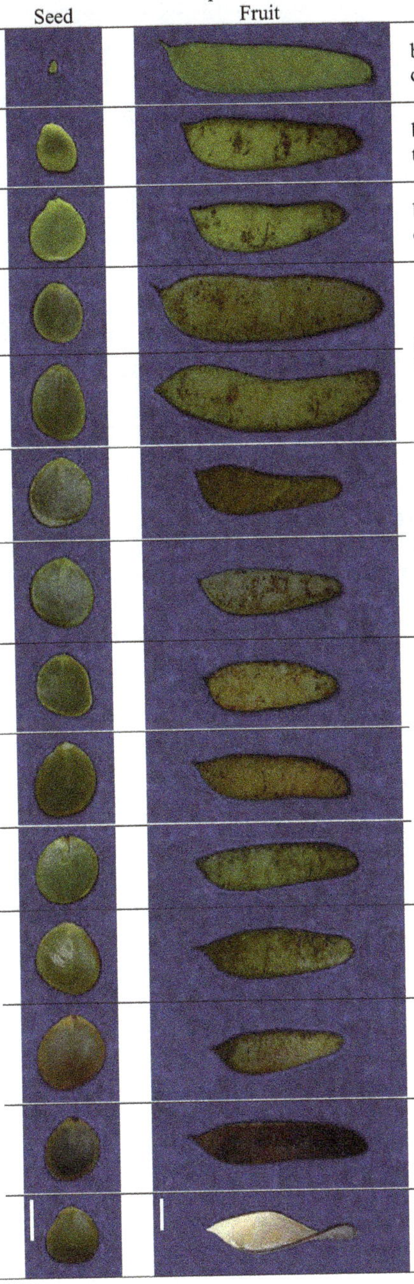

Stage	DAA	Visual aspects		Characteristics	
		Seed	Fruit	Seed	Fruit
S1	100			bright green, thin and fragile coat	light green and very flexible
S2	145 to 175			bright green, cotyledons smaller than the coat	light green, with brown spots, very flexible
S3				bright green, cotyledons of size close to the coat	
S4	200			green, coat thin and fragile, very flexible	
S5	215				
S6	230			green, rough coat and fragile, very flexible	green, with brown spots, flexible
S7	245				
S8	260			green, rough coat and fragile, flexible	
S9	285				
S10	300				
S11				green with brown spots, rough coat, flexible	green, with brown spots, slightly flexible
S12	307 to 330			light brown and opaque, rigid rough coat, slightly flexible	
S13				brown and opaque, and rigid rough coat	pale brown or brown, rigid
Dispersed				green to brown, opaque, and rigid rough coat	brown or grey, rigid, often twisted.

Figure 1. Morphological characteristics of *P. pluviosa* fruits and seeds in 13 stages of maturation. The pods were harvested directly from the branches of marked flowers and the estimated time to reach each stage was calculated after two periods of observation: Sep/09 to Aug/10 (2009/10) and Sep/10 to Jul/11 (2010/11). Bars: seeds= 1 cm, fruits= 2 cm.

Despite the long time needed to complete seed development and maturation, the major changes were observed just at the end of the process when the fruits become brown at the same time they become rigid (S13), and could indicate last stage before dehiscence. After seed dispersion, the fruits are usually twisted, dry and detached from the mother plant. The seed coat of *P. pluviosa* was green until 300 DAA, when turned completely to brown (Figure 1). It was possible to find some dispersed seeds with green color. For a practical purpose aiming to identify seed maturity, the color of the fruits could be grouped into five main stages (S1, S2 to S3, S4 to S9, S10 to S12 and S13), as shown in Figure 1.

The size of the seeds showed a different pattern of development when compared to fruit dimensions (Figure 2). Length, width and thickness of seeds reached maximum values at the end of maturation (Figures 2B, D, F), precisely

at stage 12, and could indicate that the point of maximum seed dry matter accumulation is close to shedding. On the other hand, there was no significant variation in fruit size during almost all the maturation process (Figures 2A, C), except thickness, which slightly increased until 300 DAA (Figure 2E), and could be related to seed dry matter accumulation (Figure 3B). The decrease in fruit thickness and in seed size at late maturation (Figures 2B, D, F) could be related to water loss (Figure 3A), corresponding to the natural desiccation at the end of seed maturation.

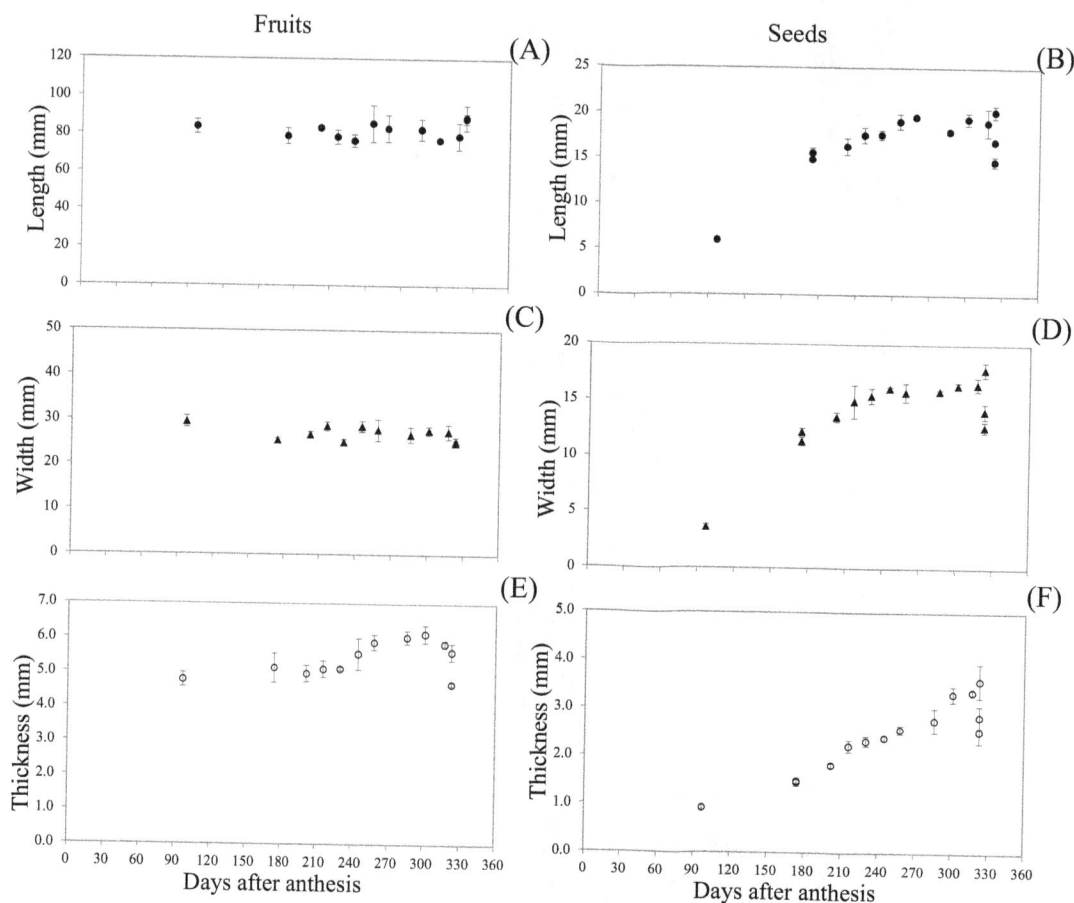

Figure 2. Length, width and thickness of *P. pluviosa* fruits (A, C and E) and seeds (B, D and F) collected from Sep/10 to Jul/11 (2010/11) in different stages at Rubião Jr. *campus*, Botucatu (SP, Brazil). Data are mean ±SD of three replicates of five fruits or fifteen seeds.

Indeed, water content gradually decreased from the beginning (*ca.* 3.5 g. g⁻¹) to the end (*ca.* 0.6 g. g⁻¹) of maturation, mainly in the last stages (Figure 3A). It is interesting to note that, despite the reduced water content, the water potential remained unchanged during most part of the maturation process (Figure 3A). In recently dispersed seeds (RDS), the water content and water potential reached the lowest values obtained in this study (0.11 and 0.09 g. g⁻¹, -64.5 and -73.4 MPa, respectively, Figure 3A). Dry matter

accumulation, similar in both analyzed periods, increased gradually until the last month of analysis. However, seeds of 2010/11 had higher dry matter than 2009/10, mainly at stage 12, being 0.376 and 0.261 g. seed⁻¹, respectively (Figure 3B). These could be associated with differences in environmental conditions, mainly rainfall and temperature as also observed for other species (Daws et al., 2004; Martins et al., 2009). In *Eugenia pyriformis*, for example, seed dry matter accumulation was influenced by temperature variations between two consecutive years (Lamarca et al., 2013). In *P. pluviosa*, besides temperature, rainfall distribution also seemed to affect dry mass accumulation once the highest range of temperature

and more defined rainfall frequency were registered in 2010/11 (Figure 4B). However, the evaluation of temperature and rainfall variations in *P. pluviosa* seed development needs more investigation with an experimental approach.

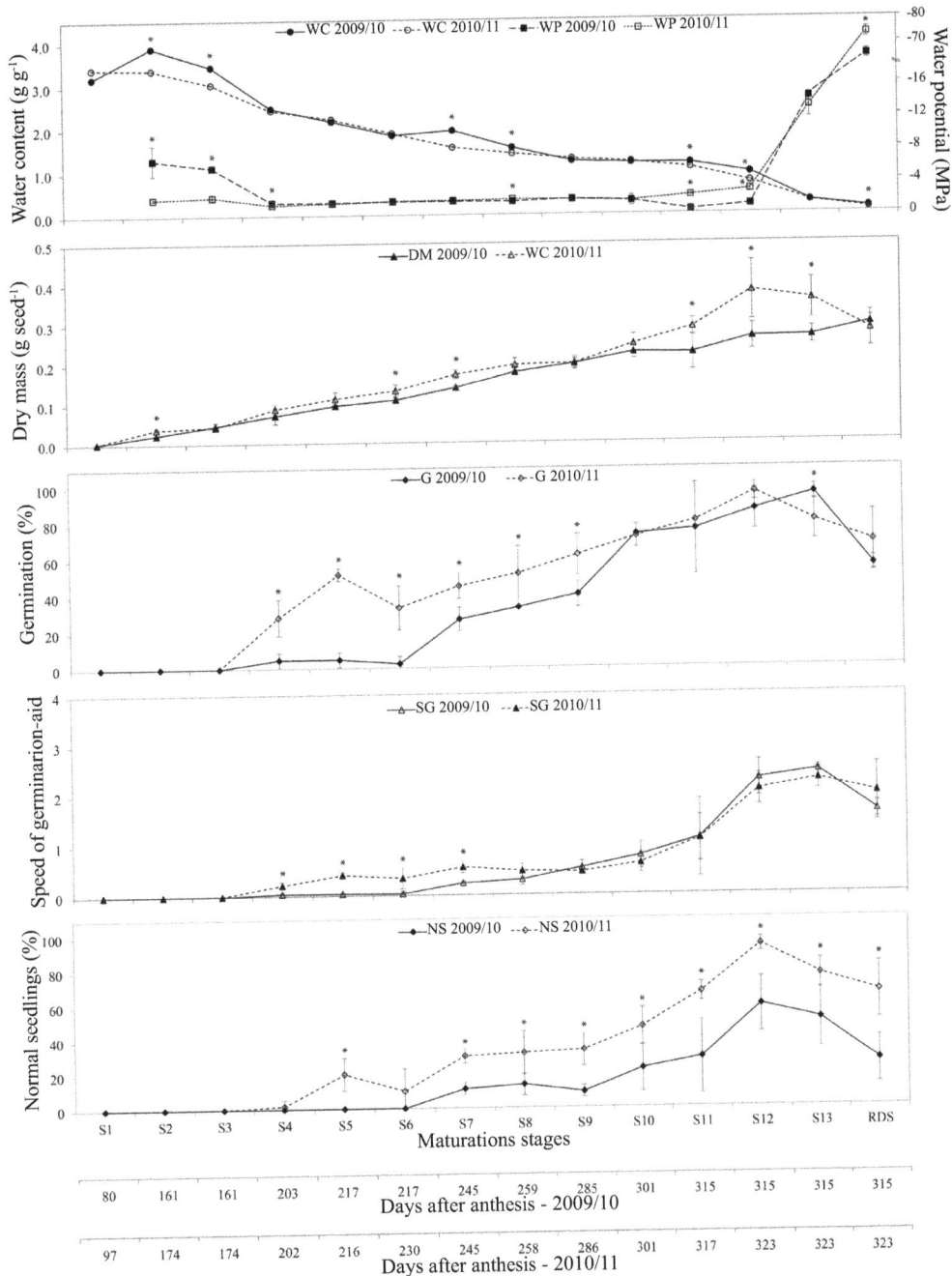

Figure 3. Physical and physiological characterization (mean ±SD) of *P. pluviosa* seeds at each stage of maturation in two consecutive cycles: Sep/09 to Aug/10 (2009/10) and Sep/10 to Jul/11 (2010/11). Water content (WC - g. g^{-1}) and water potential (WP - MPa) - (A); Dry matter (DM - g. seed^{-1}) - (B); Germination (G - %) - (C); Speed of germination-aid (SG) - (D); Normal seedlings (NS - %) - (E). Asterisks indicate significant difference among the maturation periods (*t* test, $P < 0.05$, $n = 4$).

The higher percentage of *P. pluviosa* seeds capable to germinate in the second period of analysis (2010/11) seems to be related with optimal environmental conditions during seed development (Figure 3C), when higher temperature ranges were registered, especially after 200 DAA (Figure 4). Increase of air temperature during seed development increased germinability of various species (Fenner, 1991; Martins et al., 2009), while under lower temperature (Figure 4A) less dry matter is accumulated and seeds have lower capacity to produce normal seedlings, as also reported by Kermode (1990) for a number of species. However, seeds obtained at 300 DAA in both cycles had the same germination percentage (around 70%) and reached maximal values (100%) at stages 12 and 13 (Figures 3C, D). The main difference between the two maturation cycles regards to seed capability to produce normal seedlings, which was higher in 2010/11 (Figure 3E).

Seeds of *P. pluviosa* are dispersed in the driest period of the year (July to August, Figure 4) in the southeast region of Brazil, which is consistent with seed tolerance to water stress, as suggested by Daws et al. (2004); Dussert et al. (2000) and Pritchard et al. (2004) for other species. Moreover, minimal temperatures at the time of seed dispersal in the regions of natural occurrence of this species are typically around 10 °C (INMET, 2015). Despite their short storability at room temperature (8 months - Figliolia et al., 2001), seeds of *P. pluviosa* could be viable in natural environment until the next rainy period (December to January, Figure 4).

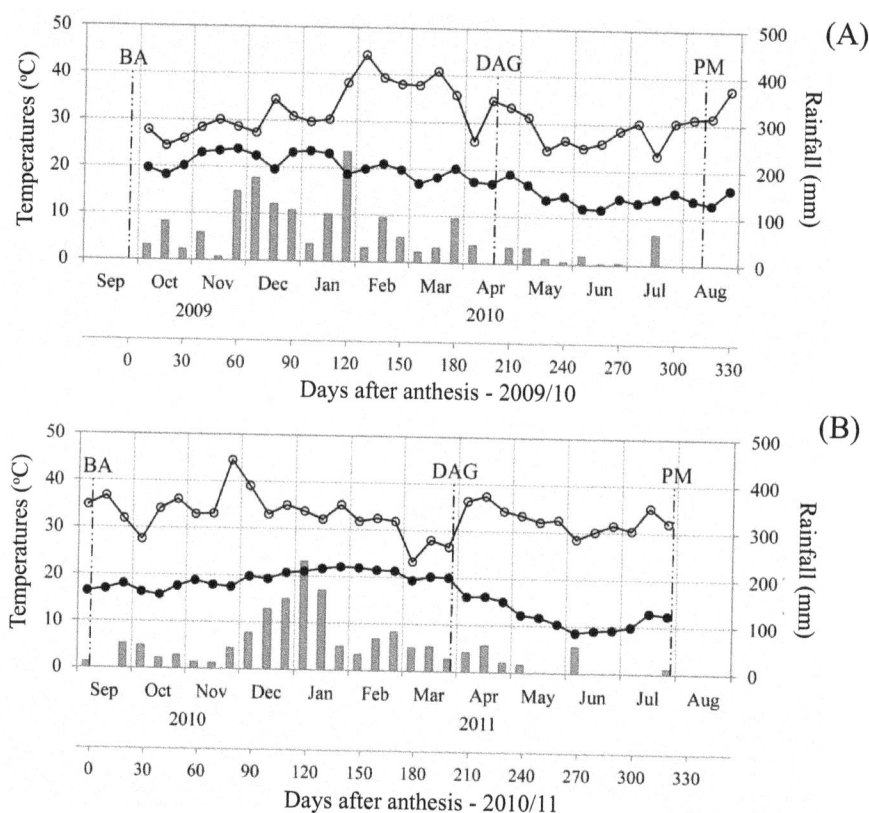

Figure 4. Meteorological data from the study site at Botucatu (SP, Brazil), in two complete periods of *P. pluviosa* seeds maturation: Sep/09 to Aug/10 (2009/10) - (A); Sep/10 to Jul/11 (2010/11) - (B). Maximum (○) and minimum (●) temperatures and rainfall (columns). The beginning of the anthesis (BA), development of ability to germinate (DAG), and physiological maturity point (PM) are indicated to each analyzed period.

Poincianella pluviosa RDS, as shown in this study, are not the most suitable material for storage and seedling production as those from stage 13, since RDS had lower rates of germination and seedling development (Figures 3C, E). Such seeds do not present physical dormancy (Ferraz-Grande and Takaki, 2006) as mentioned before and deteriorate a week after the germination tests, indicating biological degradation after shedding. Additionally, the RDS lot contained green seeds possibly due to asynchronous maturation of *P. pluviosa* seeds, similarly to what was reported for the flowering and fruiting processes of this species (Lewis and Gibbs, 1999). Therefore, some immature seeds can be dispersed with the mature ones, resulting in a heterogeneous lot with lower germination and seedling production. The implication of dispersal of immature seeds in neo-tropical forests has been discussed, emphasizing that the degree of immaturity in which seeds are detached from the mother plant is crucial for desiccation tolerance and seed longevity in such environments (Barbedo et al., 2013).

Seeds of *P. pluviosa* contained the highest dry mass (*i.e.*, physiological maturity, Bewley et al., 2013) between 315-323 DAA (stage 12), immediately before the natural desiccation. Some authors consider that the increase in seed quality, including seed longevity, can occur between periods of maximum accumulation of dry matter and seed dispersal (Hay and Probert ,1995; Probert et al., 2007). According to this concept, seeds of *P. pluviosa* should be harvested at stage 13, when natural desiccation already occurred and their water content was close to 0.26 g. g^{-1} (Figure 3A).

While tolerant to desiccation, seeds of *P. pluviosa* do not exhibit some other characteristics typically found in classic models of tolerant seeds. They accumulate cyclitols instead of raffinose family oligosaccharides (Silva, JPN unpublished data), both related to membrane stabilization in the dry state (Obendorf, 1997). On the other hand, the maturation process of *P. pluviosa* is so extensive that exceeds the reported viability of such seeds during storage at room temperature (Figliolia et al., 2001).

Altogether, our findings highlight that *P. pluviosa* seeds have the ability to develop normal seedlings very early during the long maturation cycle. This behavior could propitiate maintaining seeds able to produce new plantlets even during the maturation process, possibly allowing seedling establishment in different periods and climatic conditions. Future work will enable the determination of the viability of immature seeds and seedlings in environmental conditions and under stressed treatments, such as drought. The information provided here contributes to a better understanding of seed behavior allowing the possibility to obtain high quality seeds of this important neotropical species.

Conclusions

The complete maturation process of *P. pluviosa* seeds was attained 315-330 days after anthesis (DAA). The maximum dry mass and seed vigor were reached close to 320 DAA, although high production of seedlings was obtained before physiological maturity.

Acknowledgements

The authors thank the Administration of the Instituto de Biociências, Rubião Jr *Campus*, for the permission to harvest seeds and Juliana Iassia Gimenez for helping to collect fruits. This research work is part of JPN Silva PhD thesis supported by a fellowship from Coordenação de Aperfeiçoamento de Pessoal de Nível Superior - CAPES. CJ Barbedo is a research fellow of Conselho Nacional de Desenvolvimento Científico e Tecnológico - CNPq. This research work was financially supported by CNPq n° 478298/2011.

References

BARBEDO, C.J.; BILIA, D.A.C.; FIGUEIREDO-RIBEIRO, R.C.L. Tolerância à dessecação e armazenamento de sementes de *Caesalpinia echinata* L. (pau-brasil). *Brazilian Journal of Botany*, v.25, p.431-439, 2002. http://www.scielo.br/pdf/rbb/v25n4/a07v25n4.pdf

BARBEDO, C.J.; CENTENO, D.C.; FIGUEIREDO-RIBEIRO, R.C.L. Do recalcitrant seeds really exist? *Hoehnea*, v.40, n.4, p.583-595, 2013. http://www.scielo.br/scielo.php?script=sci_arttext&pid=S2236-89062013000400001

BEWLEY, J.D.; BRADFORD, K.J.; HILHORST, H.W.M.; NONOGAKI, H. *Seeds:* Physiology of Development, Germination and Dormancy. New York: Springer, 2013. 392p. http://www.springer.com/new+%26+forthcoming+titles+%28default%29/book/978-1-4614-4692-7

BONJOVANI, M.R.; BARBEDO, C.J. Sementes recalcitrantes: intolerantes a baixas temperaturas? Embriões recalcitrantes de *Inga vera* Willd. subsp. affinis (DC.) T.D.Penn. toleram temperatura sub-zero. *Brazilian Journal of Botany*, v.32, p.345-356, 2008. http://dx.doi.org/10.1590/S0100-84042008000200017

BUTLER, L.H.; HAY, F.R.; ELLIS, R.H.; SMITH, R.D. Post-abscission, pre-dispersal seeds of *Digitalis purpurea* remain in a developmental state that is not terminated by desiccation *ex planta*. *Annals of Botany*, v.103, p.785-794, 2009. http://aob.oxfordjournals.org/content/early/2009/01/09/aob.mcn254.full

CEPAGRI. Centro de Pesquisas Meteorológicas e Climáticas Aplicadas à Agricultura. Clima dos Municípios Paulistas, Campinas. http://www.cpa.unicamp.br/outras-informacoes/clima_muni_463.html. Accessed on: Feb 20th, 2015.

CHAPIN, M.H. Flowering and fruiting phenology in certain palms. *Palms*, v.43, p.161-165, 1999. http://www.cabdirect.org/abstracts/20000310066.html

CORTE, V.B.; BORGES, E.E.L.; PONTES, C.A.; LEITES, I.T.A.; VENTRELLA, M.C.; MATHIAS, A.A. Mobilização de reservas durante a germinação das sementes e crescimento das plântulas de *Caesalpinia peltophoroides* Benth. (Leguminosa e- Caesalpinoideae). *Revista Árvore*, v.30, p.941-949, 2006. http://www.scielo.br/scielo.php?pid=s0100-67622006000600009&script=sci_arttext

CORVELLO, W.B.V.; VILLELA, F.A.; NEDEL, J.L.; PESKE, S.T. Maturação fisiológica de sementes de cedro (*Cedrela fissilis* Vell.). *Revista Brasileira de Sementes*, v.21, p.23-27, 1999. http://www.abrates.org.br/revista/artigos/1999/v21n2/artigo04.pdf

DAWS, M.I.; LYDALL, E.; CHMIELARZ, P.; LEPRINCE, O.; MATTHEWS, S.; THANOS, C.A.; PRITCHARD, H.W. Developmental heat sum influences recalcitrant seed traits in *Aesculus hippocastanum* across Europe. *New Phytologist*, v.162, p.157-166, 2004. http://onlinelibrary.wiley.com/doi/10.1111/j.1469-8137.2004.01012.x/full

DUSSERT, S.; CHABRILLANGE, N.; ENGELMANN, F.; ANTHONY, F.; LOUARN, J.; HAMON, S. Relationship between seed desiccation sensitivity, seed water content at maturity and climatic characteristics of native environments of nine *Coffea* L. species. *Seed Science Research*, v.10, p.293-300, 2000. http://journals.cambridge.org/action/displayAbstract?fromPage=online&aid=695464

EIRA, M.T.S.; SILVA, E.A.A.; CASTRO, R.D.; DUSSERT, S.; WALTERS, C.; BEWLEY, D.; HILHORST, H.W.M. Coffee seed physiology. *Brazilian Journal of Plant Physiology*, v.18, p.149-163, 2006. http://dx.doi.org/10.1590/S1677-04202006000100011

ELLIS, R.H.; PIETA FILHO, C. The development of seed quality in spring and winter cultivars of barley and wheat. *Seed Science Research*, v.2, p.9-15, 1992. http://journals.cambridge.org/action/displayAbstract?fromPage=online&aid=1351384

FENNER, M. The effects of the parent environment on seed germinability. *Seed Science Research*, v.1, p.75-84, 1991. http://dx.doi.org/10.1017/S0960258500000696

FERRAZ-GRANDE, F.G.A.; TAKAKI, M. Efeitos da luz, temperatura e estresse de água na germinação de sementes de *Caesalpinia peltophoroides* Benth. (Caesalpinoideae). *Bragantia*, v.65, p.37-42, 2006. http://www.scielo.br/scielo.php?script=sci_arttext&pid=S0006-87052006000100006

FIGLIOLIA, M.B.; SILVA, A.; AGUIAR, I.B.; PERECIN, D. Efeito do acondicionamento e do ambiente de armazenamento na conservação de sementes de sibipiruna. *Revista Brasileira de Horticultura Ornamental*, v.7, p.57-62, 2001. http://rbho.emnuvens.com.br/rbho/article/view/78

HAY, F.R.; PROBERT, R.J. Seed maturity and the effects of different drying conditions on desiccation tolerance and seed longevity in foxglove (*Digitalis purpurea* L.). *Annals of Botany*, v.76, p.639-647, 1995. http://aob.oxfordjournals.org/content/76/6/639

ISTA. International Seed Testing Association. International rules for seed testing. *Seed Science and Technology*, v.13, p.356-513, 2015. https://www.seedtest.org/en/international-rules-_content--1--1083.html

INMET. Instituto Nacional de Meteorologia. Ministério da Agricultura, Pecuária e Abastecimento, Brasília. http://www.inmet.gov.br/portal/index.php?r=bdmep/bdmep. Acessed on: Feb 20th, 2015.

KERMODE, A.R. Regulatory mechanisms involved in the transition from seed development to germination. *Critical Reviews in Plant Science*, v.9, p.155-195, 1990. http://www.tandfonline.com/doi/abs/10.1080/07352689009382286

LAMARCA, E.V.; PRATAVIERA, J.S.; BORGES, I.F.; DELGADO, L.F.; TEIXEIRA, C.C.; CAMARGO, M.B.P.; FARIA, J.M.R.; BARBEDO, C.J. Maturation of *Eugenia pyriformis* seeds under different hydric and thermal conditions. *Anais da Academia Brasileira de Ciências*, v.85, p.223-233, 2013. http://dx.doi.org/10.1590/S0001-37652013005000006

LEWIS, G.P. *Caesalpinia*, a revision of the Poincianella-Erythrostemon group. Kew: Royal Botanic Gardens, 1998. 233p.

LEWIS, G.P. *Poincianella* in Lista de Espécies da Flora do Brasil. Jardim Botânico do Rio de Janeiro. http://floradobrasil.jbrj.gov.br/jabot/floradobrasil/FB109778. Acessed on: Feb 20th, 2015.

LEWIS, G.P.; GIBBS, P. Reproductive biology of *Caesalpinia calycina* and *C. pluviosa* (Leguminosae) of the caatinga of north-eastern Brazil. *Plant Systematics and Evolution*, v.217, p.43-53, 1999. http://link.springer.com/article/10.1007%2FBF00984921

MAGUIRE, J.D. Speed of germination-aid in selection and evaluation for seedling emergence vigor. *Crop Science*, v.2, p.176-177, 1962. https://www.crops.org/publications/cs/abstracts/2/2/CS0020020176

MARTINS, C.C.; BOVI, M.L.A.; NAKAGAWA, J.; MACHADO, C.G. Secagem e armazenamento de sementes de juçara. *Revista Árvore*, v.33, p.635-642, 2009. http://www.scielo.br/scielo.php?script=sci_arttext&pid=S0100-67622009000400006

MICHEL, B.E.; KAUFMANN, M.R. The osmotic potential of polyethylene glycol 6000. *Plant Physiology*, v.51, p.914-916, 1973. http://www.plantphysiol.org/content/51/5/914.abstract

NEWTON, R.J.; BOND, W.J.; FARRANT, J.M. Seed development, morphology and quality in selected species of the nut-fruited Restionaceae. *South African Journal of Botany*, v.68, p.226-230, 2002. http://www.ajol.info/index.php/sajb/article/view/20524

OBENDORF, R.L. Oligosaccharides and galactosyl cyclitols in seed desiccation tolerance. *Seed Science Research*, v.7, p.63-74, 1997. http://journals.cambridge.org/action/displayAbstract?fromPage=online&aid=1293144

PÉREZ, H.E.; HILL, L.M.; WALTERS, C. An analysis of embryo development in palm: interactions between dry matter accumulation and water relations in *Pritchardia remota* (Arecaceae). *Seed Science Research*, v.22, p.97-111, 2012. http://dx.doi.org/10.1017/S0960258511000523

PONTES, C.A.; CORTE, V.B.; BORGES, E.E.L.; SILVA, A.G.; BORGES, R.C.G. Influência da temperatura de armazenamento na qualidade das sementes de *Caesalpinia peltophoroides* Benth. (sibipiruna). *Revista Árvore*, v.30, p.43-48, 2006. http://www.scielo.br/scielo.php?script=sci_arttext&pid=S0100-67622006000100006

PRITCHARD, H.W.; DAWS, M.I.; FLETCHER, B.J.; GAMÉNÉ, C.S.; MSANGA, H.P.; OMONDI, W. Ecological correlates of seed desiccation tolerance in tropical African dryland trees. *American Journal of Botany*, v.91, p.863-870, 2004. http://www.ncbi.nlm.nih.gov/pubmed/21653442

PROBERT, R.J.; ADAMS, J.; CONEYBEER, J.; CRAWFORD, A.; HAY, F. Seed quality for conservation is critically affected by pre-storage factors. *Australian Journal of Botany*, v.55, p.326-335, 2007. http://dx.doi.org/10.1071/BT06046

QUEIROZ, L.P. *Leguminosas da Caatinga*. Feira de Santana. Universidade Estadual de Feira de Santana, 2009. 914p. http://www.cabdirect.org/abstracts/20113044322.html

SALLON, S.; SOLOWEY, E.; COHEN, Y.; KORCHINSKY, R.; EGLI, M.; WOODHATCH, I.; SIMCHONI, O.; KISLEV, M. Germination, genetics, and growth of an ancient date seed. *Science*, v.320, p.1464, 2008. http://www.sciencemag.org/content/320/5882/1464.short

SANTANA, D.G.; RANAL, M.A. *Análise da germinação: um enfoque estatístico*. 1. ed. Brasília. Universidade de Brasília, 2004. 248p. http://www.editora.unb.br/lstDetalhaProduto.aspx?pid=26

SOUZA, C.D.; MARINHO, C.R.; TEIXEIRA, S.P. Ontogeny resolves gland classification in two Caesalpinoid legumes. *Trees*, v.27, p.801-813, 2013. http://link.springer.com/article/10.1007%2Fs00468-012-0835-z

5

Isoenzyme activity in maize hybrid seeds harvested with different moisture contents

Thaís Francielle Ferreira[1*], Valquíria de Fátima Ferreira[1], João Almir Oliveira[1], Marcos Vinícios de Carvalho[1], Leonardo de Souza Miguel[1]

ABSTRACT – The analysis of isoenzyme activity is an important monitoring and characterization tool of the physiological quality of seeds and to understand the deterioration. The purpose of this work was to study the isoenzyme expression allied to the quality of maize hybrid seeds harvested at different moisture levels and subjected to chemical treatment. A completely randomized experimental design was used with four replicates, in a 3x2 factorial arrangement with three moisture levels (45%, 40% and 35%), and two forms of seeds tillage (with and without treatment). Seeds from maize hybrids, semi-hard BM 810 and dented BM 3061, were used. Seeds were manually gathered on ears. Chemical treatment was performed with commercial products Maxin® + K-obiol® + Actellic®. Seed quality was assessed by moisture test, incidence of mechanicals damage, first count of germination, germination, emergence, emergence speed index, mean emergence time, accelerated aging, and electrical conductivity. Isoenzyme expressions were assessed by means of the systems superoxide dismutase (SOD), catalase (CAT), esterase (EST), alcohol dehydrogenase (ADH), malate dehydrogenase (MDH), peroxidase (PO) and α-amilase. Isoenzyme expressions are different, depending on moisture levels at harvest, the hybrid maize and seeds quality. Seeds treatment does not interfere in their isoenzymes expression.

Index terms: *Zea mays,* isoenzyme expression, physiological quality.

Atividade isoenzimática em sementes de milho híbridos colhidas com diferentes teores de água e tratadas

RESUMO – A análise da atividade enzimática é uma importante ferramenta no monitoramento e caracterização da qualidade fisiológica de sementes e no entendimento da deterioração. Objetivou-se com este trabalho estudar a expressão de isoenzimas aliada a qualidade de sementes de milho híbrido colhidas com diferentes teores de água e submetidas ao tratamento químico. O delineamento utilizado foi o inteiramente casualizado, com quatro repetições, em esquema fatorial 3x2, sendo três teores de água (45%, 40% e 35%), e duas formas de manejo das sementes (tratadas e não tratadas). Foram utilizadas sementes de milho dos híbridos: BM 810, semiduro e BM 3061, dentado, colhidas manualmente em espigas. O tratamento químico foi realizado com os produtos comerciais Maxin® + K-obiol® + Actellic®. A qualidade das sementes foi avaliada pelos testes de umidade, incidência de danos mecânicos, primeira contagem de germinação, germinação, emergência, índice de velocidade de emergência, tempo médio de emergência, envelhecimento acelerado e condutividade elétrica. As expressões isoenzimáticas foram avaliadas por meio dos sistemas superóxido dismutase (SOD), catalase (CAT), esterase (EST), álcool desidrogenase (ADH), malato desidrogenase (MDH), peroxidase (PO) e α-amilase. A expressão de isoenzimas é divergente dependendo do teor de água na colheita, do híbrido e da qualidade das sementes. O tratamento de sementes não interfere na expressão das isoenzimas.

Termos para indexação: *Zea mays*, expressão isoenzimática, qualidade fisiológica.

Introduction

The companies producing maize seeds every year have improved the cultivation system to obtain high quality seeds. This search triggers a high investment in technologies that enable a better use of time in relation to early harvesting.

The early harvest of maize seeds ensures higher quality due to less exposure to adverse environmental conditions, better use of planting areas, the possibility to vacate them earlier, besides enabling the planning of the drying processes, providing better utilization of the production and processing infrastructure (Ferreira et al., 2013).

[1]Departamento de Agricultura, UFLA, Caixa Postal 3037, 37200-000 – Lavras, MG, Brasil.
* Corresponding author <franthata@yahoo.com.br>

However, the seeds harvested close to physiological maturity, a period during which the seeds have maximum vigor level, show high water content and this involves the improvement of postharvest techniques so that there is not a reduction in seeds quality for, on a molecular level, a number of mechanisms contribute to the deterioration.

The enzymes involved in the deterioration, such as esterase (EST), malate dehydrogenase (MDH), alcohol dehydrogenase (ADH), catalase (CAT) and peroxidase (PO) have great potential as molecular markers to monitor and characterize the seeds physiological quality (Veiga et al., 2010), besides providing an understanding of the causes of reduced vigor and viability (Galvão et al., 2014).

The delayed harvest leads to increased decay, and this is evident by the reduction of the peroxidase enzyme activity, which provides antioxidant protection of seeds (Tunes et al., 2014), and by the increased ADH enzyme activity, which increases anaerobic respiration (Caixeta et al., 2014).

Therefore, with the early harvest and the high moisture content of the seeds, there may be a significant increase in respiratory rate and deterioration if the processes subsequent to harvesting are not properly conducted (Galvão et al., 2014).

According to Caixeta et al. (2014), on seeds processed and stored for eight months, there is increased activity of the malate dehydrogenase enzyme, because the deteriorating process is steeper.

The loss of peroxidase enzyme activity due to increased deterioration can make seeds more sensitive to the effect of oxygen and free radicals on membrane unsaturated fatty acids, which will cause degeneration of these ones and compromise seed vigor (Martins et al., 2011).

The joint analysis of several enzyme systems enables the verification of changes that occur inside the seeds when subjected to some kind of treatment that influences quality and productivity (Tunes et al., 2014).

Studies of isoenzyme expression in maize seeds harvested with different moisture contents and changes in enzyme activity due to seeds chemical treatment are scarce. Thus, the aim of this research was to study the activity of isoenzymes superoxide dismutase (SOD), catalase (CAT), esterase (EST), alcohol dehydrogenase (ADH), malate dehydrogenase (MDH), peroxidase (PO) and α-amylase at the expense of the hybrid maize seed quality, treated and untreated, harvested with different moisture contents.

Material and Methods

Hybrid seeds used were BM 810 and BM 3061, classified as semi-hard and toothed, produced by the company Biomatrix in the city of Paracatu, located North-West of the Brazilian state of Minas Gerais, whose geodetic coordinates are 17° 13′ 19″ S of latitude, 46° 52′ 30″ W of longitude and 1,008 meters of altitude.

At random points in the defined experimental area, manual harvesting of maize ears was done when the seeds had 45%, 40% and 35% water content. The multi-grain apparatus Grain Analysis Computer 2100 was used to ascertain the moisture of the seeds. After harvest, the maize ears went through mechanical grain husking in machine to take the stover of the brand CWA in the rotation of 312 rpm and drying in a stationary dryer at 35 °C until the seeds reached 22% water content and 42 °C until they reached 12% water content. After drying, the seeds were mechanically threshed and treated with Maxim® + K-obiol® + Actellic®, being 13.75 mL of Maxim XL, 0.45 mL of K-obiol, 0.45 mL of Actellic, 2.5 mL of dye, 100 mL of water (turned to 100 Kg). The treatments are specified in Table 1.

Table 1. Description of the treatments obtained by the different moisture contents at the time of harvest of maize (45%, 40% and 35%) with and without chemical treatment (T and UT), hybrids 1 and 2, water content (WC) .

Treatments obtained	Hybrid	Harvest moisture (M)	Treatment Chemical	(WC %)
1	BM 810	45%	T	9.7
2	BM 810	40%	T	10.5
3	BM 810	35%	T	11.3
4	BM 810	45%	UT	9.1
5	BM 810	40%	UT	10.5
6	BM 810	35%	UT	11.1
7	BM 3061	45%	T	9.7
8	BM 3061	40%	T	10.1
9	BM 3061	35%	T	10.1
10	BM 3061	45%	UT	10.2
11	BM 3061	40%	UT	10.4
12	BM 3061	35%	UT	10.1

The seeds were submitted to manual classification in sieves at Central Laboratory of Seeds of the Department of Agriculture of the Federal University of Lavras, MG, Brazil. For hybrid BM 810, the seeds used were the ones retained in the oblong sieve 18/64 and for hybrid BM 3061, the ones retained in the circular sieve 20/64, and these sieves were selected due to the higher amount of retained seeds.

The determinations performed for the assessment of the seeds quality were: *Moisture content*, done by the oven method, described in the Rules for Seed Testing, with results expressed as mean percentage per treatment (Table 1) (Brasil, 2009); *Incidence of mechanical damage*, where seeds were immersed in the dye solution *amaranth®* at 0.1%, for 2 minutes, then rinsed under running water and assessment according to the methodology described by Oliveira et al. (1998); *Germination*, according to Brasil (2009), using 50 seeds for each of the four replications and assessment held at the fourth (*first count of germination*) and on the seventh day after sowing, computing the percentage of normal seedlings; *Accelerated aging*, with the use of gerbox-type plastic boxes adapted with hanging aluminum screen – in each germination box were added 40 mL of water and a single layer of seeds on the entire screen. They were then kept in a B.O.D. (Biochemical Oxygen Demand)-type germination chamber at 42 °C for 96 hours (Marcos-Filho, 1999) and after this period, the seeds were submitted to the germination test; *Seedling emergence*: the seeds were sown in plastic trays containing soil + sand as substrate, in the 2:1 ratio, moistened to 60% of the holding capacity. The trays were kept in the chamber at the temperature of 25 °C and a photoperiod of 12 hours, with daily assessments of emergence of normal seedlings and a final score at 14 days after sowing. The final emergence percentage (% E), the mean emergence time (MET) and the emergence speed index (ESI) were considered (Maguire, 1962); *Electrical conductivity:* it was performed according to Vieira and Krzyzanowski (1999), with the aid of a conductivity meter Digimed CD-21 and the results expressed in $\mu S.\ cm^{-1}.g^{-1}$; for the seeds treated, the conductivity value was subtracted from the blank test without seeds, only with the treatment product diluted in water so as to exclude the interference of seed treatment products in the values obtained.

The enzyme activities were also analyzed: α-amylase, catalase, esterase, peroxidase, superoxide dismutase, malate dehydrogenase and alcohol dehydrogenase. For each system were used two samples of 50 seeds from each treatment. The seeds were soaked in the presence of antioxidant PVP and liquid nitrogen and subsequently stored at -86 °C. For the extraction of enzymes catalase, esterase, peroxidase, superoxide dismutase, malate dehydrogenase and alcohol dehydrogenase was used buffer Tris HCl 0.2 M pH 8.0 + 0.1% of β-mercaptoethanol at the ratio of 250 μL per 100 mg of seeds.

For the extraction of α-amylase enzyme, the seeds were germinated on paper roll for a period of 70 hours. After this period, plumule and root seeds were discarded and the remainder was macerated in mortar on ice, in the presence of liquid nitrogen. For extraction, 200 mg of the powder of germinated seeds were resuspended in 600 μL of the extraction buffer (Tris-HCl 0.2 M, pH 8.0 + 0.4% of PVP). The material was homogenized in a stirrer and kept in a refrigerator overnight, followed by centrifugation at 16,000 x g for 60 minutes at 4 °C.

The electrophoretic technique was performed in polyacrylamide gels system at 7.5% (separating gel) and 4.5% (concentrating gel). For the system α-amylase was added 0.5% of soluble starch in the polyacrylamide gel. The gel/electrode system used was Tris-glycine pH 8.9. 15 uL of the sample supernatant were applied and the technique was conducted at 150 V for 4 hours.

Gels were developed for enzymes catalase, esterase, malate dehydrogenase, alcohol dehydrogenase, superoxide dismutase, peroxidase α-amylase (Alfenas et al., 2006).

A completely randomized design was used in a 3 x 2 factorial arrangement, whose factors were the seed moisture content at harvest (45%, 40% and 35%), seed treatment (treated and untreated) with four replicates per treatment. The hybrids were analyzed separately. For the isoenzymes expression, visual analysis of the expression bands was performed.

To compare the averages, Tukey test at 5% probability was used, by software Sisvar (Ferreira, 2011). For testing water content and enzymes, statistical analyses of the data were not performed.

Results and Discussion

In the physiological tests of first count of germination, germination, seedling emergence, emergence speed index and electrical conductivity there was significance only for the harvest moisture factor, i.e., the seed treatment did not interfere in the results of both hybrids. There was interaction of harvest moisture and seeds treatment only for the accelerated aging test of hybrids BM 810 and BM 3061.

For hybrid BM 810, the seeds that were harvested at 45% water content showed a higher percentage of mechanical damage, particularly those classified into more serious damage (grades 3 and 2) and the seeds harvested at 40% and 35% water content were classified undamaged (grade 0) (Figure 1).

The germination of the seeds harvested at 35% water content was lower than the germination of the seeds that were harvested at 40% and 45% water content. The vigor of the seeds harvested at 45% water content was higher than the vigor of the seeds harvested at 35% and 40% water content in the tests of first count of germination, seedling emergence and electrical conductivity. The emergence speed index was the same, regardless of the water content in which the seeds were harvested (Table 2). The vigor of the treated and untreated seeds assessed by the accelerated aging test was higher in seeds at 35% water content. For the seed harvested at 40% and 45% moisture content, the treatment showed better performance in the accelerated

aging test (Table 4).

The seeds of hybrid BM 3061 were more resistant to severe mechanical damage, although slight damage has been assessed (grade 1) (Figure 2). It is observed that the higher the water content of the seeds when harvested, the higher the incidence of damage considered serious (grade 3). The incidence of serious damage to seeds harvested at 35% and 40% water content did not differ. This high incidence of mechanical damage was caused above all by the steps of husking and threshing which, being mechanized, cause injuries in seeds.

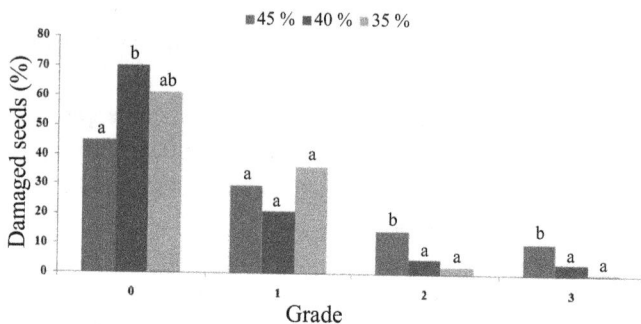

Figure 1. Incidence of mechanical damage in maize seeds hybrid BM 810 – semi-hard, harvested with different moisture contents (45%, 40% and 35%) and classified into four levels (0, 1, 2 and 3) in accordance with the intensity of the damage. Means followed by the same letter for each grade do not differ by Tukey test at 5% probability.

Table 2. Average values of first count of germination (FC), germination (G), emergence (E), emergence speed index (ESI) and electrical conductivity (EC) of maize seeds hybrid BM 810, harvested with different moisture content (M).

M (%)	FC (%)	G (%)	E (%)	ESI	EC (μS.cm^{-1}.g^{-1})
35	98 a	89 b	98 a	11.17 a	26.24 a
40	96 a	97 a	97 ab	11.19 a	42.30 b
45	85 b	98 a	92 b	11.81 a	66.96 c
CV (%)	4.35	3.63	3.90	6.72	19.14

Means followed by the same letter in the column do not differ by Tukey test at 5% probability.

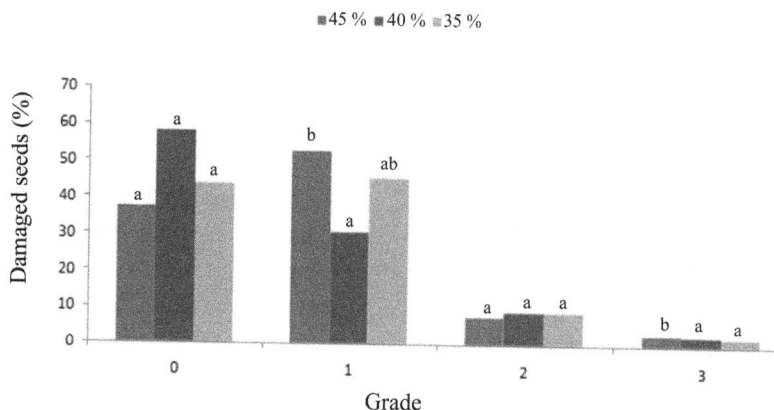

Figure 2. Incidence of mechanical damage in maize seeds hybrid BM 3061 – toothed, harvested with different moisture contents and classified into different grades. Means followed by the same letter for each grade do not differ by Tukey test at 5% probability.

The quality of the seeds harvested at 40% water content was higher than those harvested at 35% and 45% water content in the first count of germination, germination and emergence speed index. The electrical conductivity of the seeds at 45% water content was superior to others (Table 3). By the accelerated aging test, the vigor of treated seeds overcame the vigor of untreated seeds in seeds harvested at 35% and 40% water content. In treated seeds, there was no vigor difference by the accelerated aging test, taking into consideration the water content in which the seeds were harvested (Table 4). The poor quality of seeds harvested at 45% water content is related to their susceptibility to mechanical damage during the husking and threshing mechanical processes. The damage caused in the seeds is a gateway to organisms that are harmful to quality and it is due to this fact that the vigor of untreated seeds was lower in the accelerated aging test when compared to the treated seeds

Table 3.   Average values of first count of germination (FC), germination (G), emergence (E), emergence speed index (ESI) and electrical conductivity (EC) of maize seeds hybrid BM 3061, harvested with different moisture content (M).

M (%)	FC (%)	G (%)	E (%)	ESI	EC (μS cm^{-1}.g^{-1})
35	98 ab	99 ab	99 a	11.77 ab	17.05 c
40	100 a	100 a	100 a	12.13 a	20.78 b
45	96 b	98 b	98 a	11.73 b	29.81 a
CV (%)	1.72	1.39	1.61	2.41	9.73

Means followed by the same letter in the column do not differ by the Tukey test at 5% probability.

Table 4.   Mean values of normal seedlings germinated after accelerated aging of hybrid maize seeds, treated and untreated, harvested with different moisture content (M) for each hybrid.

M (%)	BM 810		BM 3061	
	T	UT	T	UT
35	100 aA	97 aA	98 aA	59 bB
40	91 bA	81 bB	93 aA	83 aB
45	84 bA	64 cB	94 aB	67 bA
CV (%)	4.56		10.20	

Means followed by the same lowercase letter in the column and uppercase letter on the row do not differ by Tukey test at 5% probability.

There was a higher intensity of the bands disclosed for enzymes superoxide dismutase (SOD) and catalase (CAT) in seeds of hybrid 1 (BM 810) when harvested at 45% water content. Unlike hybrid 2 (BM 3061), in which lower intensity of the bands of the enzymes in question was observed (Figure 3). The intensity of the bands disclosed for SOD was higher in BM 810 in the seeds harvested at 45% water content. There was a higher expression of enzyme catalase for these seeds that showed the highest incidence

of mechanical damage and thus further deterioration, when compared with the ones at 40% and 35% water content.

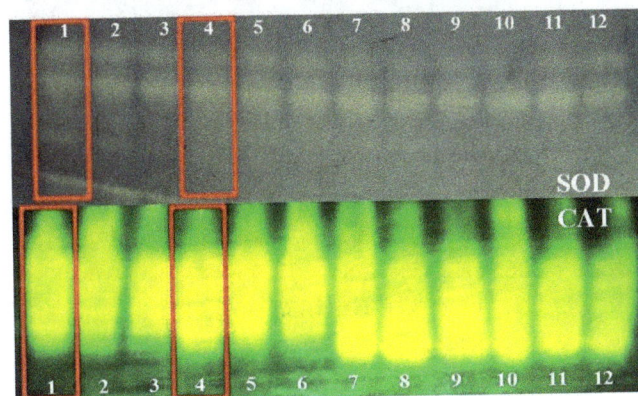

Figure 3. Enzymatic patterns of maize seeds harvested with different moisture contents (45%, 40% and 35%) treated (T) and untreated (UT). Hybrid 1 – BM 810 (1 – 45% T, 2 – 40% T, 3 – 35% T, 4 – 45% UT, 5 – 40% UT and 6 – 35% UT); Hybrid 2 – BM 3061 (7 – 45% T, 8 – 40% T, 9 – 35% T, 10 – 45% UT, 11 – 40% UT and 12 – 35% UT), disclosed for superoxide dismutase (SOD) and catalase (CAT).

With the increase of injuries in the seeds, the synthesis induction processes and activity of enzymes and hormones may be affected, reducing the activity of important enzymes in the respiratory process and removing free radicals, reducing the seeds physiological quality (Galvão et al., 2014). Marcos-Filho (2005) states that the increase of the enzyme catalase activity indicates the evolution of the deterioration due to the need for more intense action of the participating enzymes of the antioxidant complex.

These results corroborate the ones by Veiga et al. (2010) who stated that among the causes of decay events are changes in enzyme activity, which enable monitoring and characterizing the seeds quality.

It is observed in Figure 4 an increased expression of peroxidase enzyme in seeds harvested at 35% water content. The enzyme activity increases when the seeds water content is reduced to create protection mechanisms. Divergent results were found by Galvão et al. (2014), who observed that the delayed harvest reduced the peroxidase enzyme activity, indicating the occurrence of further deterioration of the seeds with the seeds delayed harvest.

The balance between the generation and removal of radicals during drying and storage of seeds, relates to its longevity (Martins et al., 2011), suggesting that the difference observed between the seeds harvested with different water content is related to the seeds age.

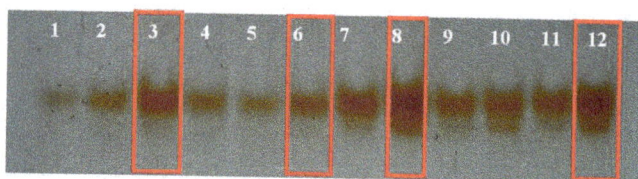

Figure 4. Enzymatic patterns of maize seeds harvested with different moisture contents (45%, 40% and 35%) treated (T) and untreated (UT). Hybrid 1 – BM 810 (1 – 45% T, 2 – 40% T, 3 – 35% T, 4 – 45% UT, 5 – 40% UT and 6 – 35% UT); Hybrid 2 – BM 3061 (7 – 45% T, 8 – 40% T, 9 – 35% T, 10 – 45% UT, 11 – 40% UT and 12 – 35% UT), disclosed for peroxidase.

The membrane systems of seeds harvested at 45% water content had a greater effect of exposure to oxygen due to the higher content of mechanical damage that they suffered during the process of husking, resulting in a lower expression of peroxidase enzyme, which can be also proven by the results of physiological tests where their poor performance is noted in both hybrids studied (Martins et al., 2011).

Most of the enzyme peroxidase activity in hybrid BM 3061 was observed when seeds were harvested at 40% water content and this may be related to their better quality and higher vigor, which can be proven by the germination tests and emergence speed index (Table 3).

In addition to protective enzymes, there are the deteriorative enzymes. Among them is esterase, which promotes hydrolysis of esters, where these reactions are directly related to the lipids metabolism. As an example, membrane phospholipids. Esterase promotes the destabilization of the lipid bilayer, accentuating the deterioration process (Vieira et al., 2006). Greater expression of this enzyme was observed in seeds harvested at 45% moisture content for both hybrids studied, which makes it clear that the seeds immaturity and the mechanical damage suffered during the processing and drying procedures contributed to the high activity of this enzyme (Figure 5).

For enzymes alcohol dehydrogenase (ADH) and malate dehydrogenase (MDH), there was a greater intensity of bands on seeds harvested at 45% water content (Figure 6). The absence of oxygen promotes the beginning of the fermentation metabolism by induction of ADH, wherein acetaldehyde is reduced to ethanol by nicotinamide adenine dinucleotide (NAD). According to Veiga et al. (2010) this enzyme is important since it converts acetaldehyde into ethanol, a compound with less toxicity, and reduces the speed of the deterioration process. Thus, the seeds are less susceptible to the

deleterious effects of acetaldehyde with the highest activity of ADH (Carvalho et al., 2014).

It was observed that the changes occurring in the ADH enzyme activity seem to be due more to the mechanical damage that impairs the seeds metabolism, where higher enzyme activity was observed in both hybrids in the seeds harvested at 45% water content in both treated and untreated seeds.

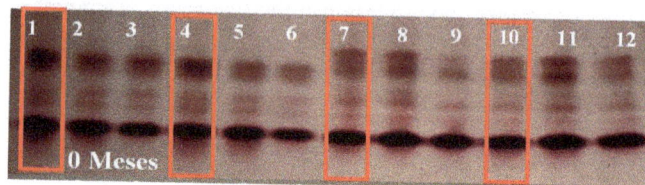

Figure 5. Enzymatic patterns of maize seeds harvested with different moisture contents (45%, 40% and 35%) treated (T) and untreated (UT). Hybrid 1 – BM 810 (1 – 45% T, 2 – 40% T, 3 – 35% T, 4 – 45% UT, 5 – 40% UT and 6 – 35% UT); Hybrid 2 – BM 3061 (7 – 45% T, 8 – 40% T, 9 – 35% T, 10 – 45% UT, 11 – 40% UT and 12 – 35% UT), disclosed for esterase.

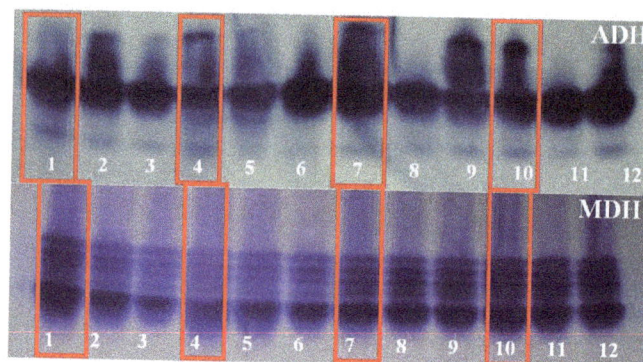

Figure 6. Enzymatic patterns of maize seeds harvested with different moisture contents (45%, 40% and 35%) treated (T) and untreated (UT). Hybrid 1 – BM 810 (1– 45% T, 2 – 40% T, 3 – 35% T, 4 – 45% UT, 5– 40% UT and 6 35% UT); Hybrid 2 – BM 3061 (7 – 45% T, 8 – 40% T, 9 – 35% T, 10 – 45% UT, 11 – 40% UT and 12 – 35% UT), disclosed for ADH – alcohol dehydrogenase and MDH – malate dehydrogenase.

Carvalho et al. (2014), when studying the expression of malate dehydrogenase (MDH) in soybean seeds, noted that the main differences were found with the advancement of the storage period at six and eight months and that in these storage periods the seeds stored in cold chamber had higher

MDH activity compared to the ones stored in conventional warehouse, due to the higher stress suffered in uncontrolled conditions, particularly at eight months of storage. Vieira et al. (2013) found decreased activity of MDH, from six months of storage at 10 °C and 25 °C, but with greater effect at nine and twelve months of storage at 25 °C.

The highest expression of MDH enzyme in seeds harvested at 45% water content can be related to the physiological quality, since this fact is directly related to the incidence of mechanical damage, which was higher in these seeds. This fact is due to the damage to mitochondrial membranes, and this organelle is the one that is more susceptible to peroxidation. The increased activity may have occurred because of increased respiration in the seeds that were in the deteriorating process, since the enzymes involved in respiration can be activated in lower quality seeds (Tunes et al., 2014).

With respect to α-amylase, lower expression was observed in seeds harvested at 45% moisture content in both hybrids studied (Figure 7). The development of α-amylase activity is an important event that can be detected during early seed germination, and its main role is to provide substrates for seedling use until it becomes photosynthetically self-sufficient (Caixeta et al., 2014).

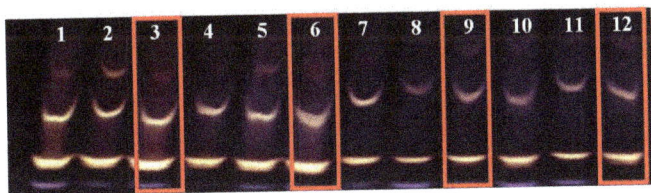

Figure 7. Enzymatic patterns of maize seeds harvested with different moisture contents (45%, 40% and 35%) treated (T) and untreated (UT). Hybrid 1 – BM 810 (1 – 45% T, 2 – 40% T, 3 – 35% T, 4 – 45% UT, 5 – 40% UT and 6 – 35% UT); Hybrid 2– BM 3061 (7 – 45% T, 8 – 40% T, 9 – 35% T, 10 – 45% UT, 11 – 40% UT and 12 – 35% UT), disclosed for α–amylase.

Seeds harvested at 35% moisture content showed higher enzyme activity, because as the seeds lose water during the maturation process, they gain greater tolerance to desiccation and thus become tolerant to high drying temperature, which, according to Rosa et al. (2005), makes these seeds present a higher synthesis of the α-amylase enzyme than intolerant seeds.

Conclusions

The isoenzymes expression varies according to the hybrid and the seeds quality.

There is an increase in the activity of α-amylase and peroxidase enzymes, and decreased activity of enzymes superoxide dismutase (SOD), catalase (CAT), esterase (EST), alcohol dehydrogenase (ADH) and malate dehydrogenase (MDH) as the seeds water content at harvest is reduced.

Seeds treatment does not interfere with the isoenzymes expression.

References

ALFENAS, A.C. *Eletroforese e marcadores bioquímicos em plantas e microorganismos.* Viçosa, MG: UFV, 2006. 627 p.

BRASIL. Ministério da Agricultura, Pecuária e Abastecimento. *Regras para análise de sementes.* Ministério da Agricultura, Pecuária e Abastecimento. Secretaria de Defesa Agropecuária. Brasília: MAPA/ACS, 2009. 395 p. http://www.agricultura.gov.br/arq_editor/file/2946_regras_analise__sementes.pdf

CAIXETA, F.; VON PINHO, E.V.R; GUIMARÃES, R.M; PEREIRA, P.H.A.R.; CATÃO, H.C.R.M.; CLEMENTE, A.C.S.. Determinação do ponto de colheita na produção de sementes de pimenta malagueta e alterações bioquímicas durante o armazenamento e a germinação. *Científica*, v.42, n.2, p.187-197, 2014. http://www.cientifica.org.br/index.php/cientifica/article/view/537

CARVALHO, E.R.; MAVAIEIE, D.P.R.; OLIVEIRA, J.A.; CARVALHO, M.V.; VIEIRA, A.R. Alterações isoenzimáticas em sementes de cultivares de soja em diferentes condições de armazenamento. *Pesquisa Agropecuária Brasileira*, v.49, n.1, p.967-976, 2014. http://www.scielo.br/pdf/pab/v49n12/0100-204X-pab-49-12-00967.pdf

FERREIRA, D.F. Sisvar: a computer statistical analysis system. *Ciência e Agrotecnologia*, v.35, p.1039-1042, 2011. http://www.scielo.br/pdf/cagro/v35n6/a01v35n6.pdf

FERREIRA, V.F.; OLIVEIRA, J.A.; FERREIRA, T.F.; REIS, L.V.; ANDRADE, V.; COSTA-NETO, J. Quality of maize seeds harvested and husked at high moisture levels. *Journal of Seed Science*, v.35, n.3, p.276-277, 2013.http://www.scielo.br/pdf/jss/v35n3/01.pdf

GALVÃO, J.C.C.; CONCEIÇÃO, P.M.; ARAÚJO, E.F.; KARSTEN, J.; FINGER, F.L. Alterações fisiológicas e enzimáticas em sementes de milho submetidas a diferentes épocas de colheita e métodos de debulha. *Revista Brasileira de Milho e Sorgo*, v.13, n.1, p.14-23, 2014.http://rbms.cnpms.embrapa.br/index.php/ojs/article/viewArticle/428

MAGUIRE, J.D. Speed of germination and in selection and evaluation for seedling emergence and vigor. *Crop Science*, v.2, n.2, p.176-177, 1962. https://dl.sciencesocieties.org/publications/cs/abstracts/2/2/CS0020020176

MARCOS-FILHO, J. Teste de envelhecimento acelerado. In: KRZYZANOWSKI, F.C.; VIEIRA, R.D.; FRANÇA-NETO, J.B. (Eds.). *Vigor de sementes*: conceitos e testes. Londrina: ABRATES, cap.3, p.3.1-3.24, 1999.

MARCOS-FILHO, J. *Fisiologia de sementes de plantas cultivadas.* Piracicaba: FEALQ, 2005. 495 p.

MARTINS, C.C.; NAKAGAWA, J.; RAMOS, P.R.R. Isoenzimas no monitoramento da deterioração de sementes de *Euterpe espiritosantensis* Fernandes. *Revista Árvore*, v.35, n.1, p.85-90, 2011. http://www.scielo.br/pdf/rarv/v35n1/10.pdf

OLIVEIRA, J.A.; CARVALHO, M.L.M.; VIEIRA, M.G.G.C.; SILVA, E.A.A. Utilização de corantes na verificação de incidência de danos mecânicos em sementes de milho. *Revista Brasileira de Sementes*, v.20, n.2, p.125-128, 1998. http://www.abrates.org.br/revista/artigos/1998/v20n2/artigo21.pdf

ROSA, S.D.V.F., VON PINHO, E.V.R.; VIEIRA, E.S.; VEIGA, R.D.; VEIGA, A.D. Enzimas removedoras de radicais livres e proteínas lea associadas à tolerância de sementes de milho à alta temperatura de secagem. *Revista Brasileira de Sementes*, v.27, n.2, p.91-101, 2005. http://www.scielo.br/pdf/rbs/v27n2/a14v27n2.pdf

TUNES, L.V.M.; FONSECA, D.A.R.; MENEGHELLO, G.E; REIS, B.B.; BRASIL, V.D.; RUFINO, C.A.; VILLELA, F.A. Qualidade fisiológica, sanitária e enzimática de sementes de arroz irrigado recobertas com silício. *Revista Ceres*, v.61, n.5, p.675-685, 2014. http://www.scielo.br/pdf/rceres/v61n5/11.pdf

VEIGA, A.D.; VON PINHO, E.V.R.; VEIGA, A.D.; PEREIRA, P.H.A.R.; OLIVEIRA, K.C.; VON PINHO, R.G. Influência do potássio e da calagem na composição química, qualidade fisiológica e na atividade enzimática de sementes de soja. *Ciência e Agrotecnologia*, v.34, p.953-960, 2010. http://www.scielo.br/pdf/cagro/v34n4/v34n4a22.pdf

VIEIRA, B.G.T.L.; BARBOSA, G.F.; BARBOSA, R.M.; VIEIRA, R.D. Structural changes in soybean seed coat due to harvest time and storage. *Journal of Food, Agriculture and Environment*, v.11, p.625-628, 2013. http://www.world-food.net

VIEIRA, R.D.; KRZYZANOWSKI, F.C. Teste de condutividade elétrica. In: KRZYANOWSKI, F.C.; VIEIRA, R.D.; FRANÇA-NETO, J.B. (Ed.). *Vigor de sementes:* conceitos e testes. Londrina: ABRATES, 1999. p. 1-26. PMid:10374823.

VIEIRA, M.G.G.C.; VON PINHO, E.V.R.; SALGADO, K.C.P.C. Técnicas moleculares em sementes. *Informe Agropecuário*, v. 27, n. 232, p. 88-96, 2006.

6

Seeds treatment times in the establishment and yield performance of soybean crops

Cristian Rafael Brzezinski[1*], Ademir Assis Henning[2], Julia Abati[1], Fernando Augusto Henning[2], José de Barros França-Neto[2], Francisco Carlos Krzyzanowski[2], Claudemir Zucareli[1]

ABSTRACT – The objective was to assess the early treatment effect of soybean seeds and pre-sowing with different combinations of chemicals on the establishment of plants and crop yield performance. The design was in randomized blocks in a 2x7 factorial arrangement, with two times for seed treatment and seven treatments (six chemical treatments and an untreated control). The treatments were: 1) fipronil + pyraclostrobin + thiophanate methyl; 2) imidacloprid + thiodicarb + carbendazin + thiram; 3) abamectin + thiamethoxam + fludioxonil + mefenoxam + thiabendazole; 4) carbendazin + thiram; 5) fludioxonil + mefenoxam + thiabendazole; 6) carboxin + thiram; and 7) untreated control (water only). The assessments were: seedling emergence, final stand, plant height and insertion of first pod, number of pods per plant, seeds per pod and per plant, thousand-seed weight and grain yield. Early treatment of soybean seeds (240 days prior to sowing) hinders the establishment of the crop, the thousand-seed weight and grain yield in relation to the pre-sowing treatment. Chemical treatments tested containing fungicides and insecticides associated favor the establishment of the crop, but do not alter the soybean yield performance.

Index terms: *Glycine max* (L.) Merrill, fungicides, insecticides, seedling emergence, grain yield.

Épocas de tratamento de sementes no estabelecimento e desempenho produtivo da cultura da soja

RESUMO – O objetivo foi avaliar o efeito do tratamento de sementes de soja antecipado e em pré-semeadura com diferentes combinações de produtos químicos, sobre o estabelecimento de plantas e desempenho produtivo da cultura. O delineamento foi de blocos ao acaso, em esquema fatorial 2x7, sendo, duas épocas de tratamento de sementes e sete tratamentos (seis tratamentos químicos e uma testemunha sem tratamento). Os tratamentos foram: 1) fipronil + piraclostrobina + tiofanato metílico; 2) imidacloprido + tiodicarbe + carbendazin + thiram; 3) abamectina + tiametoxan + fludioxonil + mefenoxam + thiabendazole; 4) carbendazin + thiram; 5) fludioxonil + mefenoxam + thiabendazole; 6) carboxin + thiram; e 7) testemunha sem tratamento (somente água). As avaliações foram: emergência de plântulas, estande final, altura de plantas e inserção da primeira vagem, número de vagens por planta, sementes por vagens e por planta, massa de mil sementes e produtividade de grãos. O tratamento de sementes de soja antecipado (240 dias antes da semeadura) prejudica o estabelecimento da cultura, a massa de mil sementes e a produtividade de grãos, em relação ao tratamento em pré-semeadura. Os tratamentos químicos testados contendo fungicidas e inseticidas associados favorecem o estabelecimento da cultura, porém não alteram o desempenho produtivo da soja.

Termos para indexação: *Glycine max* (L.) Merrill, fungicidas, inseticidas, emergência de plântulas, produtividade de grãos.

Introduction

In the field, soybean plants are subject to biotic and abiotic stresses that can negatively affect their development and thus the grain yield. These production losses occur primarily through the occurrence of diseases, pests and nematodes (Bradley, 2008).

The most important diseases in soybean crops are caused by fungi, which can be transmitted by seeds or already be present in the soil at planting. These pathogens, associated or not to soil pests, harm germination and seedling establishment, thereby reducing the stand and crop yield (Lucca Filho, 2003; Mertz et al., 2009). Nematodes can also affect the soybean crop by infection of the roots, causing stand failure, lower

[1]Departamento de Agronomia, UEL, Caixa Postal 6001, 86051-990 - Londrina, PR, Brasil.

[2] Embrapa Soja, Caixa Postal 231, 86001-970 – Londrina, PR, Brasil.
*Corresponding author <cristian_brzezinski@yahoo.com.br>

development and death of seedlings, affecting the final crop yield (Araujo et al., 2012; Dias et al., 2010).

Several tillage techniques are adopted so that these factors can cause the least possible damage to the soybean crop, and chemical seed treatment stands out. This process consists of applying compounds capable of protecting seeds against these deleterious effects, helping to control these diseases in the initial period of crop establishment, favoring the seedling emergence and development (Balardin et al., 2011).

The seeds chemical treatment was usually performed in pre-sowing, both in the farmer's property and in the resale itself. However, with the technological advancement of agriculture, seed companies are adopting techniques that can optimize logistics and maximize crop yield, such as the industrial seed treatment (IST). In this process, seeds are treated in the processing line itself and subsequently bagged and stored until the sowing time.

IST associates the use of innovative equipment and techniques such as the use of new formulations containing fungicides, insecticides and nematicides in the same treatment, as well as its carrier, which can maximize efficiency of the products, help protect operators and avoid environmental contamination.

Despite the benefits, industrial treatment may have some limiting factors, such as the possible effects that the active ingredients of the chemical products have on the seeds during storage, and later in the field. According to Munkvold et al. (2006) insecticides active ingredients may in some circumstances be harmful to seeds. Vanin et al. (2011) have found that treating sorghum seeds with the active ingredient acephate has caused reductions in germination percentage and seedling emergence due to phytotoxicity. Dan et al. (2013) have found reductions in the emergence of seedlings derived from soybean seeds treated with insecticide thiamethoxan during storage.

Given the above, and knowing that seed treatment has become a practice incorporated into agricultural production, particularly for soybean and maize, it is essential to study new formulations combinations that are being used in the IST. As well as the effect of storage of treated seed on the establishment of plantations and the development of soybean in the field.

The objective of this study was to assess the early treatment effect of soybean seeds and pre-sowing with different combinations of chemicals on the establishment of plants and crop yield performance.

Material and Methods

The experiments were conducted during the 2012/2013 harvest, in two locations with different soil and climatic

characteristics. The first area was located in the Brazilian city of Faxinal, PR (location 1), which is at a latitude 23° 56' 38.26" S; longitude 51° 14' 04.03" W, with an altitude around 1056 m. The climate, according to Köppen climate classification, is Cfa, mesothermal humid subtropical with an average temperature in the coldest month below 15 °C and maximal annual average temperature at 23 °C. The second experimental area was in the Brazilian city of Boa Esperança, PR (location 2), with latitude 24° 17' 30.40" S and longitude 52° 45' 43.33" W, and altitude around 618 m. The climate, according to Köppen climate classification, is Cfa, with an average temperature in the coldest month below 15 to 18 °C and maximum annual average temperature from 27 to 29 °C (IAPAR, 2014).

The maximum and minimum daily temperature data and rainfall during the growing period for the two experimental areas are presented in Figure 1.

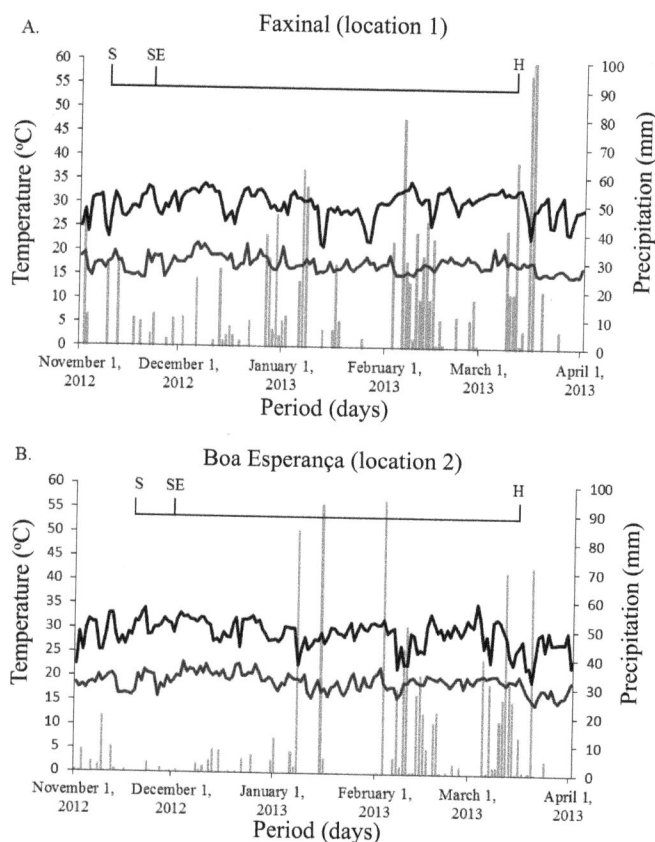

Figure 1. Maximum and minimum daily temperatures (°C) and rainfall (mm) in the Brazilian cities of Faxinal (location 1) (A) and Boa Esperança (location 2) (B) for the soybean crop development period. S: sowing; SE: seedling emergence; and H: harvest.

Soybean cultivars assessed were the BRS 360 RR, presenting early maturity of 105-120 days with relative maturity group 6.2, of indeterminate growth and edaphoclimatic region of adaptation 2. And cultivar BRS 284 of early cycle with relative maturity group 6.3, of indeterminate growth and regions of adaptation 1, 2 and 3 (Carneiro et al., 2013).

The seeds used in the experiment were subjected to two times of treatment with different products and formulations, and time 1: early seeds treatment (seeds treated and stored in a non-controlled temperature environment for 240 days in order to simulate the conditions and the maximum storage period used in the industrial processing of soybean seeds); time 2: treated seeds just before sowing (pre-sowing). The spray solution volume used was 600 mL.100 kg^{-1} of seeds, with the indicated dosage of the product and water (product + water). Seed treatment was made in polyethylene bags. The products were added to disposable syringes and the bags were vigorously shaken for homogeneous distribution of the spray solution over the seeds. The chemical treatments used are presented in Table 1.

Table 1. Active ingredients, commercial products, types and doses used for seed treatment of soybean cultivars BRS 360 RR and BRS 284.

Treatments	Active ingredient (a.i.)	Trade name	Type[1]	Commercial product dose[2]	Water dose[3]
1	fipronil + pyraclostrobin + thiophanate-methyl	Standak Top®	I + F + F	200	400
2	imidacloprid + thiodicarb + carbendazin + thiram	Cropstar® + Derosal Plus®	I + I + F + F	300+200	100
3	abamectin + thiamethoxam + fludioxonil + mefenoxam + thiabendazole	Avicta Completo (Avicta 500 FS® + Cruiser® 350 FS + Maxim Advanced®)	N + I + F + F + F	200+125+100	175
4	carbendazin + thiram	Derosal Plus®	F + F	200	400
5	fludioxonil + mefenoxam + thiabendazole	Maxim Advanced®	F + F + F	100	500
6	carboxin + thiram	Vitavax-Thiram 200 SC®	F + F	250	350
7	Control without treatment	-	-	-	600

[1] Type: I: insecticide; F: fungicide and N: nematicide
[2] Commercial product dose: mL.100 kg^{-1} of seeds
[3] Water dose: mL.100 kg^{-1} of seeds
Spray solution volume: 600 mL.100 kg^{-1} of seeds

The experimental design was randomized blocks in a 2x7 factorial arrangement, with four replications, being the factors: two treatment times (early treatment of seeds and in pre-sowing); and seven seed treatments (six chemical treatments and an untreated control – only with water).

The experimental plot consisted of four rows, six meters long, with row spacing of 0.5 m. The two central rows were considered as the plot floor area, leaving 0.5 m at the ends (borders). Sowing was performed mechanically with a density of 16 seeds per linear meter in the planting furrow. Fertilization and cultivation were the same as those used in the commercial fields where the experiments were conducted.

To determine the establishment of plants and yield performance of soybean cultivars, the following assessments were held:

Seedling emergence and plant final stand: the number of plants present in the four rows of each plot was recorded. For the variable seedling emergence, the assessment was conducted on the twelfth day after sowing, and for the final stand at harvest. The results were expressed as percentages.

Plants heights: assessed in ten plants randomly measured from the floor area of each plot, in full maturity stage (95% of pods with ripe coloring – R8) from the soil surface to the apical end of the main stem. The results were expressed in centimeters.

First pod insertion height: determined by assessing ten plants randomly collected in the floor area of each experimental plot, measuring the distance from the plant neck to the insertion of the first pod. The results were expressed in centimeters.

Number of pods per plant: obtained by the ratio between the number of pods and the total number of sample plants, in this case, ten plants per plot.

Number of seeds per pod: determined by the ratio between the total number of seeds and the total number of pods of ten plants randomly taken from the floor area of the plot.

Number of seeds per plant: determined by the ratio between the total number of seeds and the total number of plants, in this case, ten plants randomly taken from the floor area of the plot.

Thousand-seed weight: obtained by means of eight subsamples of 100 seeds per replication. Based on the weight of the subsamples, determined by an analytical balance, the mean, variance, standard deviation and coefficient of variation were calculated for subsequently obtaining the thousand-seed weight (Brasil, 2009). The result was expressed in grams.

Grain yield: obtained by weighing the seeds harvested in the floor area of each experimental plot, with humidity corrected to 13.0% (wet basis) and transformed into kg.ha^{-1}.

The data were submitted to normality tests (Shapiro-Wilk) and homoscedasticity (Hartley) and by means of these ones it was verified that there was no need for transformation. Analysis of variance was carried out and means were compared by Tukey test at 5% probability, separately for each cultivar and location. Analyses were performed using the computer program Sistema para Análise de Variância – SISVAR (System for Analysis of Variance) (Ferreira, 2011).

Results and Discussion

By means of the data obtained, it was found that there was no interaction between the factors studied. Thus, only the significant single effects were presented for treatment times and seeds chemical treatment.

Early treatment of seeds (time 1) for both cultivars and crop locations has damaged seedling emergence compared to treatment in pre-sowing (time 2) (Table 2). This result may be associated with reduced seed physiological quality (germination and vigor) during the storage period, marked by a possible phytotoxic effect of the active ingredients of the chemicals used in the treatments. Similar results were found by Krohn and Malavasi (2004) and Pereira et al. (2011), upon noticing that soybean seeds treated with fungicides and stored for longer than four months and six months, respectively, had lower emergence compared to the other treatment times.

Table 2. Yield performance of two soybean cultivars (BRS 360 RR and BRS 284) yield in the Brazilian cities of Faxinal (location 1) and Boa Esperança (location 2), according to two seed treatment times: treated and stored seeds (time 1) and seeds treated in pre-sowing (time 2).

Faxinal (location 1)					
-------------- BRS 360 RR --------------					
Times	SE (%)				M1000 (g)
1	92 b				170.9 b
2	96 a				250.9 a
CV (%)	1.64				4.37
---------------- BRS 284 ----------------					
Times	SE (%)	FS (%)	NPP	NSP	M1000 (g)
1	57 b	53 b	68.2 a	149.4 a	162.4 b
2	80 a	73 a	61.1 b	132.0 b	174.6 a
CV (%)	7.36	7.23	14.32	15.80	4.46
Boa Esperança (location 2)					
---------------- BRS 360 RR ----------------					
Times	SE (%)	FS (%)	PIH (cm)	NSPo	NSP	PRO (kg.ha^{-1})
1	57 b	56 b	9.87 a	1.83 a	122.96 a	2505.5 b
2	65 a	62 a	8.33 b	1.41 b	95.62 b	2684.7 a
CV (%)	11.13	9.21	30.40	19.26	20.39	13.99
---------------- BRS 284 ----------------					
Times	SE (%)	FS (%)	PIH (cm)	NPP	NSPo	NSP
1	41 b	37 b	8.14 a	104.49 a	1.70 a	176.4 a
2	55 a	46 a	7.30 b	89.58 b	1.47 b	131.6 b
CV (%)	12.81	13.58	20.93	19.11	17.39	22.79

Means within each column followed by the same letter do not differ by Tukey test (p $\leq$ 0.05).
SE: seedling emergence; FS: plants final stand; PIH: insertion height of the first pod; NPP: number of pods per plant; NSPo: number of seeds per pod; NSP: number of seeds per plant; M1000: thousand-seed weight and PRO: grain yield.

Regarding the effect of chemical treatments on seedling emergence, it was observed that treatments 1 (fipronil + pyraclostrobin + thiophanate-methyl) and 3 (abamectin + thiamethoxam + fludioxonil + mefenoxam + thiabendazole) propitiated highest percentages of emergence for both cultivars sowed at location 2 (Table 3). This result demonstrates the importance of using the new combinations (insecticide + fungicide or insecticide + fungicide + nematicide) in seed treatment. Mainly due to the beneficial effect promoted when controlling pathogens and insect pests, especially when sowing coincides with adverse conditions of temperature and rainfall distribution, as happened at location 2 (Figure 1B).

Table 3. Seedling emergence (SE), final plant stand (FS) and thousand-seed weight (M1000) for two soybean cultivars (BRS 360 RR and BRS 284) grown in the Brazilian city of Boa Esperança (location 2), with different chemical treatments.

| --------------- BRS 360 RR --------------- | | | |
Treatments[1]	SE (%)	FS (%)	M1000 (g)
1	70 a	68 a	124.7 ab
2	63 ab	63 ab	125.1 ab
3	71 a	68 a	112.1 b
4	59 b	57 bc	114.6 ab
5	54 b	50 c	125.6 a
6	57 b	50 c	124.8 ab
7 (control)	53 b	50 c	126.3 a
CV (%)	11.13	9.21	8.07

| --------------- BRS 284 --------------- | | |
Treatments	SE (%)	FS (%)
1	57 a	50 a
2	51 ab	47 ab
3	58 a	54 a
4	49 ab	41 bc
5	45 b	39 bc
6	46 b	38 c
7 (control)	32 c	26 d
CV (%)	12.81	13.58

Means within each column followed by the same letter do not differ by Tukey test (p ≤ 0.05).
[1]Treatments: 1: fipronil + pyraclostrobin + thiophanate-methyl; 2: imidacloprid + thiodicarb + carbendazim + thiram; 3: abamectin + thiamethoxam + fludioxonil + mefenoxam + thiabendazole; 4: carbendazim + thiram; 5: fludioxonil + mefenoxam + thiabendazole; 6: carboxin + thiram; 7: control (no treatment).

From the results, it was also possible to see that, for the experiment conducted in location 1, the lack of a significant effect of chemical treatment for the variable seedling emergence was probably due to the favorable conditions of humidity and temperature during the crop establishment (Figure 1A). According to Goulart (2005), the effect of the treatment becomes somewhat evident when soybeans are planted in ideal conditions of temperature and soil moisture due to the rapid germination and seedling emergence providing an escape in relation to the attack of soil fungi. However, when seeding is performed in water deficit conditions, as in the second location, the protective effect of treatment of soybean seeds becomes more evident. Balardin et al. (2011) and Conceição et al. (2014) working with soybeans, and Abati et al. (2014) with wheat, have achieved a similar result, when observing that treating seeds with fungicides, insecticides and polymers alone or combined have resulted in higher emergence values in the field. According to Pereira et al. (1993), the treatment protective effect on the seeds can last for a period of four to 12 days; however, this effect depends on a combination of several factors such as the product formula and mode of action, the treatment quality and the seeds physical and physiological quality.

The results found on the assessment of the plants final stand corroborate those obtained in seedling emergence (Tables 2 and 3), both for seed treatment times and for the chemical treatments. Thus it demonstrates that there was no significant reduction in stand along the development of the crop in terms of environment biotic and abiotic factors.

For the variable insertion height of the first pod, there was an effect of seed treatment times only for the experiment conducted in location 2 (Table 2). From the averages presented, higher insertion height of the first pod for the plants grown from seeds submitted to the early treatment (time 1) was seen in both cultivars. However, despite this significant difference, there was no need to change the height of the harvester cutting boom because these were 9.87 and 8.33 cm for cultivar BRS 360 RR and 8.14 and 7.30 cm for cultivar BRS 284, for seeds treated and stored and treated in pre-sowing, respectively.

As for the yield components, there was an isolated effect of seed treatment times for the number of pods per plant, of seeds per pod and seeds per plant of cultivar BRS 284, produced in both crop locations, and for the number of seeds per pod and seeds per plant of cultivar BRS 360 RR produced in location 2 (Table 2). It was observed that the early treatment of seeds (time 1) led to the development of plants with a larger number of pods, seeds per pod and seeds per plant, compared to the treatment performed in pre-sowing (time 2). These differences were found due to the reduced plant population observed in the assessment of seedling emergence. This result is closely related to the balance between the production of flowers per plant and the ratio of these who develop until the pod, since the number of flowers per plant is determined by the number of flowers per node and the number of nodes per plant. At higher densities, there is greater competition for light and reduced availability of photoassimilates, causing the plant to decrease the number of branches and produce a smaller number of nodes. In these nodes, the reproductive buds develop. Thus, decreasing the number of branches reduces the number of potential nodes, and consequently, the number of pods and their components (Bord and Settimi, 1986; Jiang and Egli, 1993; Mauad et al., 2010).

For the thousand-seed weight of cultivar BRS 284 assessed in the experiment conducted in location 1, the treatment in pre-sowing (time 2) had yielded increased thousand-seed weight produced in relation to the early treatment (time 1) (Table 2). This difference may be due to reduced seed quality during storage, more sharply for chemically treated seeds. Consequently, the crop establishment was damaged, and thus, under smaller populations, soybean plants tend to adapt to these conditions by means of changes in plant morphology and yield components, producing a higher number of pods per plant and a higher number of seeds per pod. As for populations that are appropriate to plants, they distribute better their photoassimilates for grain filling, favoring the increase of their average mass. Similar results were obtained by Mauad et al. (2010), in work performed in order to assess the effect of sowing density in the soybean yield components.

When comparing the chemical treatments, it was observed for location 1 that the plants in treatment 3 (abamectin + thiamethoxam + fludioxonil + mefenoxam + thiabendazole) produced seeds with lower mass compared to control for cultivar BRS 360 RR (Table 3). However, this difference probably did not occur due to the active ingredients of the products used in the treatment, but by the reduction of the establishment of plants associated with soybeans recovery plasticity, observed from the seedling emergence, as shown above.

For grain yield, there was a single effect of seed treatment times only for cultivar BRS 360 RR produced in location 2 (Table 2). It was observed that the treatment in pre-sowing (time 2) provided greater grain yield compared to the early treatment (time 1), with an average yield of 2684 and 2505 kg.ha^{-1}, respectively. However, these values were found below the national average obtained for the same harvest, which was 2854 kg.ha^{-1} (CONAB, 2014). This was due to the water restriction period observed in the experimental site during the crop establishment (Figure 1B), which hindered the plants development and, consequently, the yield components (number of seeds per pod and number of seeds per plant).

Based on the results obtained, it is possible to see the importance of an appropriate choice of treatment time and seed chemical treatments compared to the establishment of plants in the field, especially under adverse conditions of temperature and humidity, as a precursor for obtaining high yields of grains. Thus, for the tested treatment times, treatment in pre-sowing favored the establishment of plants in the field and grain yield in relation to early treatment (240 days). However, the importance of new studies testing more treatment times should be emphasized, mainly related to reduced storage time of treated seeds in order to avoid possible phytotoxic effects of the products active ingredients on the seeds that later harm the establishment and development of plants in the field.

Conclusions

Early treatment of soybean seeds (240 days prior to sowing) hinders the establishment of the crop, the thousand-seed weight and grain yield in relation to the pre-sowing treatment.

Chemical treatments tested containing fungicides and insecticides associated favor the establishment of the crop, but do not alter the soybean yield performance.

Acknowledgments

To CAPES (Coordenação de Aperfeiçoamento de Pessoal de Nível Superior; Brazilian Coordination of Improvement of Higher Education), for granting the scholarship to the first author. To Universidade Estadual de Londrina and the company Empresa Brasileira de Pesquisa Agropecuária, Centro Nacional de Pesquisa de Soja (Embrapa Soja) for the structure and financial support for the development of the work.

References

ABATI, J.; ZUCARELI, C.; FOLONI, J.S.S; HENNING, F.A.; BRZEZINSKI, C.R.; HENNING, A.A. Treatment with fungicides and insecticides on the physiological quality and health oh wheat seeds. *Journal of Seed Science*, v.36, n.4, p.392-398, 2014. http://www.scielo.br/pdf/jss/v36n4/a02v36n4.pdf

ARAUJO, F.F.; BRAGANTE, R.J.; BRAGANTE, C.E. Controle genético, químico e biológico de meloidoginose na cultura da soja. *Pesquisa Agropecuária Tropical*, v.42, n.2, p.220-224, 2012. http://www.scielo.br/pdf/pat/v42n2/13.pdf

BALARDIN, R.S.; SILVA, F.D.L.; DEBONA, D.; CORTE, G.D.; FAVERA, D.D.; TORMEN, N.R. Tratamento de sementes com fungicidas e inseticidas como redutores dos efeitos do estresse hídrico em plantas de soja. *Ciência Rural*, v.41, n.7, p.1120-1126, 2011. http://www.scielo.br/pdf/cr/v41n7/a5711cr4207.pdf

BORD, J.E.; SETTIMI, J.R. Photoperiod effect before and after flowering on branch development in determinate soybean. *Agronomy Journal*, v.78, n.6, p.995-1002, 1986. http://www.researchgate.net/publication/250102478_Photoperiod_Effect_Before_and_After_Flowering_on_Branch_Development_in_Determinate_soybean

BRADLEY, C.A. Effect of fungicide seed treatments on stand establishment, seedling disease, and yield of soybean in North Dakota. *Plant Disease*, v.92, n.1, p.120-125, 2008. http://apsjournals.apsnet.org/doi/pdf/10.1094/PDIS-92-1-0120

BRASIL. Ministério da Agricultura, Pecuária e Abastecimento. *Regras para análise de sementes*. Ministério da Agricultura, Pecuária e Abastecimento. Secretaria de Defesa Agropecuária. Brasília: MAPA/ACS, 2009. 395p. http://www.agricultura.gov.br/arq_editor/file/2946_regras_analise__sementes.pdf

CARNEIRO, G.E.S.; PÍPOLO, A.E.; MELO, C.L.P.; LIMA, D.; MIRANDA, L.C.; PETEK, M.R.; BORGES, R.S.; GOMIDE, F.B.; DALBOSCO, M.; DENGLER, R.U. *Cultivares de soja – Macrorregiões 1, 2, 3*. Centro-sul do Brasil. Embrapa Soja, Londrina, PR, 2013. 56p.

CONAB. Companhia Nacional de Abastecimento. *Acompanhamento da safra brasileira – Grãos*. Disponível em: http://www.conab.gov.br/OlalaCMS/uploads/arquivos/14_06_10_12_12_37_boletim_graos_junho_2014.pdf Accessed on: July 12th, 2014.

CONCEIÇÃO, G.M.; BARBIERI, A.P.P.; LÚCIO, A.D.; MARTIN, T.N.; MERTZ, L.M.; MATTIONI, N.M.; LORENTZ, L.H. Desempenho de plântulas e produtividade de soja submetida a diferentes tratamentos químicos nas sementes. *Bioscience Journal*, v.30, n.6, p.1711-1720, 2014. http://ainfo.cnptia.embrapa.br/digital/bitstream/item/113098/1/Desempenho-de-plantulas-e-produtividade-de-soja-submetida-a-diferentes-tratamentos-quimicos-nas-sementes.pdf

DAN, L.G.M.; BRACCINI, A.L.; BARROSO, A.L.L.; DAN, H.A.; PICCININ, G.G.; VORONIAK, J.M. Physiological potential of soybean seeds treated with thiamethoxam and submitted to storage. *Agricultural Sciences*, v.4, n.11, p.19-25, 2013. http://www.scirp.org/journal/PaperInformation.aspx?PaperID=40173

DIAS, W.P.; GARCIA, A.; SILVA, J.F.V.; CARNEIRO, G.E.S.N. *Nematoides em Soja*: Identificação e controle. Londrina-PR: Embrapa Soja, 2010. 8p. (Circular Técnica, 76). http://www.cnpso.embrapa.br/download/CT76_eletronica.pdf

FERREIRA, D.F. Sisvar: A computer statistical analysis system. *Ciência e Agrotecnologia*, v.35, n.6, p.1039-1042, 2011. http:/www.scielo.br/pdf/cagro/v35n6/a01v35n6.pdf

GOULART, A.C.P. *Importância do tratamento de sementes de soja com fungicidas em condições de déficit hídrico do solo*. Embrapa Agropecuária Oeste. 2005. 6p (Comunicado Técnico, 106). http://ainfo.cnptia.embrapa.br/digital/bitstream/item/24686/1/COT2005106.pdf

IAPAR. Instituto Agronômico do Paraná. *Cartas climáticas do Estado do Paraná*. http://200.201.27.14/Sma/Cartas_Climaticas/Classificacao_Climatica.htm. Accessed on: May 21th, 2014.

JIANG, H.; EGLI, D.B. Shade induced change in flower and pod number and flower and fruit abscission in soybean. *Agronomy Journal*, v.85, n.2, p.221-225, 1993. https://www.agronomy.org/publications/aj/abstracts/85/2/AJ0850020221?access=0&view=pdf

KROHN, G.N.; MALAVASI, M.M. Qualidade fisiológica de sementes de soja tratadas com fungicidas durante e após o armazenamento. *Revista Brasileira de Sementes*, v.26, n.2, p.91-97, 2004. http://www.scielo.br/pdf/rbs/v26n2/24494.pdf

LUCCA FILHO, O.A. *Patologia de Sementes*. In: PESKE, S.T., TOSENTHAL, M.D., ROTA, G.R. (Eds.) Sementes: Fundamentos Científicos e Tecnológicos. Ed. Universitária. 2003. p.225-282.

MAUAD, M.; SILVA, T.L.B.; ALMEIDA NETO, A.I.; ABREU, V.G. Influência da densidade de semeadura sobre características agronômicas na cultura da soja. *Revista Agrarian*, v.3, n.9, p. 175-181, 2010. http://www.periodicos.ufgd.edu.br/index.php/agrarian/article/view/75/649

MERTZ, L.M.; HENNING, F.A.; ZIMMER, P.D. Bioprotetores e fungicidas químicos no tratamento de sementes de soja. *Revista Ciência Rural*, v.39, n.1, p.13-18, 2009. http://www.scielo.br/pdf/cr/v39n1/a03v39n1.pdf

MUNKVOLD, G.; SWEETS, L.; WINTERSTEEN, W. *Iowa commercial pesticide applicator manual* – Category 4. Ames: Iowa State University, 2006. 39p.

PEREIRA, L.A.G.; COSTA, N.P.; ALMEIDA, M.R.; FRANÇA-NETO, J.B.; GIGLIOLI, J.L.; HENNING, A.A. Tratamento de sementes de soja com fungicida e/ ou antibiótico, sob condição de semeadura em solo com baixa disponibilidade hídrica. *Revista Brasileira de Sementes*, v.15, n.2, p.241-246, 1993. http://www.abrates.org.br/revista/artigos/1993/v15n2/artigo17.pdf

PEREIRA, C.E.; OLIVEIRA, J.A.; GUIMARÃES, R.M.; VIEIRA, A.R.; EVANGELISTA, J.R.E.; OLIVEIRA, G.E. Tratamento fungicida e peliculização de sementes de soja submetidas ao armazenamento. *Ciência e Agrotecnologia*, v.35, n.1, p.158-164, 2011. http://www.scielo.br/pdf/cagro/v35n1/a20v35n1.pdf

VANIN, A.; SILVA, A.G.; FERNANDES, C.P.C.; FERREIRA, W.S.; RATTES, J.F. Tratamento de sementes de sorgo com inseticidas. *Revista Brasileira de Sementes*, v.33, n. 2, p. 299-309, 2011. http://www.scielo.br/pdf/rbs/v33n2/12.pdf

7

Micropilar and embryonic events during hydration of *Melanoxylon brauna* Schott seeds

Eduardo Euclydes de Lima e Borges[1], Glauciana da Mata Ataíde[2*], Antônio César Batista Matos[1]

ABSTRACT – Germination is a complex process that involves molecules properties that make up the cell walls, hydrolytic enzymes that break the bonds between the polymers and action of reactive oxygen substance. *Melanoxylon brauna* is a forest species of high economic value. In order to evaluate the physiological and biochemical changes that occur in the embryonic axis during germination, fresh matter, length, activities of the enzymes pectin methylesterase, polygalacturonase, superoxide dismutase, catalase, peroxidase and hydrogen peroxide levels were quantified in the embryonic axis. Furthermore, in the micropyle area the composition of carbohydrates and micropyle physical resistance were evaluated with and without drying. During soaking, if there are increases in fresh matter and length of the embryonic axis, there is the same trend of polygalacturonase and pectin methylesterase enzymes. The hydrogen peroxide content was reduced during the soaking, as well as the puncture force of the micropylar area. It is concluded that the seed coat and the cotyledons are responsible for 90% of the water soaked by the seeds. The events in the micropyle and embryonic axis occur independently in the first 16 hours. The weakening of the micropyle features an elastic step and a plastic one. Enzymes pectin methylesterase and polygalacturonase act in cellular expansion of the embryonic axis.

Index terms: puncture force, hydrogen peroxide, cell wall, germination.

Eventos micropilar e embrionário na hidratação de sementes de *Melanoxylon brauna* Schott

RESUMO – A germinação é um processo complexo que envolve propriedades das moléculas que compõem a parede celular, enzimas hidrolíticas que quebram ligações entre polímeros da pectina e da hemicelulose e ação das substâncias reativas de oxigênio. *Melanoxylon brauna* é uma espécie florestal de alto valor econômico. Com o objetivo de avaliar alterações fisiológicas e bioquímicas no eixo embrionário durante a germinação, foram quantificados a massa fresca, o comprimento, as atividades das enzimas pectina metilesterase, poligalacturonase, superóxido dismutase, catalase, peroxidase e os teores de peróxido de hidrogênio. Foram avaliados na região micropilar a composição dos carboidratos e a resistência da micrópila seca e sem secagem. Durante a embebição, ocorrem aumentos na massa fresca e comprimento dos eixos embrionários, mesma tendência das enzimas poligalacturonase e pectina metilesterase, que participam da expansão celular. O conteúdo de peróxido de hidrogênio reduziu durante a embebição, assim como a força de ruptura da região micropilar. Conclui-se que o tegumento e os cotilédones são responsáveis por 90% da água embebida pelas sementes. Os eventos na micrópila e eixo ocorrem independentes nas primeiras 16 horas. O enfraquecimento da micrópila apresenta uma etapa elástica e outra plástica. As enzimas pectina metilesterase e a poligalacturonase atuam na expansão celular do eixo embrionário.

Termos para indexação: força de ruptura, peróxido hidrogênio, parede celular, germinação.

Introduction

Germination is the result of expansion of the embryonic axis due to the embryo potential growth and cell wall weakening of the micropyle, which is the limiting component of the radicle growth. According to Nonogaki (2014), germination will occur when the active embryo radicle is able to break the strength exerted by tissue wraps. Different enzymes are involved in the germination process, both in the embryonic axis and in the micropyle.

The seeds micropylar endosperm resistance decreases by the action of enzymes during germination, which weaken the cell wall and allow root protrusion. Among these enzymes, the activity of pectin methylesterase was

[1]Departamento de Engenharia Florestal, UFV, 36570-000 – Viçosa, MG, Brasil.

[2]Departamento de Silvicultura, UFRRJ, 23895-000 – Seropédica, RJ, Brasil.
*Corresponding author <glaucianadamata@yahoo.com.br>

detected by Scheler et al. (2015) in *Lepidium sativum* seeds, with the penetration of the radicle in the endosperm being dependent of cells cohesion loss in place, resulting from the action of the enzyme. Moreover, the growth of the embryonic axis of *Dalbergia nigra* seeds by polygalacturonase was observed, which facilitates expansion of the axis by effect of hydration, according to Ataíde et al. (2013). Reactive oxygen species (ROS) are considered toxic molecules formed during normal metabolic functions and induced when plants are exposed to environmental stimuli (Éaux and Toledano, 2007). However, they also act as signaling molecules in response to different stresses during germination (Gomes and Garcia, 2013; Tenhaken, 2015) or even germination under normal conditions, as occurs in *Dalbergia nigra* seeds (Matos et al., 2014). Reactions catalyzed by peroxidase also result in stimulation of sprouting, especially in the weakening of the cell wall, since they produce the hydroxyl radical (Richards et al., 2015).

To avoid irreversible cell damage, the antioxidant system enzymes come into play when ROS levels exceed certain levels, as verified for *Dalbergia nigra* seeds (Matos et al., 2014). The first line of defense in plants against oxidative stress is the superoxide dismutase (SOD), an enzyme that catalyzes the conversion of superoxide radicals ($O_2^{\bullet-}$) at H_2O_2 (Alscher et al., 2002). The H_2O_2 formed in turn can be removed from the cells by enzymes such as peroxidase (POX), and catalase (CAT) which, in general, use it as a substrate, reducing it to water.

Among the forest native species of ecological and economic importance for use in plantations, there is the *Melanoxylon brauna*, popularly known as braúna, naturally occurring in the Atlantic Forest in Northeast and Southeast Brazil, being the species known for the quality and great economic value of its wood (Lorenzi, 2009). Currently, it is in the Brazilian List of Endangered Flora Species (MMA, 2008). Despite the great economic and environmental value and studies involving aspects related to the seed species physiology (Flores et al., 2014a; Flores et al., 2014b; Borges et al., 2015), more scientific research about the spread of *M. brauna* is needed, particularly with an emphasis on dehydration and germination steps.

Thus, it is proposed to quantify the physiological and biochemical changes which occur in the embryonic axis and in the micropyle during the germination of *M. brauna* seeds, such as the antioxidant system enzymes activity, polygalacturonase and pectin methylesterase and the mechanical resistance of the micropyle, with a view to understanding the germination process of the species.

Material and Methods

Melanoxylon brauna seeds were collected from ripe fruit in five trees from natural regeneration in the Brazilian district of Cataguases, Minas Gerais state (23K 0721180 and 7701850, UTM, 609 m of altitude), and taken to Forest Seed Analysis Laboratory at University Federal of Viçosa. After drying in the sun, at up to about 12-15% moisture, the seeds were processed and mixed, forming a single lot. The seeds were stored in a refrigerator (about 5 °C) in cardboard drums until the beginning of the experiments.

The fresh matter of seeds parts were quantified (integument, cotyledons and embryos) and of isolated embryos during soaking. To this end, whole seeds were kept on two sheets of germitest-type paper moistened with water at 30 °C, with constant light from four 40 W daylight-type fluorescent lamps, and sampled at times zero, 2, 4, 16 and 24 hours, when the seeds were dissected in integument, cotyledons and embryonic axis, superficially dried on paper towels and each part weighed in scales of down to a hundredth precision. In the fresh matter evaluations of the embryonic axes isolated from these seeds, they were removed with a scalpel and placed to soak under the same temperature and light conditions above outlined, and the samples were withdrawn at times zero, 2, 4, 16 and 24 hours.

The determinations done in isolated embryos at times zero, 2, 4, 16 and 24 hours were:

Fresh matter – Five replications of ten embryonic axes were used by means of weighing the sample on a precision balance.

Length of the embryonic axes – To calculate the average length of the embryonic axis, these were individually measured using a photographic enlarger and a millimeter ruler. Five replicates of ten embryonic axes were used.

Pectin methylesterase enzyme – The extraction of the pectin methylesterase enzyme was performed according to the method described by Pinto et al. (2011), with modifications. Four replications of 0.1 g of embryonic axes removed at times zero, 2, 4, 16 and 24 hours of soaking were used, which were homogenized in 4.0 mL of solution of NaCl 1.0 M, pH 7.5, containing 1.0% (m/v) of insoluble polyvinylpolypyrrolidone (PVPP). This solution was centrifuged at 15,000 g for 30 minutes in a centrifuge refrigerated at 4 °C, and the supernatant (enzyme extract) was used for determination of enzyme activity. The activity was quantified according to Grsic-Rausch and Rausch (2004), being considered that a pectin methylesterase unit corresponds to 1 µM of NADPH formed per minute in pH 7.5 and 25 °C.

Polygalacturonase Enzyme – The enzyme extracts were obtained as described by Guimarães et al. (2001). Four replications of 0.1 g of embryonic axes removed at times zero, 2, 4, 16 and 24 hours of soaking were used. The activity assessment was based on the dosage reducing sugars produced according to the method of DNS (3,5-Dinitrosalicylic acid), according to Miller (1959). The enzyme unit was defined as the amount of protein required to produce the equivalent of 1.0 µmol of glucose per minute.

Hydrogen peroxide (H_2O_2) – The extraction and production of hydrogen peroxide were conducted according to Gay and Gebicki (2000). Three replicates of 0.1 g for each sample withdrawal time were used.

Superoxide dismutase (SOD) and catalase (CAT) enzymes – The crude enzyme extract used were obtained from the method described by Flores et al. (2014a). Three replicates of 0.1 g for each sample withdrawal time were used for both enzymes. The superoxide dismutase activity was determined according to Del Longo et al. (1993) and Beauchamp and Fridovich (1971). The activity of catalase was determined according to Hodges et al. (1997). The enzyme activity was calculated using the molar extinction coefficient of 36 M.cm^{-1}, according to Anderson et al. (1995).

Peroxidase enzyme – Enzymatic extracts were obtained according to Peixoto et al. (1999). The activity was determined according to Kar and Mishra (1976) using the molar extinction coefficient of 2.47 mM^{-1}.cm^{-1} (Chance and Maehley, 1995) and expressed in µmol^{-1}.min^{-1}.mg^{-1} of protein. Three replicates of 0.1 g per hour of sample withdrawal were used.

The protein contents of the enzymatic extracts were determined by the method by Bradford (1976), using BSA (bovine serum albumin) as a standard.

To quantify changes in seeds micropyle, whole seeds samples were germinated under the same conditions described above, removing the seeds micropylar area for the following assessments:

Micropyle resistance in whole seeds – The resistance of the micropyle area was quantified by the puncture force in whole seeds in the soaking times of 0 (control), 24, 48 and 72 hours of soaking after removal of the embryonic axis in each of these times.

Resistance of the isolated micropyle, wet and after drying in the soaking times of 0, 2, 4, 16, 24, 48 and 72 hours – In these times, the puncture force was measured in the wet seeds and after drying at room temperature. Independent samples were used for the measurement of wet and dry seeds. The micropyle resistance was also quantified after subjecting them to a temperature of 80 °C for 24 hours in order to kill the seed and subsequent denaturation of the enzymes present in the micropyle, and the puncture force was measured after this period, whether or not followed by drying.

All quantifications of alterations in the cell wall strength were done by the strength for tissues rupture using a texture analyzer (Stable Micro Systems Texture Analyzer). The micropyle area (about 2 mm) of the dry or hydrated seeds was cut using a multi-tool kit with a circular saw. Then it was positioned on the probe and drilled. The required force to puncture the endosperm expressed in Newton (N) was used as a strength parameter of the micropylar endosperm to the embryo elongation during germination. The resistance data were obtained from the average of five replicates of 10 individual seeds, taken at random after each aforementioned range.

Extraction of the cell wall and carbohydrates quantification– The micropyle cell wall was purified according to Borges et al. (2000), with modifications: 300 mg of micropylar material were homogenized in a tris-HCl buffer, 50 mM, pH 8.0, at 5 °C. After filtration in a nylon filter, two more washings and filtering were done. The residue was washed in deionized water at 5 °C and kept in suspension, being considered as a crude extract of the cell wall.

The extraction of carbohydrates of the micropyle area was performed as described by Carpita and Gilbeaut (1993), and alditol acetates were prepared for monosaccharide analysis in gas chromatography, according to Englyst and Cummings (1984). Four replications were done in the soaking times of 0, 2, 4, 16, 24, 48 and 72 hours.

Results and Discussion

The hydration of *M. brauna* seed embryos began slowly after four hours in water, while the cotyledons and the integument started in the first two hours of soaking, continuously reaching higher values for 24 hours (Figure 1). Two aspects must be highlighted in the hydration of the three compartments. In the case of cotyledons, the size of both is clearly superior to the embryo, besides having reserves with hygroscopic properties, such as starch, soluble sugars and proteins, and thus, being capable to hydrate more efficiently.

The integument, by having a high concentration of galactomannan (unpublished data) in a thin inner layer, acts in the same way, since this reserve tissue is highly hygroscopic. Due to the reduced size of the embryonic axis with respect to the integument and the cotyledons and being wrapped by both, it can be stated that the root protrusion is determined in part by the greater time to achieve water content needed to swell and increase the potential growth, acting against the integument tissues that are opposed to this expansion.

This is corroborated by the data in Figure 2, which represent the increase in fresh matter and length of isolated embryos. Please note that the value achieved in 24 hours for fresh matter is similar to that obtained for whole seeds embryos that were soaked for 72 hours. This increase in fresh matter is followed by the increase in the average length of embryos, following the three-phase standard proposed by Bewley et al. (2013).

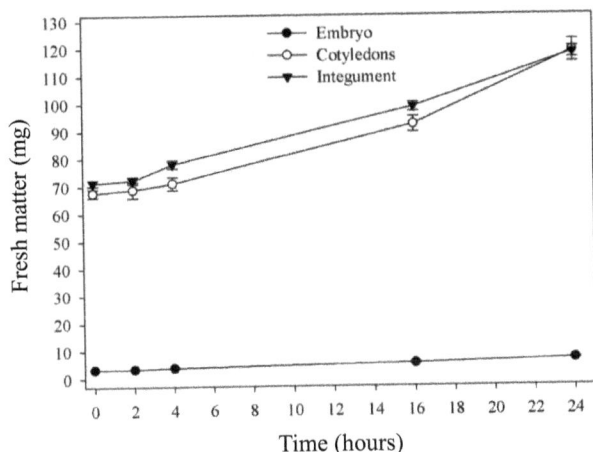

Figure 1. Fresh matter of the parts of seeds of *Melanoxylon brauna* (embryonic axis, cotyledons and integument) during soaking at a temperature of 30 °C. +– Standard error.

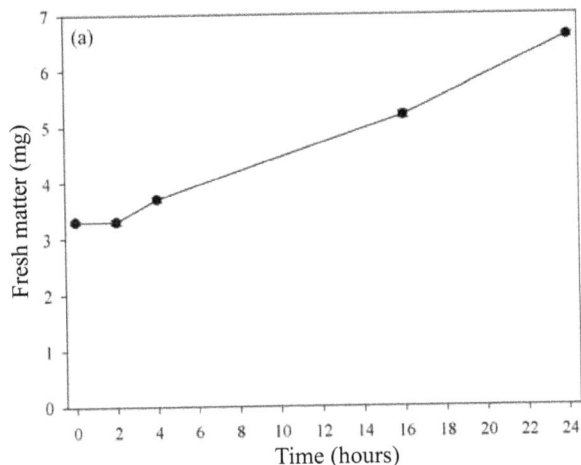

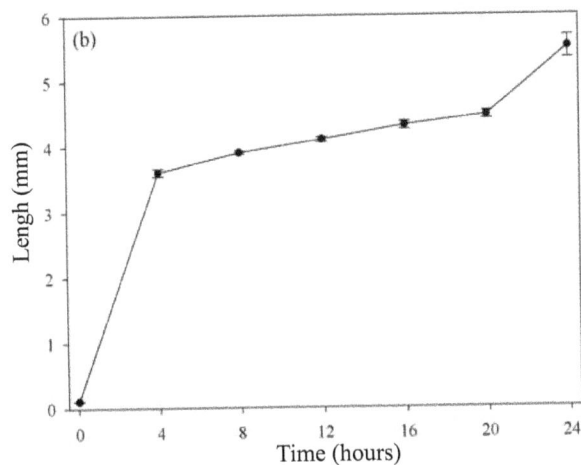

Figure 2. Fresh matter (a) and medium length (b) of embryonic axes isolated from *Melanoxylon brauna* seeds during soaking at a temperature of 30 °C. +– Standard error.

Thus, it is clear that once having access to water, axis expansion starts. Possibly by the activation of enzyme activities that weaken the connections that maintain the cell wall cohesion of the embryonic axis cells in the soaking exponential phase, with small increases in expansion of the embryonic axis in the stationary phase. These stages correspond to soaking phases I and II, respectively, according to Bewley et al. (2013). Alternatively, physical expansion of the wall may have occurred due to the polymers mechanical properties such as xyloglucan, for example, which comprises it (Park and Cosgrove, 2012). Other elements are involved for axis new growth to occur, corresponding to soaking phase III, which corresponds to the phase of root protrusion (Bewley et al., 2013).

By the results obtained with enzymes pectin methylesterase (Figure 3a) and polygalacturonase (Figure 3b) in embryonic axis, it is possible to see that both are pre-formed and their activities have greatly increased in 16 hours and are more pronounced in 24 hours. The polygalacturonase activity is high at the beginning of hydration, which may imply that demethylation and calcium bridges formation have already taken place during the seed formation phase. The expansion in the second phase would be conditioned to the presence of both enzymes. Together they would act more strongly in the weakening or extensibility of cell wall from four hours, more specifically in pectin, allowing its expansion with the water inlet. Thus, enzyme participation in the initial expansion of the embryonic axis is clear.

Factors that increase the wall extensibility influence hydration, which in turn allows cell growth. Pectin is formed by homogalacturonan or rhamnogalacturonans I and II. According to Peaucelle et al. (2012), the degree of methyl esterification of the homogalacturonan defines porosity, elasticity and compressibility of the wall. The action of pectin methylesterase allows increased wall susceptibility to the action of polygalacturonase, resulting in changes in the wall biomechanical properties (Wakabayashi et al., 2003). Thus both act, allowing cell expansion upon hydration.

Moreover, the observed increase in fresh matter of the isolated embryonic axes in the first four hours (Figure 2a) also appears to be related to the presence of reactive oxygen substances. A high concentration of hydrogen peroxide up to four hours of hydration was observed, stabilizing at 16 and with a marked reduction in 24 hours, when it reached minimum values (Figure 4).

Hydrogen peroxide breaks bonds between polysaccharides in non-enzymatic reactions, causing weakening of the wall (Schopfer, 2001). Thus the weakening of the cell wall by non-enzyme reactions may have reduced the wall potential, allowing cell expansion by the hydration resulting from the water potential difference between the medium and the axis cells in phase I. This possibility is in accordance with

the proposal that the hydration in phase I is purely physical, according to several authors cited by Weitbrecht et al. (2011).

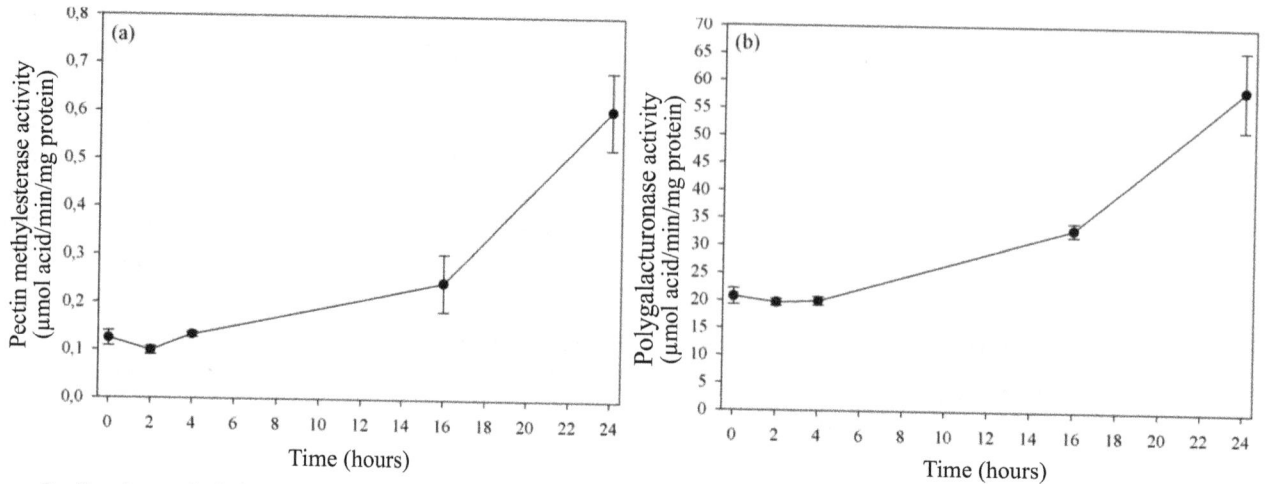

Figure 3. Pectin methylesterase activity (a) and polygalacturonase (b) in embryonic axes isolated from *Melanoxylon brauna* seeds during soaking at a temperature of 30 °C. +– Standard error.

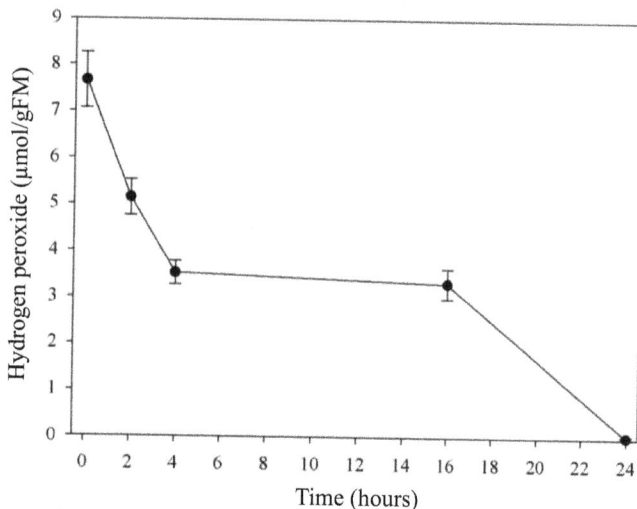

Figure 4. Hydrogen peroxide content (µmol/g FM) in embryonic axes isolated from *Melanoxylon brauna* seeds during soaking at a temperature of 30 °C. +– Standard error.

The conformation of the two curves (Figure 2a and Figure 4) is clear, where it is possible to observe inverse behaviors between them. The concentration of hydrogen peroxide also has the three-phase conformation with a decrease in 2 hours, followed by a stabilization phase between 4 and 16 hours, and decreasing again in 24 hours. The reduction in the peroxide concentration over the hydration period has eliminated the chance of possible damage caused by the presence of the reactive oxygen substance.

The production of hydrogen peroxide can have a different source from that usually cited (mitochondrion or peroxisome),

being produced by the action of NADPH peroxidase which, according to Bedard et al. (2007), is the enzyme producing ROS in plants and animals. Moreover, the production and maintenance of hydrogen peroxide can also be by the action of NADPH oxidase, according to Karmer et al. (2010). According to them, hydrogen peroxide production outside the cell could return in the form of hydrogen peroxide, formed by the effect of extracellular pH 5.0.

In Figure 5a, it is seen that the enzyme superoxide dismutase activity was stable in the first 16 hours and increased by 24 hours, and it was not possible to implicate it in the first phase of action in the reduction of reactive oxygen species such as superoxide. It is possible that this concentration is low at this stage, with no need for dismutation by the enzyme. It is noteworthy that the presence of the substance under the conditions of this study not detected.

On the other hand, the enzyme catalase (Figure 5b) had high activity at first and then decreased until 16 hours of soaking, returning to the same level of two hours. Unlike the latter, peroxidase (Figure 5c) showed low activity in the first four hours, with successive increases in up to 24 hours. It is noticed that both enzymes have fairly similar activities in the control embryos, with more pronounced decrease of catalase, which approached the peroxidase values in up to four hours of hydration. Subsequently, there was a substantial increase in the latter, which can indicate greater effectiveness of peroxidase to eliminate excess hydrogen peroxide produced by the increase in metabolic activity during hydration.

Considering that the production of ROS occurs during seed

formation due to the synthesis metabolism (Liu et al., 2014), catalase activity would act to reduce the level of these substances during germination so they do not cause damage to the cells, but keeping them in a concentration that would act in the wall flexibility. Thus, hydration occurring in phase I would be possible by the Fenton reaction. Subsequently, with the hydration increases and metabolic activation, the increase of ROS induced the increase of peroxidase

activity. It is noteworthy that, according to Weitbrecht et al. (2011), phase III is typically metabolic and therefore the production of ROS is high, especially by the mitochondrion, one of the producers of these substances (Vanlerberghe, 2013). Thus, without the ROS reduction in phase III there would be inactivation of enzymes catalase and peroxidase, without which the flexibility of the wall and subsequent increase in size would not have occurred.

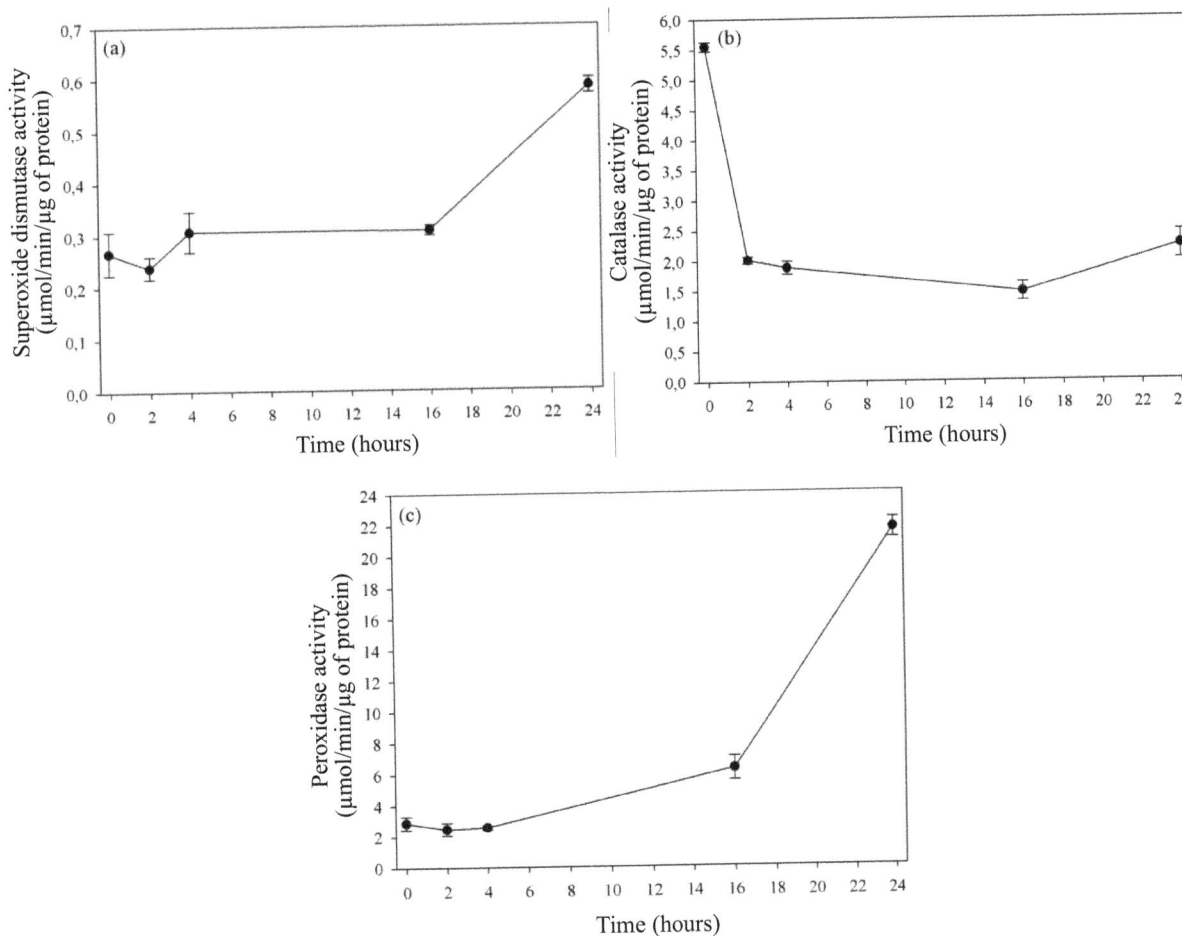

Figure 5. Activity of enzymes superoxide dismutase (a), catalase (b), and peroxidase (c) in embryonic axes isolated from *Melanoxylon brauna* seeds during soaking at a temperature of 30 °C. +– Standard error.

In assessing the change in the micropyle wall strength in whole seeds, shown in Figure 6a, it was possible to observe its marked reduction in 24 hours, in the same manner as in the micropyle isolated in two hours (Figure 6b). The puncture force values found during these times near 1 N are similar. In Figure 6b it is seen that the weakening of the wall occurs in two hours and remains relatively constant for 72 hours. Such similarity could not be otherwise, unless the embryo inhibited any activity at that location, which does not seem to be the case.

Drying the micropyle resulted in increasing its resistance. For permanent deformation of the wall to occur, there would

be the need for changes in both components. The wall elastic element could be credited to the mechanical action of the polymers (Suslov et al., 2010), thus keeping the wall expansion property. As for the plastic change, it would result from the action of enzymes attached to the wall or exuded from the radicle to that location, which would degrade the wall component, such as, for example, of pectin.

After submission to the temperature of 80 °C and subsequent drying of the micropyle, the puncture force increased compared to the force required to puncture the wet seeds, as shown in Figure 6c. Thus, the possible action of soluble enzymes in the micropyle is discarded in view of their deterioration by heat.

Thus, the presence of proteins in the cell wall can be an actual fact, in case of its synthesis in a similar manner to *Arabidopsis* seeds, which, according to Dekkers et al. (2013),

the expansin gene is expressed exclusively in the micropylar endosperm. According to the authors, although there are variations in cell wall architecture among different species, the weakening of the wall by specific proteins is widely conserved in the seed germination mechanism.

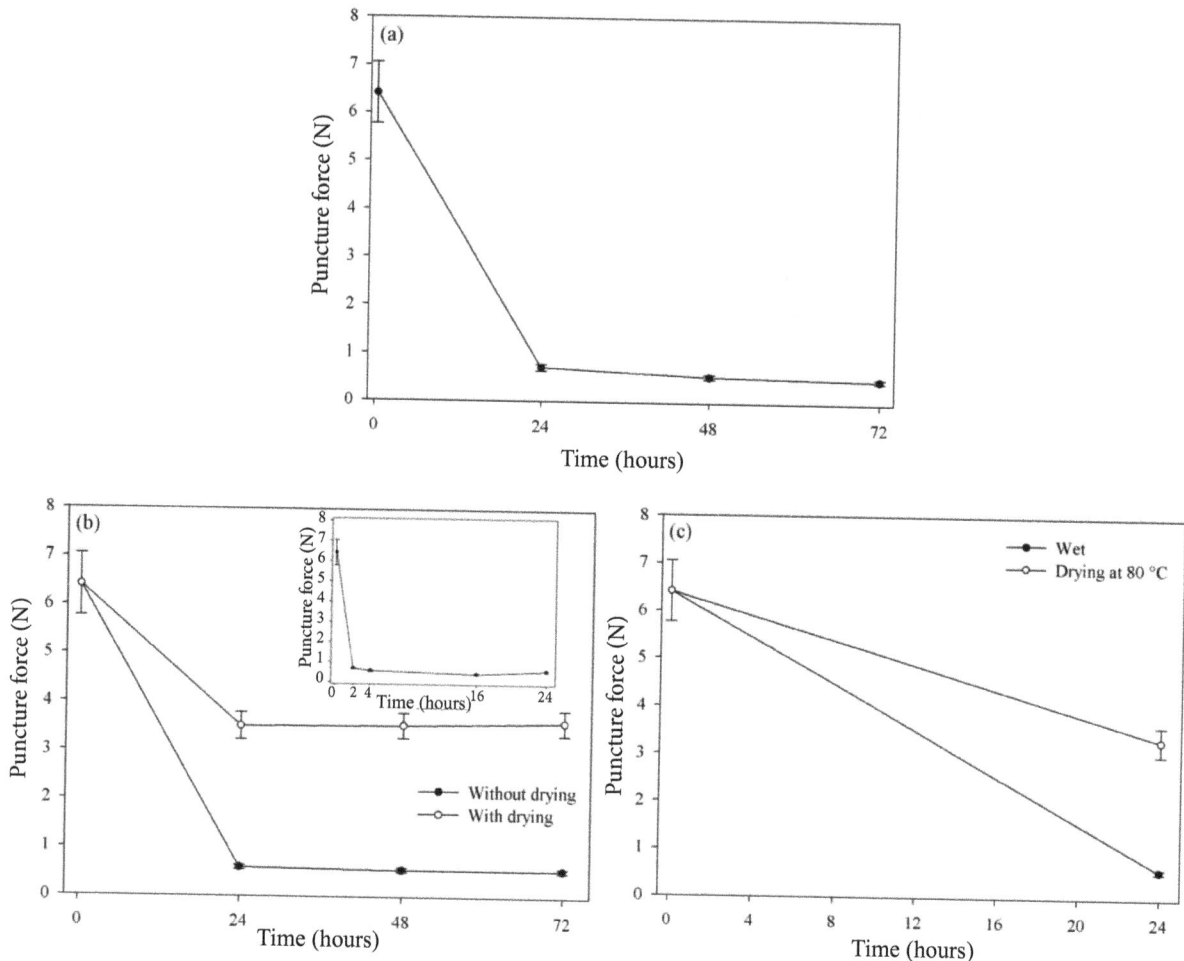

Figure 6. Required puncture force to rupture (N) the micropylar area of whole seeds (a) the isolated micropyle (b) and after subjection to a temperature of 80 °C (c) in *Melanoxylon brauna* seeds during soaking at a temperature of 30 °C. +– Standard error. The detail in Figure 6b is the puncture force in the micropyle isolated within the first 24 hours.

Considering the results obtained in the initial hydration phase, expansin would be one of the candidates to work in the elastic component of the micropyle wall. According to Cosgrove (2000), one of the features of expansin is the immediate extension of the wall, which begins within seconds after the extract containing expansin being placed in contact with the substrate, besides not causing continuous wall weakening.

The analysis of the micropyle cell wall composition (Figure 7) shows the possible formation of galactomannan, a cell wall reserve material, and rhamnogalacturonan, characteristic polysaccharide of pectin, responsible for wall stability by forming bonds with calcium or boron (Peaucelle

et al., 2012). Apparently, the presence of rhamnogalacturonan, consisting in rhamnose and galactose, is small, which allows the expansion of the wall due to increased turgor being easier. Moreover, the absence of xylose and glucose eliminates the possibility of xyloglucan-cellulose intersection, which would give greater resistance to the turgor wall.

It is evident that the micropyle cell wall is not structured as in other composite tissues of cellulose-hemicellulose and pectin. There were no changes in the sugars concentrations, assuming that there was no participation of hydrolytic enzymes, with the initial expansion of the wall being due to reduction in strength.

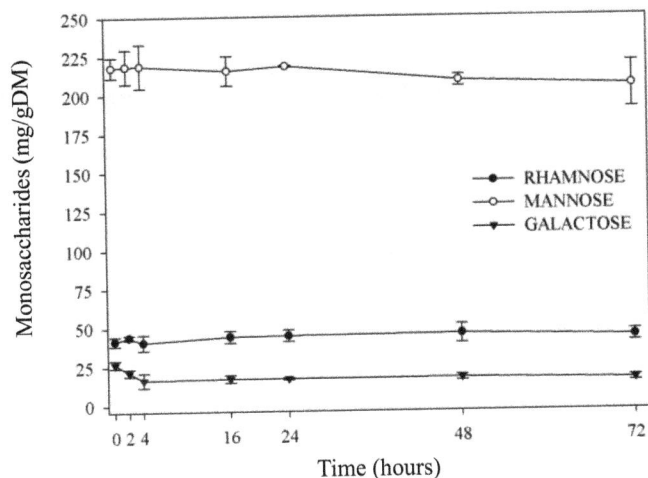

Figure 7. Cell wall of monosaccharides of the micropylar area of *Melanoxylon brauna* seeds during soaking at a temperature of 30 °C. +– Standard error.

According to Prietruszka and Lewicka (2007), the growth results from increase and irreversible deformation of the cell wall, while the elastic extension is not permanent and is reversed when the force is removed. According to the authors, the polymers, as the cell wall components are inert and have a stable elastic behavior at a wide temperature range, suggest that the elastic behavior is purely physical. Thus, the micropylar wall plastic component resistance of the *M. brauna* seeds is overcome by metabolic action, such as, for example, enzymes attached to the micropyle wall or exuded from the embryo axis.

Conclusions

The integument and cotyledons are responsible for 90% of the water soaked by the *Melanoxylon brauna* seeds within 72 hours;

The events in the embryonic axis and in the micropyle occur independently in the first 16 hours;

The weakening of the micropyle occurs in two stages, one plastic and another elastic;

Enzymes polygalacturonase and pectin methylesterase are involved in cell expansion of the embryonic axis;

Germination of *M. brauna* seeds comprises the change in plasticity of the micropyle wall by a physical effect (phase I), followed by permanent alteration of the wall (plastic element) by the action of enzymes or proteins (phases II-III).

Acknowledgments

The authors thank CNPq [Conselho Nacional de Desenvolvimento Científico e Tecnológico (National Counsel of Technological and Scientific Development)] for the productivity scholarship for the first author and Professor Lúcio Alberto de Miranda Gomide, of Universidade Federal de Viçosa, for making available the texture analyzer.

References

ALSCHER, R.G.; ERTURK, N.; HEATH, L.S. Role of superoxide dismutase (SODs) in controlling oxidative stress. *Journal of Experimental Botany*, v.53, p.1331-1341, 2002.http://jxb-oxfordjournals-org.ez35.periodicos.capes.gov.br/content/53/372/1331.full.pdf+html

ANDERSON, M.D.; PRASAD, T.K.; STEWART, C.R. Changes in isozyme profiles of catalase, peroxidase, and glutathione reductase during acclimation to chilling in mesocotyls of maize seedlings. *Plant Physiology*, v.109, p.1247-1257, 1995. http://www.ncbi.nlm.nih.gov/pubmed/12228666

ATAÍDE, G.M.; BORGES, E.E.L.; GONÇALVES, J.F.C.; GUIMARÃES, V.M.; FLORES, A.V.; BICALHO, E.M. Activities of alfa-galactosidase and poligalacturonase during hydratation of *Dalbergia nigra* (Vell.) Fr All. Ex Benth) seed at different temperatures. *Journal of Seed Science*, v.35, n.1, p.92-98, 2013.http://www.scielo.br/scielo.php?pid=S2317-15372013000100013&script=sci_arttext

BEAUCHAMP, C.; FRIDOVICH, I. Superoxide dismutase improved assays and assay applicable to acrylamide gels. *Analytical Biochemistry*, v.44, p.276-287, 1971. http://www.ncbi.nlm.nih.gov/pubmed/4943714

BEDARD, K.; LARDY, B.; KRAUSE, K.H. NOX family NADPH oxidases: Not just in mammals. *Biochimie*, v.89, p.1107-1112, 2007.http://www.ncbi.nlm.nih.gov/pubmed/17400358

BEWLEY, J.D.; BRADFORD, K.J.; HILHORST, H.W.M.; NONOGAKI, H. *Seeds*: physiology of development, germination and dormancy. Nova York: Springer, 2013. 392 p.

BORGES, E.E.L.; BORGES, R.C.G.; BUCKERIDGE, M.S. Alterações nas composições de carboidratos e de ácido graxos em sementes de jacarandá-da-bahia osmocondicionadas. *Revista Brasileira de Fisiologia Vegetal*, v.12, n.1, p.10-16, 2000. http://www.cnpdia.embrapa.br/rbfv/pdfs/v12n1p10.pdf

BORGES, E.E.L.; FLORES, A.V.; ATAÍDE, G.M.; MATOS, A.C.B. Alterações fisiológicas e atividade enzimática em sementes armazenadas de *Melanoxylon brauna* Schott. *Cerne*, v.21, n.1, p.75-81, 2015. http://www.scielo.br/pdf/cerne/v21n1/2317-6342-cerne-21-01-00075.pdf

BRADFORD, M.M. A rapid and sensitive method for the quantification of microgram quantities of proteins utilizing the principle of protein-dye binding. *Analytical Biochemistry*, v.72, p.248-254, 1976. http://www.ncbi.nlm.nih.gov/pubmed/942051

CARPITA, N.G.; GILBEAUT, D.M. Structural models of primary cell walls in flowering plants: consistency of molecular structure with the physical properties of the walls during growth. *Plant Physiology*, v.3, p.1-3, 1993. http://www.ncbi.nlm.nih.gov/pubmed/8401598

CHANCE, B.; MAEHLEY, A.C. Assay of catalase and peroxidase. *Methods in Enzymology*, v.2, p.764-775, 1995.http://www.ncbi.nlm.nih.gov/pubmed/13193536

COSGROVE, D.J. Loosening of plant cell walls by expansins. *Nature*, v.407, n.2, p.321-326, 2000. http://www.nature.com/nature/journal/v407/n6802/pdf/407321a0.pdf

DEKKERS, B.J.W.; PEARCE, S.; BOLDEREN-VELDKAMP, R.P.; MARSHALL, A.; WIDERA, P.; GILBERT, J.; DROST, H.; BASSEL, G.W.; MÜLLER, K.; KING, J.R.; WOOD, A.T.A.; GROSSE, I.; QUINT, M.; KRASNOGOR, M.; LEUBNER-METZGER, G.; HOLDSWORTH, M.J.; BENTSINK, L. Transcriptional dynamics of two seed compartments with opposing roles in *Arabidopsis* seed germination. *Plant Physiology*, v.163, n.1, p.205-215, 2013. http://www.ncbi.nlm.nih.gov/pubmed/23858430

DEL LONGO, O.T.; GONZÁLEZ, C.A.; PASTORI, G.M.; TRIPPI, V.S. Antioxidant defenses under hyperoxygenic and hyperosmotic conditions in leaves of two lines of maize with differential to drought. *Plant Cell Physiology*, v.37, n.7, p.1023-1028, 1993. http://pcp.oxfordjournals.org/content/34/7/1023.full.pdf

ÉAUX, B.; TOLEDANO, M.B. Ros as signalling molecules: mechanisms that generate specificity in ROS homeostasis. *Nature Reviews Molecular Cell Biology*, v.8, p.813-824, 2007. http://www.ncbi.nlm.nih.gov/pubmed/17848967

ENGLYST, H.N.; CUMMINGS, J.H. Simplified method for the measurement of total non-starch polysaccharides by gas-liquid chromatography of constituent sugar as alditol acetates. *Analyst*, v.109, p.937-942, 1984. http://pubs.rsc.org/en/Content/ArticleLanding/1984/AN/an9840900937

FLORES, A.V.; BORGES, E.E.L.; GUIMARÃES, V.M.; GONÇALVES, J.F.C.; ATAIDE, G.M.; BARROS, D.P. Atividade enzimática durante a germinação de sementes de *Melanoxylon brauna* Schott sob diferentes temperaturas. *Cerne*, v.20, n.3, p.401-408, 2014a. http://www.redalyc.org/articulo.oa?id=74432265009

FLORES, A.V.; BORGES, E.E.L.; GUIMARÃES, V.M.; ATAIDE, G.M.; CASTRO, R.V.O. Germinação de sementes de *Melanoxylon brauna* Schott em diferentes temperaturas. *Revista Árvore*, v.38, n.6, p.1147-1154, 2014b. http://www.scielo.br/pdf/rarv/v38n6/a19v38n6.pdf

GAY, C.; GEBICKI, J.M. A critical evaluation of the effect of sorbitol on the ferric-xylenol orange hydroperoxide assay. *Analytical Biochemistry*, v.284, p.217-220, 2000. http://www.ncbi.nlm.nih.gov/pubmed/10964403

GOMES, M.G.; GARCIA, Q.S. Reactive oxygen species and seed germination. *Biologia*, v.68, p.351-357, 2013. http://link.springer.com/article/10.2478%2Fs11756-013-0161

GRSIC-RAUSCH, S.; RAUSCH, T. A coupled spectrophotometric enzyme assay for the determination of pectin methylesterase activity and its inhibition by proteinaceous inhibitors. *Analytical Biochemistry*, v.333, n.1, p.14-18, 2004. http://www.sciencedirect.com/science/article/pii/S000326970400404X

GUIMARÃES, V.M.; REZENDE, S.T.; MOREIRA, M.A.; BARROS, E.G.; FELIX, C.R. Characterization of α-galactosidases from germinating soybean seed and their use for hydrolysis of oligosaccharides. *Phytochemistry*, v.58, n.1, p.67-73, 2001. http://www.ncbi.nlm.nih.gov/pubmed/11524115

HODGES, D.M.; ANDREWS, C.J.; JOHNSON, D.A.; HAMILTON, R.I. Antioxidant enzyme responses to chilling stress in differentially sensitive inbred maize lines. *Journal of Experimental Botany*, v.48, n.310, p.1105-1113, 1997. http://jxb.oxfordjournals.org/content/48/5/1105.full.pdf

KAR, M.; MISHRA, D. Catalase, peroxidase, and polyphenolxidase activities during rice leaf senescence. *Plant Physiology*, v.57, p.315-319, 1976. http://www.ncbi.nlm.nih.gov/pmc/articles/PMC542015/

KARMER, I.; ROACH, T.; BECKETT, R.P.; WHITAKER, C.; MINIBAYEVA, F.V. Extracellular production of reactive oxygen species during seed germination and early seedling growth in *Pisum sativum. Journal of Plant Physiology*, v.167, p.805-811, 2010. http://www.sciencedirect.com/science/article/pii/S0176161710000945

LIU, N.; LIN, Z.; GUAN, L.; GAUGHAN, G.; LIN, G. Antioxidant enzymes regulate reactive oxygen species during pod elongation in *Pisum sativum* and *Brassica chinensis. PLoS ONE*, v.9, n.2, p.1-9, 2014.http://journals.plos.org/plosone/article?id=10.1371/journal.pone.0087588

LORENZI, H. Árvores brasileiras: manual de identificação e cultivo de plantas arbóreas nativas do Brasil. Nova Odessa: Plantarum, 2009. 384p.

MATOS, A.C.B.; BORGES, E.E.L.; SEKITA, M.C. Produção de espécies reativas de oxigênio em sementes de *Dalbergia nigra* sob estresse térmico. *Journal of Seed Science*, v.36, n.3, p.282-289, 2014. http://www.scielo.br/scielo.php?pid=S2317-15372014000300002&script=sci_arttext

MILLER, G.L. Use of dinitrosalicylic acid reagent for determination of reducing sugar. *Analytical Chemistry*, v.31, n.3, p.426-428, 1959. http://pubs.acs.org/doi/abs/10.1021/ac60147a030

MMA – Ministério do Meio Ambiente. Instrução Normativa n° 6, de 23 de dezembro de 2008. *Lista Oficial das Espécies da Flora Brasileira Ameaçada de Extinção*. 2008. http://www.mma.gov.br/estruturas/ascom_boletins/_arquivos/83_19092008034949.pdf

NONOGAKI, H. Seed dormancy and germination-emerging mechanism and new hypotheses. *Frontier Plant Science*, v.5, n.233, p.1-14, 2014. http://www.ncbi.nlm.nih.gov/pubmed/24904627

PARK, Y.B.; COSGROVE, D.J. Changes in cell wall biomechanical properties in the xyloglucan-deficient xxt1/xxt2 mutant of *Arabidopsis. Plant Physiology*, v.158, n.1, p.465-475, 2012. http://www.ncbi.nlm.nih.gov/pubmed/22108526

PEAUCELLE, A.; BRAYBROOK, S.; HÖFTE, H. Cell wall mechanics and growth control in plant: the role of pectins revisited. *Frontier Plant Science*, v.3, n.121, 2012. http://journal.frontiersin.org/Journal/10.3389/fpls.2012.00121/abstract

PEIXOTO, P.H.P.; CAMBRAIA, J.; SANTANA, R.; MOSQUIM, P.R.; MOREIRA, M.A. Aluminum effects on lipid peroxidation and on the activities of enzymes of oxidative metabolism in sorghum. *Revista Brasileira de Fisiologia Vegetal*, v.11, p.137-143, 1999. http://www.cnpdia.embrapa.br/rbfv/pdfs/v11n3p137.pdf

PIETRUSZKA, M.; LEWICKA, S. Effect of temperature on plant elongation and cell wall extensibility. *General Physiology and Biophysics*, v.26, n.1, p.40-47, 2007. http://www.ncbi.nlm.nih.gov/pubmed/17579253#

PINTO, L.K.A.; MARTINS, M.L.L.; RESENDE, E.D.; THIEBAUT, J.T.L. Atividade de pectina metilesterase e da β-galactosidase durante o amadurecimento do mamão cv golden. *Revista Brasileira de Fruticultura*, v.33, n.3, p.713-722, 2011. http://www.scielo.br/scielo.php?pid=S0100-29452011000300004&script=sci_arttext

RICHARDS, S.L.; WILKINS, K.A.; SWARBRECK, S.M.; ANDERSON, A.A.; HABIB, N.; SMITEH, A.G.; McANISH, M.; DAVIES, J.M. The hydroxyl radical in plants: from seed to seed. *Journal Experimental Botany*, v.66, n.1, p.37-46, 2015. http://jxb.oxfordjournals.org/content/66/1/37.full.pdf+html

SCHELER, C.; WEITBRECHT, K.; PEARCE, S.P.; HAMPSTEAD, A. BÜTTNER-MAINIK, A.; LEE, K.J.D.; VOEGELE, A.; ORACZ, K.; DEKKERS, B.J.W.; WANG, X.; WOOD, A.T.A.; BENTSINK, L.; KING, J.R.; KNOX, J.P.; HOLDSWORTH, M.J.; MÜLLER, K.; LEUBNER-METZGER, G. Promotion of testa rupture during garden cress germination involves seed compartment-specific expression and activity of pectin methylesterases. *Plant Physiology*, v.167, p.200-215, 2015. http://www.plantphysiology.org/content/167/1/200.full.pdf+html

SCHOPFER, P. Hydroxyl radical-induced cell-wall loosening in vitro and in vivo implications for the control of elongation growth. *Plant Journal*, v.28, p.678-688, 2001. http://www.ncbi.nlm.nih.gov/pubmed/11851914

SUSLOV, D.; VERBELEN, J.P.; VISSEMBERG, K. Is acid-induced extension in seed plants only protein mediated? *Plant signaling & Behavior*, v.5, n.6, p.757-759, 2010. http://www.ncbi.nlm.nih.gov/pmc/articles/PMC3001582/

TENHAKEN, R. Cell wall remodeling under abiotic stress. *Frontiers in Plant Science*, v.5, article 771, p.1-9, 2015. http://journal.frontiersin.org/article/10.3389/fpls.2014.00771/full

VANLERBERGHE, G.C. Alternative oxidase: a mitochondrial respiratory pathway to maintain metabolic and signaling homeostasis during abiotic and biotic stress in plant. *International Journal of Molecular Science*, v.14, n.4, p.6805-6847, 2013. http://www.mdpi.com/1422-0067/14/4/6805

WAKABAYASHI, K.; HOSON, T.; HUBER, D.J. Methyl de-esterification as a major factor regulating the extent of pectin depolimerization during fruit ripening: a comparasion of the action of avocado (*Persea americana*) and tomato (*Lycopersicon esculentum*). *Journal of Plant Physiology*, v.160, n.6, p.667-673, 2003. http://ac.els-cdn.com/S0176161704704514/1-s2.0-S0176161704704514-main.pdf?_tid=a17c87d0-dd61-11e4-aa23-00000aab0f26&acdnat=1428437391_a705745c8e2af0a89e5e67531ab31a41

WEITBRECHT, K.; MÜLLER, K.; LEUBNER-METZGER, G. First off the mark: early seed germination. *Journal of Experimental Botany*, v.62, n.10, p.3289–3309, 2011. http://jxb.oxfordjournals.org/content/62/10/3289.full.pdf+html

8

Accelerated aging test in niger seeds

Carla Regina Baptista Gordin[1]*, Silvana de Paula Quintão Scalon[1],
Tathiana Elisa Masetto[1]

ABSTRACT – Niger is a promising oilseed species for biodiesel production but there is no much information about the physiological potential of its seeds. Thus, the aim was to adapt the methodologies of accelerated aging test on six lots of niger seeds. The test was carried out by traditional and with saturated salt solution (20 and 40 g NaCl.100 mL^{-1}) methods at 41 and 45 °C for 24, 48, 72 and 96 hours. After the decay period, the seeds were submitted to the germination test, proceeding to an evaluation on the seventh day after sowing, counting the normal seedlings percentage. A completely randomized design with four replications of 50 seeds was used and the means were compared by Tukey's test. The accelerated aging test was correlated with seedling emergence and provided lots classification in at least two levels of vigor. For the accelerated aging test, the method with 20 g NaCl.100 mL^{-1} at 41 °C for 24 hours is recommend. The traditional method is not suitable because it provides water content variation between samples above what is tolerable.

Index terms: biodiesel, *Guizotia abyssinica* (L.f.) Cass., oilseed, vigor.

Teste de envelhecimento acelerado em sementes de niger

RESUMO – O niger é uma espécie oleaginosa de interesse para a produção de biodiesel com poucas informações sobre o potencial fisiológico de suas sementes. Portanto, objetivou-se adequar as metodologias do teste de envelhecimento acelerado para a avaliação do vigor de seis lotes de sementes de niger. Utilizaram-se os métodos tradicional e com solução saturada de sal (20 e 40 g de NaCl.100 mL^{-1}), nas temperaturas de 41 e 45 °C por 24, 48, 72 e 96 horas. Após o período de deterioração, as sementes foram submetidas ao teste de germinação e procedeu-se a avaliação, aos sete dias após a semeadura, contabilizando-se a porcentagem de plântulas normais. O delineamento experimental utilizado foi o inteiramente casualizado, com quatro repetições de 50 sementes e as médias comparadas pelo teste de Tukey. O teste de envelhecimento acelerado correlacionou-se com o teste de emergência de plântulas e proporcionou a estratificação dos lotes em pelo menos dois níveis de vigor, recomendando-se o método com 20 g NaCl.100 mL^{-1} a 41 °C por 24 horas. O método tradicional não é recomendado por proporcionar variação do teor de água entre as amostras, superior ao tolerável.

Termos para indexação: biodiesel, *Guizotia abyssinica* (L.f.) Cass., oleaginosa, vigor.

Introduction

Among the species with potential for commercial production of biodiesel, niger (*Guizota abyssinica* (L.f.) Cass.) is highlighted, characterized by high oil production (30% of the weight of the seeds), with a high content of linoleic acid. In tropical Africa, its center of origin, the oil has many uses, especially in food, and in Brazil, it is considered promising for the production of biomass when used as ground cover in autumn/winter (Getinet and Sharma, 1996; Kuo et al., 2007; Carneiro et al., 2008; Sarin et al., 2009; Solomon and Zewdu, 2009). However, although the species has economic potential, there is little information concerning the evaluation of the seeds physiological potential, as production technologies and standards for marketing them are still nonexistent.

In the evaluation of the seeds physiological quality, germination test provides essential information on the best conditions for germination in order to exploit the seeds full potential. However, it does not provide necessary information on the capacity of seeds lots to set up a stand in adverse field conditions (Ventura et al., 2012). Thus, vigor tests may be used to assess the satisfactory stand establishment under different environmental conditions, complementing the information provided by the germination test (Grey et al., 2011).

[1]Universidade Federal da Grande Dourados, Caixa Postal 533, 79804-970 – Dourados, MS, Brasil.
*Corresponding author <carlagordin@ufgd.edu.br>

Among the vigor tests, the accelerated aging test involves subjecting the seeds to high temperatures and relative humidities, simulating normal storage conditions, but with an increase in the decay rate (Moncaleano-Escandon et al., 2013).

In order to avoid uneven water absorption among the samples, which may result in a differentiated decay, affecting post-aging results, Jianhua and McDonald (1996) have proposed the replacement of water by saturated salt solutions. With this procedure, there is reduction of the medium relative humidity, ensuring that the effects of decay should be due to temperature and exposure period (Moncaleano-Escandon et al., 2013).

Thus, due to the species economic potential and the need for knowledge about assessment methods of the vigor of its seeds, this study aimed to adapt the methods of traditional accelerated aging tests and with saturated salt solutions to evaluate the niger seeds vigor.

Material and Methods

The study was conducted at the Seed Technology Laboratory of the Faculty of Agricultural Sciences at University Federal of Grande Dourados (UFGD) in 2013. Six lots of niger seeds were used, one being produced in the 2009/2010 harvest (Lot 1), in the Brazilian municipality of Primavera do Leste, MT, four produced at different times in the 2011/2012 harvest (Lots 2 to 5) on Experimental Farm of the Faculty of Agricultural Sciences at UFGD (FAECA) and the last one also produced at FAECA, in the 2012/2013 harvest (Lot 6). The seeds were kept in paper packaging and stored in a cold and dry room (15 °C and 45% RH) until the establishment of the experiments.

The lots were initially assessed for the following tests and determinations: moisture content, performed using the method of the oven at 105 ± 3 °C for 24 hours, with four replications, according to Brasil (2009); germination test conducted on Germitest® paper dampened at the equivalent of 2.5 times the dry paper mass, inside "gerbox"-type germination boxes, placed in a B.O.D (Biochemical Oxygen Demand)-type germination chamber regulated at 25 °C with continuous light (six Philips® "daylight"-type fluorescent lamps of 20 watts and photon irradiance of 32.85 µmol. $m^{-2}.s^{-1}$), using four replications of 50 seeds (Gordin et al., 2012). Evaluations were performed seven days after sowing, recording the germination percentage, taking into account the formation of normal seedlings (developed shoot and root system); germination first count held together with the germination test, counting the number of normal seedlings obtained on the third

day after sowing, according to results obtained in preliminary tests; germination speed index, according to Maguire (1962); mean germination time, according to the formula proposed by Edmond and Drapalla (1958); and seedling length and dry matter by randomly choosing ten normal seedlings, measured with a digital caliper and dried in an oven at 65 °C for 72 hours, followed by weighing on a precision scale.

Seedling emergence under controlled conditions was obtained from seedings in trays filled with dystrophic red latosol, wrapped in a greenhouse coated with Sombrite®, with 30% dimming in the experimental area at FCA/ UFGD. The temperature and average relative humidity in the experiment conduction period were 31 °C and 58%. The speed index and the mean emergence time were recorded and, at 15 days after planting, the emergence percentage and dry matter length and seedling assessments were held, obtained in the same way as for the germination test; the initial stand was held in conjunction with the field emergence testing, registering the number of seedlings emerged on the third day after sowing, according to results obtained in pretests.

The traditional accelerated aging test was performed for each lot in individual chambers made of wire mesh suspended inside, where an even layer of niger seeds (1 g) was distributed. Within each individual compartment were added 40 mL of distilled water, being 100% relative humidity (Jianhua and McDonald, 1996) and the boxes were placed in the greenhouse at 41 and 45 °C for 24, 48, 72 and 96 hours. After these periods, the seeds were immersed for five minutes in 2% sodium hypochlorite, washed in distilled water and subjected to the germination test (Gordin et al., 2012), recording the germination percentage at seven days after sowing. The moisture content of the seeds was also determined (Brasil, 2009) before and after the aging period. For the accelerated aging test with a saturated solution, the same methodology as for the traditional test was used, replacing the distilled water by 40 mL of saturated sodium chloride (NaCl) in the concentrations of 20 and 40 g.100 mL^{-1}, corresponding to 76 and 55% of relative humidity, respectively (Jianhua and McDonald, 1996).

A completely randomized design with four replications of 50 seeds was used. The data were submitted to tests of normality and homogeneity of variance and then to ANOVA and, if significant, the averages were compared by Tukey's test at 5% probability, by means of the computer program SISVAR® (Ferreira, 2011). Later, the simple Pearson correlation coefficients (r) were calculated among the initial characterization and accelerated aging testing and field emergence testing, determining the significance of the r values by the t-test at 5% probability.

Results and Discussion

Lots of niger seeds differed as to seed physiological quality in all traits assessed, highlighting lot 6 as having more vigor than the others, except for the percentage of germination and seedling length held during the emergence test, which was not sensitive in detecting the differences between the lots (Table 1).

By the first count of germination, lots 1, 2 and 4 had intermediate vigor, noting that lots 3 and 5 showed the lowest and 6 the highest vigor compared to the others. By the germination speed index, lot 6 was also considered superior compared to the others, as well as by the mean germination time, which also identified lots 1, 2, 3 and 4 with high vigor compared to lot 5. The high vigor of lot 6 was also identified by the seedlings length and dry matter (both in the germination and emergence tests), which also found lots 1, 2, 4 and 5 with high vigor compared to the others (Table 1).

Table 1. Seed germination and seedlings emergence of niger (*Guizotia abyssinica*) seeds from different lots.

Lots	Germination						Emergence					
	G	FC	GSI	MGT	TL	DM	E	IS	ESI	MET	TL	DM
	(%)			days	mm	g.seedling^{-1}	(%)			days	mm	g.seedling^{-1}
1	86 a[1]	51 bc	6.19 b	2.40 ab	53.9 ab	0.0028 ab	33 b	9 b	1.33 c	5.23 b	65.4 a	0.0042 ab
2	81 a	48 bc	9.25 b	1.98 a	26.4 b	0.0033 a	38 b	26 b	2.53 bc	4.13 ab	69.0 a	0.0037 ab
3	85 a	29 c	5.15 b	2.20 ab	46.8 b	0.0023 b	30 b	20 b	1.48 c	4.18 ab	73.1 a	0.0028 b
4	84 a	53 bc	8.46 b	2.09 ab	57.6 ab	0.0030 a	44 ab	27 b	3.41 b	3.48 a	65.8 a	0.0052 a
5	84 a	59 ab	5.31 b	2.61 b	48.2 ab	0.0030 a	41 ab	18 b	2.44 bc	5.00 ab	72.2 a	0.0044 ab
6	91 a	79 a	20.67 a	1.94 a	79.1 a	0.0030 a	71 a	55 a	7.35 a	3.80 ab	78.2 a	0.0043 ab
C. V. (%)	6.1	11.3	20.5	11.8	27.2	10.3	15.2	22.6	25.3	16.8	13.4	22.0

[1]Means followed by the same letter in the column do not differ among themselves by the Tukey's test at 5% probability. (G) germination; (FC) germination first count; (GSI) germination speed index; (MGT) mean germination time; (TL) seedlings total length; (DM) seedlings dry matter; (E) emergence; (IS) initial stand; (ESI) emergence speed index; (MET) mean emergence time; (TL) seedlings total length and (DM) seedlings dry matter.

In the seedling emergence test, the seeds of lot 6 showed up with more vigor than in the other lots by the percentage of emergence, initial stand and emergence speed index, and on the last test it was possible to see that lots 1 and 3 had lower emergence speed, being classified as less vigorous compared to lots 2, 4 and 5, of intermediate vigor. Lots 2-6 did not differ as to the mean germination time, with high vigor with respect to lot 1 (Table 1).

Often the germination test is not sensitive enough to detect differences between lots, as observed in this study, although obtained in different growing conditions, for it provides optimal conditions for the expression of the seeds maximum physiological potential. Thus, the need to complement their results by the vigor testing in seed lots with similar germination is confirmed (Bolek, 2010).

All initial characterization tests of the lots were significantly correlated ($p < 0.05$) with seedling emergence, except the evaluation of seedlings length conducted during the germination test (Table 2). There was a negative correlation of seedlings emergence test with seedlings mean germination time and emergence speed index and a positive correlation with other tests; however, the correlations with the seedling length test in a greenhouse and seedlings dry matter conducted in both environments were considered low. Thus, the percentage assessments, speed index, germination first count, emergence speed index and initial stand are indicated for the evaluation of the niger seeds physiological quality, as they estimate the seedlings emergence.

Table 2. Simple Pearson correlation coefficients (r) estimated between germination tests and seedling field emergence, in six lots of niger (*Guizotia abyssinica)* seeds.

	Germination						Emergence				
	G	FC	GSI	MGT	TL	DM	IS	ESI	MET	TL	DM
E	0.729**	0.900*	0.949**	-0.537**	0.753ns	0.386**	0.933**	0.992**	-0.465**	0.647**	0.408**

(*) Significant at 5% probability by the t-test; (**) Significant at 1% probability by the t-test; (ns) nonsignificant at 5% probability by the t-test; (E) emergence percentage; (G) germination percentage; (FC) germination first count; (GSI) germination speed index; (MGT) mean germination time; (TL) seedlings total length; (DM) seedlings dry matter; (IS) initial stand; (ESI) emergence speed index and (MET) mean emergence time.

In Tables 3 and 4 it is possible to see the values of the seeds moisture content before and after the period of exposure to high temperatures and concentrations of NaCl. Comparing the traditional and modified procedures, it was observed, in general, that the seed water content was quite high in the traditional method, at both temperatures used.

Table 3. Moisture content of niger (*Guizotia abyssinica*) seeds before (RH) and after the aging period (AA) at the temperature of 41 °C.

Lots	RH	Traditional AA				AA with salt saturation							
						20 g NaCl.100 mL^{-1}				40 g NaCl.100 mL^{-1}			
	(%)	24 h	48 h	72 h	96 h	24 h	48 h	72 h	96 h	24 h	48 h	72 h	96 h
1	9.4	42.2	46.4	59.5	59.2	13.2	13.3	13.6	10.6	8.1	9.0	8.5	8.5
2	8.1	42.6	48.9	54.9	50.3	14.2	13.5	14.3	11.8	9.3	10.1	9.4	10.1
3	10.5	45.8	46.1	47.8	46.9	15.3	13.1	13.8	11.9	10.4	10.5	9.4	9.3
4	9.3	44.5	52.4	55.7	59.1	15.7	13.5	14.6	12.4	9.8	9.5	10.0	9.2
5	8.4	46.0	54.3	50.3	40.2	15.6	11.7	13.1	12.3	9.4	8.6	8.7	8.2
6	8.5	48.3	46.8	46.8	42.8	13.6	12.6	12.7	10.9	9.7	9.6	9.2	8.9

Table 4. Moisture content of niger (*Guizotia abyssinica*) seeds before (RH) and after the aging period (AA) at the temperature of 45 °C.

Lots	RH	Traditional AA				AA with salt saturation							
						20 g NaCl.100 mL^{-1}				40 g NaCl.100 mL^{-1}			
	(%)	24 h	48 h	72 h	96 h	24 h	48 h	72 h	96 h	24 h	48 h	72 h	96 h
1	9.4	41.4	44.4	47.2	41.0	16.5	12.0	12.5	16.5	10.2	9.0	8.9	11.4
2	8.1	45.2	52.0	40.2	40.4	15.9	13.4	14.5	17.5	10.6	10.2	10.1	13.0
3	10.5	49.9	54.8	56.7	54.4	18.3	14.0	14.5	17.6	11.2	10.2	10.1	13.4
4	9.3	50.8	60.1	60.1	54.5	16.5	13.5	13.7	17.1	10.9	10.1	10.2	13.5
5	8.4	47.2	51.6	47.3	46.5	16.0	11.8	13.4	15.5	10.2	9.0	9.0	11.5
6	8.5	40.8	42.2	46.3	43.3	17.1	12.2	14.0	16.5	10.7	10.2	9.6	12.8

According to Peng et al. (2011), increasing exposure of the seeds to artificial aging leads to the accumulation of reactive oxygen species causing lipid peroxidation and therefore less integrity and selectivity of the membranes, allowing entry of water more quickly in the cells. However, the use of a saturated salt solution promotes greater control of the aging chamber relative humidity, providing delay of the water absorption by the seeds (Jianhua and McDonald, 1996).

Thus, the seeds tend to reach hygroscopic equilibrium at higher water contents as the relative humidity increases, according to what was verified by Nery et al. (2009) and Braz et al. (2008) in seeds of forage turnip (*Raphanus sativus* L. var *oleiferus* Metzg.) and sunflower (*Helianthus annus* L.), respectively, where lower and more uniform moisture contents were obtained with the use of a saturated salt solution during the aging period, compared to those observed for the seeds aged by the traditional method, providing lower and slower water absorption by the seeds of this species.

The initial moisture content of the seeds did not vary more than two percentage points among samples, as recommended by Marcos-Filho (2005). However, after the aging period, the traditional method provided, at both temperatures, the variation of water content among the samples, exceeding the tolerable (Table 3), and the presence of microorganisms was also seen. According to Marcos-Filho (2005), these factors are undesirable for performing the test because they compromise the results fidelity, as the wetter samples are more sensitive to more intense decay, and microorganisms impair seedling germination and development, causing uncertainty regarding their normality, and hindering the test interpretation.

Importantly, these effects have been detected in a more drastic manner in typically smaller seeds, which is a criterion that fits the niger seeds, which are on average 4.54 mm long, 1.39 mm wide and 1.15 mm thick (Gordin et al., 2012). These seeds absorb water more rapidly, characterizing phase I of the seed hydration three-phase pattern and, after that period, also referred to as phase II, there is little or no water absorption, because the seeds cells can no longer expand, and it is possible to notice, already from this stage, the activation of metabolic processes required for embryo growth and early germination processes (Castro et al., 2004).

The accelerated aging tests results were significant (p < 0.05) in all methods used. In general, tests performed at the temperature of 41 °C were effective at distinguishing the lots in terms of the physiological quality in three levels of vigor, except in times of 72 hours in the traditional procedure and 96 hours with a saturated solution at a concentration of 20 g NaCl.100 mL⁻¹, which provided a distinction of lots on two vigor levels, as in the seedling emergence test. However, in all methods lot 6 was considered as having high vigor compared to the others (Table 5).

The temperature of 45 °C was harmful for the preliminary evaluation of the seeds physiological potential, and it was found that when increasing the seeds exposure time there was no germination. The deleterious effects of rising temperatures associated with stress exposure time were more intense in the traditional accelerated aging, which did not provide the germination of lots from 48 hours, and with saturated solutions these results were observed only within 96 hours of exposure to stress (Table 6).

Table 5. Initial percentages of seed germination (G) and seedling field emergence (E) of niger (*Guizotia abyssinica*) seeds and after the accelerated aging (AA) periods by the traditional procedures and with saturated salt solution (NaCl) at the temperature of 41 °C.

Lot	G	E	Traditional AA				AA with salt							
							20 g NaCl.100 mL⁻¹				40 g NaCl.100 mL⁻¹			
			24 h	48 h	72 h	96 h	24 h	48 h	72 h	96 h	24 h	48 h	72 h	96 h
							(%)							
1	86 a¹	33 b	28 c	25 b	27 a	14 bc	15 c	20 c	21 bc	26 b	18 c	14 c	17 c	13 bc
2	81 a	38 b	47 b	20 b	11 b	15 b	23 bc	20 c	17 c	17 b	20 bc	21 bc	27 bc	13 bc
3	85 a	30 b	34 bc	0 c	29 a	4 d	20 c	17 c	25 bc	15 b	14 c	15 c	23 c	20 b
4	84 a	44 ab	31 c	27 b	28 a	6 cd	33 b	40 b	36 b	25 b	31 b	31 b	37 b	16 bc
5	86 a	41 ab	23 c	17 b	17 ab	6 cd	20 c	13 c	21 bc	17 b	17 c	15 c	19 c	11 c
6	91 a	71 a	67 a	46 a	16 ab	25 a	60 a	66 a	58 a	54 a	72 a	62 a	69 a	79 a
C. V. (%)	6.1	15.2	15.9	23.9	33.3	32.6	19.9	21.3	24.3	27.9	18.5	21.5	19.3	14.2

¹Means followed by the same letter in the column do not differ among themselves by the Tukey's test at 5% probability.

Table 6. Initial percentages of seed germination (G) and seedling field emergence (E) of niger (*Guizotia abyssinica*) seeds and after the accelerated aging (AA) periods by the traditional procedures and with saturated salt solution (NaCl) at the temperature of 45 °C.

Lot	G	E	Traditional AA				AA with salt							
							20 g NaCl.100 mL⁻¹				40 g NaCl.100 mL⁻¹			
			24 h	48 h	72 h	96 h	24 h	48 h	72 h	96 h	24 h	48 h	72 h	96 h
							(%)							
1	86 a¹	33 b	11 b	0	0	0	13 c	14 c	13 c	0	13 c	17 bc	21 b	0
2	81 a	38 b	18 b	0	0	0	25 bc	22 bc	19 bc	0	21 bc	19 bc	20 b	0
3	85 a	30 b	19 b	0	0	0	22 bc	15 bc	22 bc	0	19 bc	16 c	14 b	0
4	84 a	44 ab	22 b	0	0	0	29 b	28 b	31 b	0	27 b	32 b	24 b	0
5	86 a	41 ab	0 c	0	0	0	13 c	10 c	10 c	0	13 c	18 bc	12 b	0
6	91 a	71 a	64 a	0	0	0	66 a	71 a	65 a	0	64 a	60 a	62 a	0
C. V. (%)	6.1	15.2	35.4	0	0	0	25.0	23.4	23.8	0	23.3	25.4	26.2	0

¹Means followed by the same letter in the column do not differ among themselves by the Tukey's test at 5% probability.

This indicates that from a certain amount of exposure to aging treatment the seeds lose vigor, become more sensitive to stress during germination and subsequently are unable to promote the damages reparation, losing the ability to germinate (Rajjou and Debeaujon, 2008; Samarah and Al-Kofahi, 2008). In studies carried out by Bittencourt and Vieira (2006) and Mendes et al. (2010), maize (*Zea mays* L.) and castor (*Ricinus communis* L.) seeds germination reduction was observed after accelerated aging, with increasing temperature decay from 42 to 45 °C, in the first case, and from 41 to 45 °C, in the second one. The test completion on *Jatropha curcas* (common names include Barbados nut, purging nut, physic nut, or JCL (abbreviation of *Jatropha curcas* Linnaeus)) seeds by Pereira et al. (2012) has provided layering of the lots only at the temperature of 41 °C, being considered limiting the temperatures of 42 and 45 °C, while for wheat (*Triticum aestivum* L.) seeds the temperature of 45 °C was lethal (Maia et al., 2007).

Similar results were obtained by Lehner et al. (2008), where wheat (*Triticum aestivum* L.) seeds showed less sensitivity to accelerated aging in an environment with a relative humidity of 75% compared to the environment with 100% humidity. Probably, the test period was increased in the modified procedure due to the lesser degree of decay experienced by the seeds. Thus, the use of a saturated solution became important in controlling the aging chamber relative humidity, reducing the seeds decay, the variation between the samples water content and the infestation by microorganisms.

There was a significant correlation between the seedlings emergence and accelerated aging tests conducted for 24 hours by the traditional method at 41 °C. There was no correlation (r = 0) with tests conducted at 45 °C for 96 hours in the two salt concentrations, and after 48 hours in the conventional method (Table 7). The other methods were significantly correlated (p < 0.05) with seedling emergence, indicating that they are able to estimate this test, observing low and negative correlation only in the test conducted by the traditional method for 72 hours (Table 7).

However, given that the results accuracy was hampered by the samples water content variation, the traditional accelerated aging test was not recommended for the evaluation of the niger seeds physiological quality, even when it allowed the lots stratification in levels of vigor or correlated with the seedling emergence test, and testing conducted with a saturated salt solution can be used as an alternative.

Table 7. Simple Pearson correlation coefficients (r) estimated between accelerated aging tests (AA) and seedling field emergence in six lots of niger (*Guizotia abyssinica)* seeds.

| | Traditional AA | | | | AA with salt | | | | | | | |
| | | | | | 20 g NaCl.100 mL^{-1} | | | | 40 g NaCl.100 mL^{-1} | | | |
	24 h	48 h	72 h	96 h	24 h	48 h	72 h	96 h	24 h	48 h	72 h	96 h
41 °C	0.791[ns]	0.865[**]	-0.426[*]	0.748[**]	0.966[**]	0.919[*]	0.911[**]	0.920[**]	0.975[**]	0.966[**]	0.949[**]	0.912[**]
45 °C	0.863[**]	0.000	0.000	0.000	0.923[**]	0.947[*]	0.908[**]	0.000	0.943[**]	0.972[**]	0.936[**]	0.000

(*) Significant at 5% probability by the t-test; (**) Significant at 1% probability by the t-test; (ns) nonsignificant at 5% probability by the t-test and (AA) accelerated aging.

Among the studies on oil varieties, Braga Junior et al. (2011) have considered that the accelerated aging test modified with a saturated solution of 40 g NaCl.100 mL^{-1} at 40 °C for 48 hours was the most appropriate method for classification of lots of castor (*Ricinus communis* L.) seeds. On the other hand, Nery et al. (2009) have found that the accelerated aging test at 41 °C for 96 hours and with saturated solution of NaCl at 41 °C for 72 hours is effective in evaluating the physiological potential of forage turnip (*Raphanus sativus* L.) seeds. As for Amaro et al. (2014) and Lima et al. (2015), the accelerated aging test under the condition of 41 °C for 72 hours provides greater differentiation of vigor among lots of crambe abyssinica (*Crambe abyssinica* Hochst) seeds.

Thus, it appears that the methods used to conduct the test vary with the genotype observing difference in the behavior of seeds with the temperatures and times of exposure to the accelerated aging test as well as the use and the concentration of the saline solution in the same way that the rate of decay of seeds varies among species and among lots of seeds of the same species (Kibinza et al., 2011).

Conclusions

The accelerated aging test is efficient to evaluate the physiological potential of niger seeds, and the method with 20 g NaCl. 100 mL^{-1} is recommended at 41 °C for 24 hours.

The traditional method is not suitable due to providing variation of water content between the samples, exceeding what is tolerable.

References

AMARO, H.T.R.; DAVID, A.M.S.S.; SILVA NETA, I.C.; ASSIS, M.O.; ARAÚJO, E.F.; ARAÚJO, R.F. Teste de envelhecimento acelerado em sementes de crambe (*Crambe abyssinica* Hochst), cultivar FMS Brilhante. *Revista Ceres*, v.61, n.2, p.202-208, 2014. http://www.scielo.br/scielo.php?pid=S0034-737X2014000200007&script=sci_arttext

BITTENCOURT, S. R. M.; VIEIRA, R.D. Temperatura e período de exposição de sementes de milho no teste envelhecimento acelerado. *Revista Brasileira de Sementes*, v.28, n.3, p.161-168, 2006. http://www.scielo.br/pdf/rbs/v28n3/23.pdf

BOLEK, Y. Genetic variability among cotton genotypes for cold tolerance. *Field Crops Research*, v.119, n.1, p.59-67, 2010. http://www.sciencedirect.com/science/article/pii/S0378429010001620

BRAGA JUNIOR, J.M.; ROCHA, M.S.; BRUNO, R.L.A.; VIANA, J.S.; BELTRÃO, N.E.M. Teste de envelhecimento acelerado em sementes de mamona cultivar BRS – Energia. *Revista Eletrônica de Biologia*, v.4, n.1, p.88-101, 2011. http://revistas.pucsp.br/index.php/reb/article/view/2764/5847

BRASIL. Ministério da Agricultura, Pecuária e Abastecimento. *Regras para análise de sementes*. Ministério da Agricultura, Pecuária e Abastecimento. Secretaria de Defesa Agropecuária. Brasília: MAPA/ACS, 2009. 395p. http://www.agricultura.gov.br/arq_editor/file/2946_regras_analise__sementes.pdf

BRAZ, M.R.S.; BARROS, C.S.; CASTRO, F.P.; ROSSETTO, C.A.V. Testes de envelhecimento acelerado e deterioração controlada na avaliação do vigor de aquênios de girassol. *Ciência Rural*, v.38, n.7, p.1857-1863, 2008. http://www.scielo.br/pdf/cr/v38n7/a09v38n7.pdf

CARNEIRO, M.A.C.; CORDEIRO, M.A.S.; ASSIS, P.C.R.; MORAES, E.S.; PEREIRA, H.S.; PAULINO, H.B.; SOUZA, E.D. Produção de fitomassa de diferentes espécies de cobertura e suas alterações na atividade microbiana de solo de cerrado. *Bragantia*, v.67, n.2, p.455-462, 2008. http://www.scielo.br/pdf/brag/v67n2/a21v67n2.pdf

CASTRO, R.D.; BRADFORD, K.J.; HILHORST, H.W.M. Embebição e reativação do metabolismo. In: Ferreira, A.G.; Borghetti, F. (Eds.) *Germinação*: do básico ao aplicado. Porto Alegre, Artmed, 2004. p.149-162.

EDMOND, J.B.; DRAPALLA, W.J. The effects of temperature, sand and soil, and acetone on germination on okra seeds. *Proceedings of the American Society Horticultural Science*, v.71, p.428-34, 1958.

FERREIRA, D.F. Sisvar: a computer statistical analysis system. *Ciência e Agrotecnologia*, v.35, n.6, p.1039-1042, 2011. http://www.scielo.br/pdf/cagro/v35n6/a01v35n6.pdf

GETINET, A.; SHARMA, S.M. Niger (*Guizotia abyssinica* (L. f.) Cass. Promoting the conservation and use of underutilized and neglected crops 5. *Institute of Plant Genetics and Crop Plant Research*, Gatersleben/ International Plant Genetic Resources Institute, Rome, 1996. 59 p.

GORDIN, C.R.B.; MARQUES, R.F.M.; MASETTO, T.E.; SCALON, S.P.Q. Germinação, biometria de sementes e morfologia de plântulas de *Guizotia abyssinica* Cass. *Revista Brasileira de Sementes*, v.34, n.4, p.619-627, 2012. http://www.scielo.br/scielo.php?script=sci_arttext&pid=S0101-31222012000400013

GREY, T.; BEASLEY JUNIOR, J.P.; WEBSTER, T.M.; CHEN, C.Y. Peanut seed vigor evaluation using a thermal gradient. *International Journal of Agronomy*, v.2011, p.1-7, 2011. http://www.hindawi.com/journals/ija/2011/202341/

JIANHUA, Z.; McDONALD, M.B. The saturated salt accelerated aging test for small-seeded crops. *Seed Science and Technology*, v.25, p.123-131, 1996. http://agris.fao.org/agris-search/search.do?recordID=CH9700211

KIBINZA, S.; BAZIN, J.; BAILLY, C.; FARRANT, J.M.; CORBINEAU, F.; EL-MAAROUF-BOUTEAU, H. Catalase is a key enzyme in seed recovery from aging during priming. *Plant Science*, v.181, p.309-315, 2011. http://www.sciencedirect.com/science/article/pii/S016894521100166X

KUO, W.L.; CHEN, C.C.; CHANG, P.H.; CHENG, L.Y.; SHIEN, B.J.; HUANG, Y.L. Flavonoids from *Guizotia abyssinica*. *Journal of Chinese Medicine*, v.18, n.3, p.121-128, 2007. http://ejournal.nricm.edu.tw/upload/21614/18/1803-03.pdf

LEHNER, A.; MAMADOU, N.; PELS, P.; CÔME, D.; BAILLY, C.; CORBINEAU, F. Changes in soluble carbohydrates, lipid peroxidation and antioxidant enzyme activities in the embryo during aging in wheat grains. *Journal of Cereal Science*, v.47, p.555-565, 2008. http://www.sciencedirect.com/science/article/pii/S0733521007001312

LIMA, J.J.P.; FREITAS, M.N.; GUIMARÃES, R.M.; VIEIRA, A.R.; ÁVILA, M.A.B. Accelerated aging and electrical conductivity tests in crambe seeds. *Ciência e Agrotecnologia*, v.39, n.1, p.7-14, 2015. http://www.scielo.br/scielo.php?pid=S1413-70542015000100007&script=sci_arttext

MAGUIRE, J.B. Speed of germination-aid in selection and evaluation for seedling emergence vigor. *Crop Science*, v.2, n.2, p.176-177, 1962.

MAIA, A.R.; LOPES, J.C.; TEIXEIRA, C.O. Efeito do envelhecimento acelerado na avaliação da qualidade fisiológica de sementes de trigo. *Ciência e Agrotecnologia*, v.31, n.3, p.678-684, 2007. http://www.scielo.br/scielo.php?pid=S1413-70542007000300012&script=sci_arttext

MARCOS-FILHO, J. *Fisiologia de sementes de plantas cultivadas*. Piracicaba: FEALQ, 2005.495p.

MENDES, R.C.; DIAS, D.C.F.S.; PEREIRA, M.D.; DIAS, L.A.S. Testes de vigor para a avaliação do potencial fisiológico de sementes de mamona (*Ricinus communis* L.). *Ciência e Agrotecnologia*, v.34, n.1, p.114-120, 2010. http://www.scielo.br/pdf/cagro/v34n1/15.pdf

MONCALEANO-ESCANDON, J.; SILVA, B.C.F.; SILVA, S.R.S.; GRANJA, J.A.A.; ALVES, M.C.J.L.; POMPELLI, M.F. Germination responses of *Jatropha curcas* L. seeds to storage and aging. *Industrial Crops and Products*, v.44, p.684-690, 2013. http://www.sciencedirect.com/science/article/pii/S0926669012005092

NERY, M.C.; CARVALHO, M.L.M.; GUIMARÃES, R.M. Testes de vigor para avaliação da qualidade de sementes de nabo forrageiro. *Informativo Abrates*, v.19, n.1, 2009. http://www.abrates.org.br/images/stories/informativos/v19n1/artigo04.pdf

PENG, Q.; KONG, Z.; LIAO, X.; LIU, Y. Effects of accelerated aging on physiological and biochemical characteristics of waxy and non-waxy wheat seeds. *Journal of Northeast Agricultural University*, v.18, n.2, p.7-12, 2011. http://www.sciencedirect.com/science/article/pii/S1006810412600026

PEREIRA, M.D., MARTINS FILHO, S.; LAVIOLA, B.G. Envelhecimento acelerado de sementes de pinhão-manso. *Pesquisa Agropecuária Tropical*, v.42, n.1, p.119-123, 2012. http://www.scielo.br/pdf/pat/v42n1/17.pdf

RAJJOU, L.; DEBEAUJON, I. Seed longevity: Survival and maintenance of high germination ability of dry seeds. *Comptes Rendus Biologies*, v.331, p.796-805, 2008. http://www.sciencedirect.com/science/article/pii/S1631069108002011#

SAMARAH, N.H.; AL-KOFAHI, S. Relationship of seed quality tests to field emergence of artificial aged barley seeds in the Semiarid Mediterranean region. *Jordan Journal of Agricultural Sciences*, v.4, n.3, p.217-230, 2008. https://journals.ju.edu.jo/JJAS/article/viewFile/1001/994

SARIN, R.; SHARMA, M.; KHAN, A.A. Studies on *Guizotia abyssinica* L. oil: Biodiesel synthesis and process optimization. *Bioresource Technology*, v.100, p.4187-4192, 2009. http://www.ncbi.nlm.nih.gov/pubmed/19386491

SOLOMON, W.K.; ZEWDU, A.D. Moisture-dependent physical properties of niger (*Guizotia abyssinica* Cass.) seed. *Industrial crops and products*, v.29, p.165-170, 2009. http://www.researchgate.net/publication/238363949_Moisturedependent_physical_properties_of_niger_%28_Guizotia_abyssinica_Cass.%29_seed

VENTURA, L.; DONÀ, M.; MACOVEI, A.; CARBONERA, D.; BUTTAFAVA, A.; MONDONI, A.; ROSSI, G.; BALESTRAZZI, A. Understanding the molecular pathways associated with seed vigor. *Plant Physiology and Biochemistry*, v.60, p.196-206, 2012. http://www.ncbi.nlm.nih.gov/pubmed/22995217

9

Germination inhibits the growth of new roots and seedlings in *Eugenia uni lora* and *Eugenia brasiliensis*

Talita Silveira Amador[1], Claudio José Barbedo[1*]

ABSTRACT – Seeds of *Eugenia* species can produce new roots and whole plants even when much of its reserves is removed. However, new roots and seedlings rarely are formed spontaneously, and after cutting, each seed fragment usually produces only one new seedling, suggesting some control of the formation of several seedlings. It is possible, therefore, that germination leads to the production of inhibitory substances avoiding the development of new embryonic tissues. In the present work we have analyzed the potential of germinating seeds of *Eugenia uniflora* and *Eugenia brasiliensis* to inhibit new roots and seedling growth. Seeds were germinated after totally or partially fractionated. This last one was also totally fractionated after the development of a seedling, and the halves were also germinated. The results showed that the germination of the *E. uniflora* and *E. brasiliensis* seeds have inhibited the formation of new roots and seedlings.

Index terms: Myrtaceae, cutting seeds, recalcitrant seeds.

Potencial de inibição da formação de raízes e plântulas em sementes germinantes de pitangueira (*Eugenia uniflora*) e grumixameira (*E. brasiliensis*)

RESUMO – Sementes de espécies de *Eugenia* têm potencial para gerar novas raízes e até plantas inteiras mesmo após a remoção de grande parte de suas reservas. Contudo, a formação de novas raízes e plântulas raramente ocorre de forma espontânea e, quando as sementes são fracionadas, cada fragmento normalmente produz apenas uma nova plântula, sugerindo algum autocontrole na formação de várias plântulas. É possível, portanto, que uma vez iniciada a germinação, a semente produza substâncias inibitórias à diferenciação de novos tecidos embrionários. No presente trabalho analisou-se, em sementes de *Eugenia uniflora* e *Eugenia brasiliensis*, o potencial de inibição do crescimento de raízes e plântulas a partir do início da primeira germinação. Sementes dessas espécies foram submetidas a fracionamento total ou parcial (fissura) e colocadas para germinar. Após a germinação das fissuradas, em uma parte das mesmas o fracionamento foi completado, separando-se as metades, que foram também colocadas para germinar. Os resultados mostraram que a germinação de sementes de *E. uniflora* e *E. brasiliensis* inicia processos de inibição da regeneração de novas raízes e plântulas na semente.

Termos para indexação: Myrtaceae, fracionamento de sementes, semente recalcitrante.

Introduction

Eugenia (Myrtaceae) comprises some native species in Brazil already domesticated and of great economic importance, such as *Eugenia brasiliensis* Lam. (grumixama), *E. involucrata* DC. (cereja do Rio Grande), *E. pyriformis* Camb. (uvaia) and *E. uniflora* L. (pitanga). They are species that produce fleshy fruits that are suitable for fresh consumption or for industrialization and usually have few seeds (Delgado and Barbedo, 2007; Amador and Barbedo, 2011). These show a recalcitrant behavior, with different levels of tolerance to desiccation (Delgado and Barbedo, 2007; Delgado and Barbedo, 2012).

Eugenia species seeds have the potential to generate new roots and even whole plants, even after the removal of much of their reserves (Silva et al., 2003; Prataviera et al., 2015), which is rare in nature. This feature can be used technologically to increase the potential of plantlet production. Also, there is obvious interest in understanding the factors involved in this regenerative capacity that allows the formation of new seedlings when fractionated; *Eugenia* embryo anatomy, for example, has been studied (Justo et al., 2007; Delgado et al., 2010). Interestingly, however, the formation of new roots and seedlings rarely occurs spontaneously: some kind of injury apparently is necessary for them to be formed (Silva et al., 2005; Amador and Barbedo, 2011).

[1]Instituto de Botânica, Núcleo de Pesquisa em Sementes, Caixa Postal, 68041, 04301-012 – São Paulo, SP, Brasil.
*Corresponding author <claudio.barbedo@pesquisador.cnpq.br>

Despite the existence of polyembryony in species of the Myrtaceae family (Landrum and Kawasaki, 1997), the *Eugenia* embryos have been described as monoembryonic, which appear as globular structures, wherein the difference between the cotyledon and the radicle hypocotyl axis is visible only microscopically (Gurgel and Soubihe Sobrinho, 1951; Salomão and Allem, 2001; Justo et al., 2007; Delgado et al., 2010). However, since the development of new tissues depend on the fractionation, it is possible that injuries in seeds can start a process of inducing the formation of new roots and seedlings, or can block self-inhibition of these formations in germinating seeds. This is because it has been noticed that seeds that are fractionated can generate new seedlings, but rarely more than one new formation in each fragment, suggesting some self-control in the formation of several seedlings. Rizzini (1970) has found that *E. dysenterica* DC. seeds have substances that inhibit germination and that this inhibitory potential is increased when the embryo begins to germinate. Delgado and Barbedo (2011) have also found an inhibitor effect of germination on germinating seed extracts of *E. uniflora* when applied to lettuce and bean seeds. Thus, the study of inhibitors of germination can help elucidate the hypothesis of self-inhibition, as seen in coffee seeds (Pereira et al., 2002).

In studies of organogenesis, the rate of formation and the development of somatic or zygotic embryos are strongly influenced by the chemical composition of the environment. The balance among sugars, amino acids and growth regulators, for example, has shown effects that are sometimes inducers, sometimes inhibitors of the formation of new embryos (Deo et al., 2010; Kanwar et al., 2010; Karami and Saidi, 2010; Swamy et al., 2010). Thus, in *Eugenia* seeds, the beginning of the formation of roots or shoots could change the balance of those compounds, which makes an unfavorable medium for differentiation of new seedlings. Seeds of *E. stipitata* ssp. *sororia*, which also have the ability to regenerate embryos after fractionation, had seedling development from the seed area opposite to the damaged one, where apparently was the meristematic zone (Anjos and Ferraz, 1999). However, in the seed complementary fraction, seedlings were formed on the cut surface, which shows some polarity in mobilizing hormones. This may also be related to the fact that only after splitting the halves a second seedling can be regenerated, as shown in seeds of *E. pyriformis* (Amador and Barbedo, 2011). It is possible, therefore, that once initiated the germination, the seed produce substances that are inhibitory to the differentiation of new embryonic tissues by the migration of such substances from the germinating area to the others. Thus, a seed fraction could only begin to develop new roots and seedlings after being completely separated from the other

tissues or at least sufficiently spaced from the growth area of an existing seedling. However, there is still little information allowing to verify the occurrence of this inhibition in *Eugenia* species. In the present work, in seeds of *Eugenia uniflora* and *Eugenia brasiliensis*, the potential of inhibiting the growth of roots and seedlings from the beginning of the first germination was analyzed.

Material and Methods

Fruits of *E. uniflora* and *E. brasiliensis* of two stages of ripening in each species, identified by the characteristic color of each species in the dispersion (referred to as unripe and ripe) were collected at Institute of Botany, in the Brazilian city of São Paulo, SP (23°38' S and 46°37' W) and taken to the Laboratory of the Seed Research. The seeds were manually extracted from the fruit in sieves with running water and the fruit pulp residues were removed. After washing, the seeds remained at rest on germination paper to remove the residual surface water and were stored until the beginning of the experiments in a cold chamber at 7 °C (Kohama et al., 2006), not exceeding 15 days.

Then the seeds of both maturity stages in both species were classified by size, according to their larger diameter, as small (7.0 ± 1.0 mm in *E. uniflora* and 6.5 ± 0.5 mm in *E. brasiliensis*) and large (9.8 ± 0.2 mm in *E. uniflora* and 13.0 ± 1.0 mm in *E. brasiliensis*) and evaluated for water content (expressed as a percentage, on a wet basis) by the method of oven at 103 °C for 17 hours (ISTA, 2015) with three replications of five seeds each. Samples of seeds of each size and each maturity stage were submitted to germination test in a growth chamber at 25 ± 1 °C and 95 ± 5% relative humidity, with continuous light provided by four fluorescent lamps of 40 W each. The seeds were placed on paper roll for germination, pre-moistened with tap water (Brasil, 2009), in eight replications of ten seeds each. The number of roots and seedlings produced was registered every five days, until no longer issuing new roots or epicotyls for 30 consecutive days. The protrusion of the primary root with at least 1.0 cm was used for the calculation of germinative seeds and the production of normal seedlings for calculating germination.

The remainder of the seeds, still separated by species, degree of maturity and size, was divided into two subgroups, one of which was placed to germinate in plastic trays with fine-grained vermiculite, and irrigated with water whenever necessary. The seeds were removed from the substrate as the roots reached 1 cm in length, and then stored in chambers at 7 °C within perforated polyethylene bags to obtain the number required for the experiments. The seeds of this subgroup were called pre-germinated seeds (PGS), while the others were called non-germinated seeds (NGS).

Seeds of each subgroup of each size and each maturity stage were submitted to three treatments of incision (with a scalpel): control (kept whole), complete or partial lengthwise incision. In the NGS, the incision passed from the hilum center and in the PGS the incision was made so that one side remained with all the protruding root. In the full incision, hereinafter referred to as fractionation, two fractions were obtained and placed to germinate side by side. In PGS, the fraction containing the root was named fraction R, and its opposite, fraction S. In the NGS, the name was arbitrary. In the partial incision, hereinafter called crack, it was held until about two-thirds of the largest diameter of the seeds, and the halves were kept linked to each other.

The cracked seeds that showed at the end of the germination tests only one primary root without noticeable development of new roots were divided into two groups. In the first group, the incision was completed until the halves were separated. The half containing the root was discarded and its complementary one was placed to germinate, and its origin control was kept and the production of roots and seedlings in the isolated fractions was assessed again. In the second group, the seeds were kept in moist vermiculite to check for possible new germinations.

A completely randomized design, in a 2 x 2 x 3 factorial arrangement (PGS or NGS versus seed size versus type of incision) for each species and each maturity stage was used. The data were submitted to analysis of variance, and means were compared by Tukey test at 5% probability, by Sisvar software.

Results and Discussion

The differences in the water contents of *E. uniflora* seeds were higher in seeds of different sizes than in the seeds of fruit with different ripening (Table 1). Throughout ripening, seed water content tends to decrease as the dry mass is accumulated (Barbedo et al., 2013; Marcos-Filho, 2015). Therefore, the results suggest that the separation of *E. uniflora* seeds by fruit maturity degree was not efficient, in this work, in producing seeds with different maturity stages, which has already been verified in other species, according to Barbedo et al. (2013). Thus, ripe fruit seeds were not necessarily riper than those of unripe fruit. Large seeds of unripe fruits may, for example, be in a more advanced degree of ripening than the small seeds of ripe fruits, because these have shown higher water content than those (Table 1). As for the *E. brasiliensis* seeds used in the study, apparently they have had greater ripening synchronization with the fruits,

because all unripe ones have shown higher water content than the ripe ones.

Table 1. Water content (%) of seeds of *Eugenia uniflora* and *E. brasiliensis* obtained in two stages of fruit ripening and separated by size in small or large.

Species	Ripening of the fruit	Size of seeds	
		Small	Large
E. uniflora	Unripe	57.5	52.7
	Ripe	57.2	52.2
E. brasiliensis	Unripe	52.7	50.2
	Ripe	47.5	47.0

In *E. uniflora* seeds, it was also found that while the values of germinative (whole) seeds have not shown difference among seeds of different sizes (Table 2, production of whole seed roots), the largest ones, of ripe fruits, have shown higher values of germination than the smaller ones (Table 2, production of shoots in non-germinated seeds). The values of germinative seeds and whole seed germination of *E. brasiliensis*, in turn, have demonstrated that the ones of ripe fruit apparently had already started a process of deterioration, as they were generally lower than those of unripe fruit (Table 3).

Seeds subjected to fractionation (Tables 2 and 3 – fractionated) have shown again the potential of the species of the genus *Eugenia* to form roots and seedlings, even when half of the reserve tissue is removed from these seeds (Silva et al., 2003; Amador and Barbedo, 2011), and even when still unripe (Teixeira and Barbedo, 2012).

Another important aspect is that the cracked small *E. uniflora* non-germinated ripe seed showed higher germination rates than the whole ones (Table 2). This suggests that the crack has induced shoot formation in seeds that would only emit the primary root when not injured. According to Amador and Barbedo (2011), in *E. pyriformis* seeds it was evident that the formation of new roots and seedlings is ruled by the balance among the promoters of this formation, resulting from the cutting injury, and inhibitors, resulting from the formation of the first root or shoot. It is possible, therefore, that in the *E. uniflora* the fractionation injury also induces the differentiation of tissues that will produce roots and seedlings. The production of roots on *E. brasiliensis* PGS was lower in some treatments (especially in small seeds) than the production on NGS (Table 3). One possible explanation for this result would be that the initial root development may have caused a very low efficiency to regenerate a new root, which reinforces the idea of a process of self-inhibition of these seeds, when starting the germination (Silva et al., 2005; Amador and Barbedo, 2011).

Table 2. Production of roots and seedlings (% relative to the initial number of seeds) per seed or fractions of *Eugenia uniflora* seeds, of small and large seeds, derived from ripe or unripe fruits, with no visible germination at the time of the incision (non-germinated) or after the protrusion of least 1 cm of root (pre-germinated), subjected to fractionation (total separation of the seed in two halves) or cracks (cut in half to 2/3 of the seed diameter without complete separation). Values represent the sum of fractions R and S.

Type of Incision	Ripe fruit		Unripe fruit	
	Large	Small	Large	Small
Production of roots				
Whole	98 bA*	95 aA	98 bA	99 aA
Fractionated	125 aA	105 aB	133 aA	104 aB
Cracked	98 bA	103 aA	102 bA	99 aA
C.V. (%)	12.64			
	Pre-germinated		Non-germinated	
	Large	Small	Large	Small
Production of shoots – seeds of ripe fruit				
Whole	95 aAa	95 aAa	89 aAa	62 bBb
Fractionated	95 aAa	91 aAa	58 bAb	59 bAb
Cracked	100 aAa	100 aAa	92 aAa	92 aAa
Production of shoots – seeds of unripe fruit				
Whole	90 aAa	64 aBb	96 aAa	92 aAa
Fractionated	64 bAb	45 bBb	98 aAa	71 bBa
Cracked	91 aAa	64 aBa	98 aAa	64 bBa
C.V. (%)	14.91			

*Means followed by the same letter (lowercase in columns, uppercase in the row and italic comparing pre-germinated with non-germinated) do not differ by Tukey test, at 5%.

Table 3. Production of roots and seedlings (% relative to the initial number of seeds) per seed or fractions of *Eugenia brasiliensis* seeds, of small and large seeds, derived from ripe or unripe fruits, with no visible germination at the time of the incision (non-germinated) or after the protrusion of least 1 cm of root (pre-germinated), subjected to fractionation (total separation of the seed in two halves) or cracks (cut in half to 2/3 of the seed diameter without complete separation). Values represent the sum of fractions R and S.

Type of Incision	Pre-germinated		Non-germinated	
	Large	Small	Large	Small
Production of roots – seeds of ripe fruit				
Whole	100 bAa*	100 aAa	88 abAa	76 aAb
Fractionated	130 aAa	100 bBa	104 aAb	81 aBb**
Cracked	100 bAa	100 aAa	71 bAb**	44 bBb**
Production of roots – seeds of unripe fruit				
Whole	100 aAa	100 aAa	100 aAa	92 aAa
Fractionated	100 aAa	111 aAa	112 aAa	51 bBb
Cracked	100 aAa	100 aAa	101 aAa	60 bBb
C.V. (%)	18.74			
Production of shoots – seeds of ripe fruit				
Whole	91 aAa	60 bBa	71 aAa	54 aBa
Fractionated	74 bBa	92 aAa	51 bAa	42 bAb
Cracked	100 aAa	100 aAa	40 bAb	30 cAb
Production of shoots – seeds of unripe fruit				
Whole	69 aBa	89 aAa	90 aAa	74 aBa
Fractionated	36 bAb	31 cAa	78 bAa	14 bBa
Cracked	71 aAa	66 bAa	80 abAa	24 bBb
C.V. (%)	21.15			

*Means followed by the same letter (lowercase in columns, uppercase in the row and italic comparing pre-germinated with non-germinated) do not differ by Tukey test, at 5%. ** They also produced roots in fraction S.

Germination of cracked seeds allowed the identification of seven different categories of post-crack germination, similar to that described for *E. pyriformis* seeds by Amador and Barbedo (2011), being shown in Figure 1A: in category I, one root or seedling are formed in one of the seed halves that was separated by the crack, which began in the area adjacent to the crack, and another root or seedling are formed in the other half; in category II, only one root or seedling are formed in the area diametrically opposed to the crack, which began in the area adjacent to the crack; in category III, only one root or seedling are formed in

one of the halves, which began distant from the crack; in category IV, one root or seedling are formed in one of the halves, which began in an area distant from the crack and a root or seedling are formed in the other half, which began in the cracked surface; in category V, one root or seedling are formed in one of the halves, which began in the area adjacent to the crack; in category VI, one root or seedling are formed in the area opposed to the crack, both initiated in the area adjacent to the crack; in category VII, two roots or seedlings are formed in one of the halves, which began in the area adjacent to the crack.

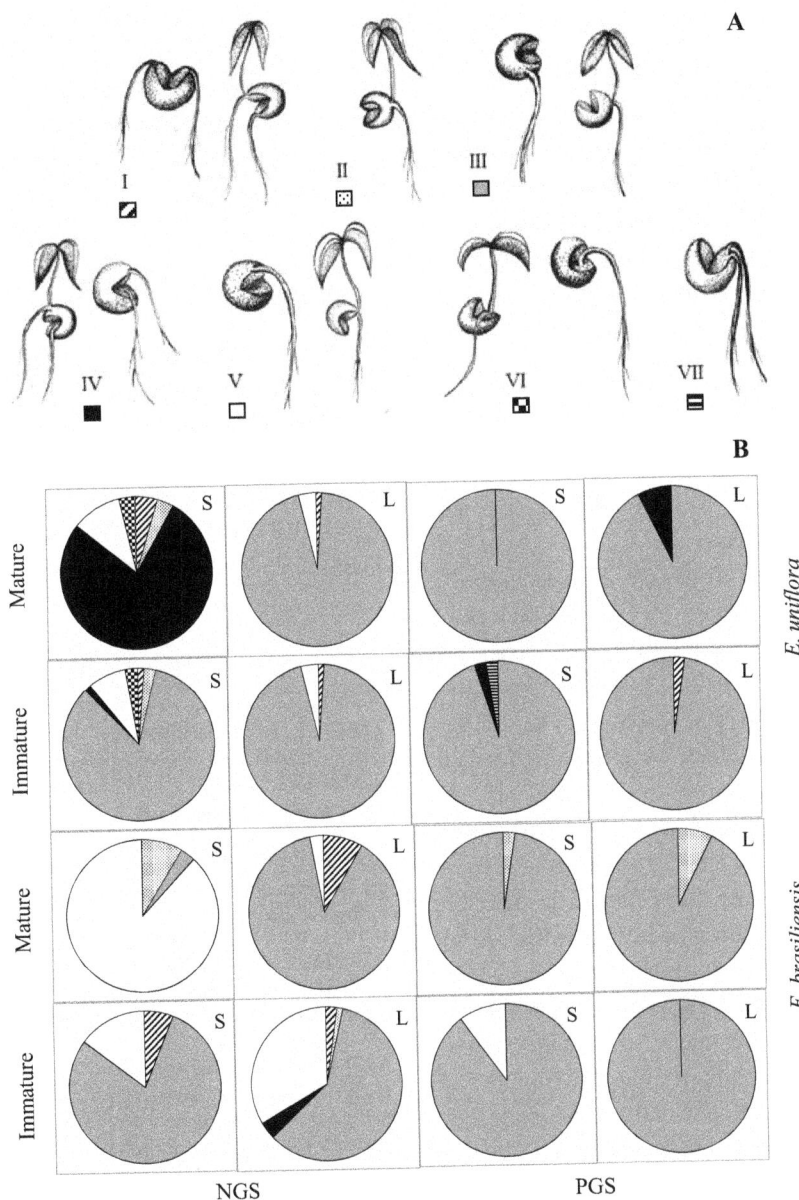

Figure 1. Models of incisions made in *Eugenia uniflora* and *Eugenia brasiliensis* seeds with their germination categories (A) and frequency distribution of the models (B) in seeds produced by unripe and ripe fruit of small (S) and large (L) sizes and in cracked seeds before (NGS) or after (PGS) germination has begun.

It was evident that category III, i.e., formation of root or seedling in an area distant from the crack, was predominant. In 13 of 16 treatments, category III represented more than 75% of all other categories (Figure 1B). Exception is for NGS of *E. uniflora* ripe fruit in which category IV (forming a second root or seedling) was predominant, and *E. brasiliensis*, in which category V (in which root or seedling are formed in the area adjacent to the crack) was predominant. In unripe *E. brasiliensis* NGS, the frequency of category V was also high, but did not reach 50%. In these last two, however, there is also the formation of only one root or seedling and, therefore, only in *E. uniflora* ripe fruit NGS the predominant category involved the formation of a second root or seedling. Interestingly, they are the same seeds in which the crack has induced the formation of seedlings in seeds that would form only root (Table 2), as discussed above. It is also in this category that the formation of the second root or seedling occurs in an area that is distant from the formation of the first root or seedling. This fact was also observed in *E. pyriformis* seeds by Amador and Barbedo (2011), who attributed the second germination to the fact that the first one occurred in a remote area, i.e., there would not be enough time for the migration of potential inhibitory substances from the first to the second germination area. Adding to the fact that in these seeds the crack may have accelerated the process of generating new seedlings, there were more favorable conditions for the factors stimulating new germination to outweigh the inhibitors, as was also discussed by Amador and Barbedo (2011).

When the fractioning of the cracked seeds of category III, which was predominant, was performed, i.e., when the crack was completed until the halves were completely separated, the fractions opposite to the fractions with root protrusion formed a new root in 28% to 75% (Table 4). The cracked seeds, of this same category, which were not fractionated, have never produced seedling or root in the other half. The lowest values were seen in NGS of small *E. uniflora* seeds, i.e., the ones with higher water content which, as noted above, were probably the unripest when harvested. It is therefore possible that the remaining reserves were not sufficient for the second germination, since most of them could have been consumed in the first germination. However, as new roots or seedlings were formed in all treatments, it is evident again that there are processes for inhibiting the development of new roots, promoted by the development of a root or seedling in the same seed, as observed in seeds of *E. pyriformis* by Amador and Barbedo (2011). These authors have considered that the production of inhibitory compounds should be continuous, cumulative and coming from the area in which there is seedling growth.

Table 4. Percentage of germinating fractions of seeds of *E. uniflora* and *E. brasiliensis*, ripe or unripe, large (L) or small (S), and pre-germinated (PGS) or non-germinated (NGS), previously cracked, which, when separated, initiated the formation of new roots.

	E. brasiliensis				*E. uniflora*			
	Mature		Immature		Mature		Immature	
	S	L	S	L	S	L	S	L
PGS	55	65	52	50	75	58	55	68
NGS	38	58	28	58	42	55	50	42

Conclusions

Seed germination of *Eugenia uniflora* and *Eugenia brasiliensis* starts the inhibition of the formation of new roots or seedlings in the seed and seed incision can block the action of these inhibitors.

Acknowledgments

The authors thank Coordenação de Aperfeiçoamento de Pessoal de Nível Superior – CAPES (Coordination of Improvement of Higher Education Personnel) for the doctorate scholarship granted to the first author and Conselho Nacional de Desenvolvimento Científico e Tecnológico – CNPq (National Counsel of Technological and Scientific Development) for the Productivity in Research scholarship granted to the second author.

References

AMADOR, T.S.; BARBEDO, C.J. Potencial de inibição da regeneração de raízes e plântulas em sementes germinantes de *Eugenia pyriformis*. *Pesquisa Agropecuária Brasileira*, v.46, p.814-821, 2011. http://dx.doi.org/10.1590/S0100-204X2011000800005

ANJOS, A.M.G.; FERRAZ, I.D.K. Morfologia, germinação e teor de água das sementes de araçá-boi (*Eugenia stipitata* ssp. *sororia*). *Acta Amazonica*, v.29, p.337-348, 1999.

BARBEDO, C.J.; CENTENO, D.C.; FIGUEIREDO-RIBEIRO, R.C.L. Do recalcitrant seeds really exist? *Hoehnea*, v.40, p.583-595, 2013. http://dx.doi.org/10.1590/S2236-89062013000400001

BRASIL. Ministério da Agricultura, Pecuária e Abastecimento. *Regras para análise de sementes*. Ministério da Agricultura, Pecuária e Abastecimento. Secretaria de Defesa Agropecuária. Brasília: MAPA/ACS, 2009. 395p. http://www.agricultura.gov.br/arq_editor/file/2946_regras_analise__sementes.pdf

DELGADO, L.F.; BARBEDO, C.J. Tolerância à dessecação de sementes de espécies de *Eugenia*. *Pesquisa Agropecuária Brasileira*, v.42, p.265-272, 2007. http://dx.doi.org/10.1590/S0100-204X2007000200016

DELGADO, L.F.; BARBEDO, C.J. Atividade inibidora da germinação em extratos de sementes *Eugenia uniflora* L. *Revista Brasileira de Sementes*, v.33, n.3, p.463-471, 2011. http://dx.doi.org/10.1590/S0101-31222011000300009

DELGADO, L.F.; BARBEDO, C.J. Water potential and viability of seeds of *Eugenia* (Myrtaceae), a tropical tree species, based upon different levels of drying. *Brazilian Archives of Biology and Technology*, v.55, p.583-590, 2012. http://dx.doi.org/10.1590/S1516-89132012000400014

DELGADO, L.F.; MELLO, J.I.O.; BARBEDO, C.J. Potential for regeneration and propagation from cut seeds of *Eugenia* (Myrtaceae) tropical tree species. *Seed Science and Technology*, v.38, p.624-634, 2010. http://dx.doi.org/10.15258/sst.2010.38.3.10

DEO, P.C.; TYAGI, A.P.; TAYLOR, M.; HARDING, R.; BECKER, D. Factors affecting somatic embryogenesis and transformation in modern plant breeding. *The South Pacific Journal of Natural and Applied Sciences*, v.28, p.27-40, 2010. http://dx.doi.org/10.1071/SP10002

GURGEL, J.T.A.; SOUBIHE SOBRINHO, J. Poliembrionia em mirtáceas frutíferas. *Bragantia*, v.11, p.141-163, 1951.

ISTA. International rules for seed testing. *Seed Science and Technology*, v.13, p.356-513, 2015. http://www.seedtest.org/en/international-rules-_content---1--1083.html

JUSTO, C.F.; ALVARENGA, A.A.; ALVES, E.; GUIMARÃES, R.M.; STRASSBURG, R.S. Efeito da secagem, do armazenamento e da germinação sobre a micromorfologia de sementes de *Eugenia pyriformis* Camb. *Acta Botanica Brasilica*, v.21, p.539-551, 2007. http://dx.doi.org/10.1590/S0102-33062007000300004

KANWAR, K.; JOSEPH, J.; DEEPIKA, R. Comparison of in vitro regeneration pathways in *Punica granatum* L. *Plant Cell, Tissue and Organ Culture*, v.100, p.199-207, 2010. http://dx.doi.org/10.1007/s11240-009-9637-4

KARAMI, O.; SAIDI, A. The molecular basis for stress-induced acquisition of somatic embryogenesis. *Molecular Biology Reports*, v.37, p.2493-2507, 2010. http://dx.doi.org/10.1007/s11033-009-9764-3

KOHAMA, S.; MALUF, A.M.; BILIA, D.A.C.; BARBEDO, C.J. Secagem e armazenamento de sementes de *Eugenia brasiliensis* Lam. (grumixameira). *Revista Brasileira de Sementes*, v.28, p.72-78, 2006. http://dx.doi.org/10.1590/S0101-31222006000100010

LANDRUM, L.R.; KAWASAKI, M.L. The genera of Myrtaceae in Brazil: an illustrated synoptic treatment and identification keys. *Brittonia*, v.49, p.508-536, 1997. http://dx.doi.org/10.2307/2807742

MARCOS-FILHO, J. *Fisiologia de sementes de plantas cultivadas*. 2.ed. Londrina, ABRATES, 2015. 660p.

PEREIRA, C.E.; VON PINHO, E.V.R.; OLIVEIRA, D.F.; KIKUTI, A.L.P. Determinação de inibidores da germinação no espermoderma de sementes de café (*Coffea arabica* L.). *Revista Brasileira de Sementes*, v.24, p.306-311, 2002. http://dx.doi.org/10.1590/S0101-31222002000100042

PRATAVIERA, J.S.; LAMARCA, E.V.; TEIXEIRA, C.C.; BARBEDO, C.J. The germination success of the cut seeds of *Eugenia pyriformis* depends on their size and origin. *Journal of Seed Science*, v.37, n.1, p.47-54, 2015. http://www.scielo.br/pdf/jss/v37n1/2317-1537-jss-37-01-00047.pdf

RIZZINI, C.T. Efeito tegumentar na germinação de *Eugenia dysenterica* DC. (Myrtaceae). *Revista Brasileira de Biologia*, v.30, p.381-402, 1970.

SALOMÃO, A.N.; ALLEM, A.C. Polyembryony in angiospermous trees of the Brazilian Cerrado and Caatinga vegetation. *Acta Botanica Brasilica*, v.15, p.369-378, 2001. http://dx.doi.org/10.1590/S0102-33062001000300007

SILVA, C.V.; BILIA, D.A.C.; MALUF, A.M.; BARBEDO, C.J. Fracionamento e germinação de sementes de uvaia (*Eugenia pyriformis* Cambess. – Myrtaceae). *Revista Brasileira de Botânica*, v.26, p.213-221, 2003. http://dx.doi.org/10.1590/S0100-84042003000200009

SILVA, C.V.; BILIA, D.A.C.; BARBEDO, C.J. Fracionamento e germinação de sementes de *Eugenia*. *Revista Brasileira de Sementes*, v.27, p.86-92, 2005. http://dx.doi.org/10.1590/S0101-31222005000100011

SWAMY, M.K.; SUDIPTA, K.M.; BALASUBRAMANYA, S.; ANURADHA, M. Effect of different carbon sources on in vitro morphogenetic response of patchouli (*Pogostemon cablin* Benth.). *Journal of Phytology*, v.2, p.11-17, 2010. http://journal-phytology.com/index.php/phyto/article/view/4377/2167

TEIXEIRA, C.C.; BARBEDO, C.J. The development of seedlings from fragments of monoembryonic seeds as an important survival strategy for *Eugenia* (Myrtaceae) tree species. *Trees, structure and function*, v.26, p.1069-1077, 2012. http://dx.doi.org/10.1007/s00468-011-0648-5.

10

Viability of *Simira gardneriana* M.R. Barbosa & Peixoto seeds by the tetrazolium test

Fabrícia Nascimento de Oliveira[1], Salvador Barros Torres[2*],
Narjara Walessa Nogueira[2], Rômulo Magno Oliveira de
Freitas[2]

ABSTRACT – Ecoregion Caatinga presents a great diversity of species with potential for exploitation. Among them 'pereiro-vermelho' (*Simira gardneriana* M.R. Barbosa & Peixoto) stands out for its importance in timber and forestry activities. Its seeds germinate slowly. Therefore, the use of tetrazolium test to estimate viability becomes essential when quick answers on seeds quality are wanted. This study has aimed to establish the best concentration of tetrazolium solution and the coloration period for assessing the viability of *Simira gardneriana* seeds. Initially, seeds were subjected to pre-wetting between paper sheets for 144 hours at 30 °C. Subsequently, the endosperm portion containing the embryo was immersed at four concentrations of tetrazolium solution (0.075, 0.1, 0.5 and 1.0%) and three coloration periods (2, 4 and 6 hours) in the dark under the temperature of 30 °C and another one at 35 °C. The percentage of viable seeds was compared with the results obtained in the germination test conducted on paper substrate at 30 °C in four replicates of 25 seeds. Tetrazolium test was efficient to estimate the viability of *S. gardneriana* seed and the concentration of 0.075% for six hours at 35 °C was the best condition.

Index terms: Rubiaceae, forest seeds, caatinga, germination, conservation.

Viabilidade de sementes *Simira gardneriana* M.R. Barbosa & Peixoto pelo teste de tetrazólio

RESUMO – A Caatinga apresenta grande diversidade de espécies com potencial de exploração, dentre estas o pereiro-vermelho (*Simira gardneriana* M.R. Barbosa & Peixoto) destaca-se pela importância nas atividades madeireira e florestal. Suas sementes germinam lentamente, com isso o uso do teste de tetrazólio para estimar a viabilidade passa a ser essencial quando se deseja respostas rápidas sobre a qualidade das sementes. Assim, objetivou-se com esse trabalho estabelecer a melhor concentração da solução de tetrazólio e o período de coloração para a avaliação da viabilidade de sementes de *S. gardneriana*. Inicialmente, as sementes foram submetidas ao pré-umedecimento entre papel por 144 horas a 30 °C. Posteriormente, a porção do endosperma contendo o embrião foi imersa em quatro concentrações da solução de tetrazólio (0,075; 0,1; 0,5 e 1,0%) e três períodos de coloração (2, 4 e 6 horas), no escuro, sob a temperatura de 30 °C e outra a 35 °C. A porcentagem de sementes viáveis foi comparada com os resultados obtidos no teste de germinação, conduzido em substrato papel, a 30 °C, em quatro repetições de 25 sementes. O teste de tetrazólio foi eficiente para estimar a viabilidade de sementes de *S. gardneriana*, sendo a concentração de 0,075% por seis horas, a 35 °C a melhor condição.

Termos para indexação: Rubiaceae, sementes florestais, caatinga, germinação, conservação.

Introduction

S. gardneriana M.R. Barbosa & Peixoto, known as pereiro-vermelho, is an endemic species of the Caatinga, which stands out for its timber, dyeing, craftsmanship and landscape values (Barbosa and Peixoto, 2000). However, the lack of studies on this species is one of the barriers which limit its use, especially with regard to handling, quality and viability of the seeds.

Although *S. gardneriana* seeds do not show dormancy, germination occurs slowly, making it difficult to obtain results when quick information is wanted. In this context, tetrazolium test has been shown to be a promising alternative in determining seeds viability and vigor of various forest species for the quality and speed in obtaining results. This test reflects the activity of dehydrogenase enzymes involved

[1]Departamento de Ciências Ambientais e Tecnológicas, Universidade Federal Rural do Semi-Árido, 59515-000 – Mossoró, RN, Brasil.

[2]Departamento de Ciências Vegetais, Universidade Federal Rural do Semi-Árido, Caixa Postal 137, 59625-900 – Mossoró, RN, Brasil.
*Corresponding author < sbtorres@ufersa.edu.br>

in the breathing process, whose seeds living and dead tissues are identified by the presence or absence of the color red, respectively (AOSA, 2009).

Recent research in forest species seeds has been developed, aiming to reduce the time required to obtain viability results, from the verification of the most appropriate methodology of tetrazolium test for each species, according to work carried out by Fava and Albuquerque (2013) on *Palicourea rigida* Kunth; Abbade and Takaki (2014) on *Tabebuia roseoalba* (Ridl.) Sandwith; Cripa et al. (2014), Kaiser et al. (2014) and Lamarca and Barbedo (2014) on species of the genus *Eugenia*; and Nogueira et al. (2014) on Pacara Earpod Tree (*Enterolobium contortisiliquum* (Vell.) Morong).

Several factors can interfere with satisfactory results in the tetrazolium test, especially those related to the criteria for interpreting the findings and methods of execution (Gaspar-Oliveira et al., 2009). Among the implementation procedures before performing this test is preconditioning, which aims the solution penetration in the tissues of interest to be evaluated (Brasil, 2009). In this sense, for seeds of *Ceiba speciosa* (A. St. – Hil.) Ravenna), Lazarotto et al. (2011) recommend immersion in water at room temperature for eight hours; Abbade and Takaki (2014) recommend for seeds of *T. roseoalba* immersion in water for 12 h at 25 °C; and for seeds of Pacara Earpod Tree (*E. contortisiliquum* (Vell.) Morong), Nogueira et al. (2014) recommend chiseling followed by soaking in water for 24 h and removal of the integument.

As regards the coloration period, the tetrazolium salt concentration and the incubation temperature directly interfere in the coloration intensity and uniformity of the seeds tissues, depending on the characteristics of each species (AOSA, 2009). The definition of these factors is important as they influence the assessment and interpretation of test results. Considering the variables involved in the tetrazolium test, there is a concern among researchers in adapting the methodology for assessment of native forest seeds. For seeds of *Tabebuia serratifolia* Vahl Nich., immersion in 0.5% tetrazolium solution for 12 hours at 30 °C is used and in seeds of *Peltophorum dubium* (Sprengel) Taubert, the solution concentration used is 0.1% and temperature of 25 °C for 150 minutes of coloration (Oliveira et al., 2005a; 2005b). For seeds of *Piptadenia moniliformis* Benth., Azerêdo et al. (2011) recommend coloration in tetrazolium solution at 0.075% for four hours at 35 °C. For seeds of *Eugenia uniflora* L., Kaiser et al. (2014) recommend the tetrazolium test at 0.5% for 2 hours at 30 °C. And for seeds of Pacara Earpod Tree, Nogueira et al. (2014) recommend seeds immersion in tetrazolium solution at 0.075% at 35 °C for three hours.

Based on the above, this work aimed to establish the best tetrazolium solution concentration and coloration period to assess the viability of *S. gardneriana* seeds.

Material and Methods

The research was developed at the Seed Testing Laboratory of the Plant Sciences Department at UFERSA, in the Brazilian city of Mossoró, RN, with seeds of *S. gardneriana* granted by the Centro de Referência para Recuperação de Áreas Degradadas da Caatinga (Reference Center for Recovery of Degraded Areas of the Caatinga) at Federal University of São Francisco Valley (UNIVASF, Universidade Federal do Vale do São Francisco), in the Brazilian city of Petrolina, PE. The seeds were collected from various matrix trees located in the Brazilian municipality of Afrânio, PE (8°30'42" S; 41°00'36" W and 540 m altitude) in December 2009.

Before conducting the experiments, the seeds were packed in paper bags and stored in cold storage (10-12 °C and 50-52% RH of the environment) for five years.

Preliminarily, seeds soaking curve was determined using four replicates of 25 seeds distributed in three sheets of paper towel moistened with water equivalent to 2.5 times the dry weight of the substrate and maintained in a B.O.D. (Biochemical Oxygen Demand) growth chamber at 30 °C. The initial fresh weight of the various seeds subsamples was recorded before soaking and after the seeds were removed, dried in absorbent paper and weighed (0.001 g) in three-hour intervals in the first 15 h and in six-hour intervals between 15 and 33 h. Then, weighings were taken at 12-hour intervals between 33 and 81 h, and finally at 24-hour intervals until the moment when stabilization of the seed weight was noticed due to the maximum water absorption. Thus, the immersion was defined as the weight increase relative to the initial weight.

The determination of the initial moisture content of the seeds was performed by the oven method at 105 °C ± 3 °C for 24 hours (Brasil, 2009) using four replications of three grams of seeds and the results expressed as a percentage (wet basis).

For the tetrazolium test, the *S. gardneriana* seeds were premoistened in paper towel and placed in a germination chamber at 30 °C for 144 hours, according to results from the soaking curve. Next, with a scalpel, seeds were longitudinally and medially cut, with the endosperm portion containing the embryo placed in 50 mL plastic cups and immersed in a solution of 2,3,5 triphenyl tetrazolium chloride at four concentrations of 0.075; 0.1; 0.5 and 1.0% for three coloration periods (2, 4 and 6 hours) in a chamber regulated at a temperature of 30 °C and the other one at 35 °C in the dark. Four replications of 25 seeds were used for each concentration combination of the tetrazolium solution and coloration period for experiments involving the incubation temperatures.

After the coloration period, the solution was drained, the material washed in running water, and the embryos were extracted from the remaining part of the endosperm and kept in water in a refrigerated environment until evaluation. Subsequently, the embryos were individually observed with a magnifying stereoscopic and evaluated for uniformity, color intensity, presence of milky white areas, tissues appearance and location of these colorations in relation to the essential areas of the embryo (hypocotyl-radicle axis and vascular area), being classified into viable and unviable, in accordance with standards specified by AOSA (Association of Official Seed Analysts) (2009) and Brasil (2009), for various agricultural and forest species: 1) viable: embryos with full coloration in light pink or bright red; radicle end without a milky white/yellowish coloration; and 2) unviable: embryos with full coloration in crimson red/intense red or milky white/yellowish; radicle end without coloration or an intense red coloration. Results were expressed as percentage of viable seeds.

To compare the results obtained in the tetrazolium test, germination test was conducted in a B.O.D. (Biochemical Oxygen Demand)-type growth chamber at 30 °C under constant white light, with four replications of 25 seeds. The seeds were distributed on paper towel rolls moistened in water equivalent to 2.5 times the mass of the dry substrate and placed inside transparent plastic bags to keep moisture. The criteria used for seed germination was the development of normal seedlings (Brasil, 2009), and assessment was carried out on the thirtieth day after sowing.

The experimental design used was a completely randomized one in a 4 x 3 + 1 factorial arrangement (4 concentrations of tetrazolium solution x 3 periods of coloration + 1 control – germination test) at a temperature of 30 °C and the other one at 35 °C in four replications of 25 seeds for each treatment. The data were submitted to normality and homogeneity test of variances to assess the need for transformation. Sequentially, analysis of variance (ANOVA) was performed by the F-test at 1% and 5% significance and means of viable seeds obtained by the tetrazolium test were compared by Tukey's test (p ≤ 0.05). The comparison between the means of viable seeds for each of the combinations in the tetrazolium test with the results of the germination test (control) were carried out by the Dunnett's test (p ≤ 0.05). Statistical analyses were performed using the software ASSISTAT version 7.6 beta (Silva and Azevedo, 2002).

Results and Discussion

S. gardneriana seeds present stabilization in water absorption from 33 hours after the start of soaking (Figure 1),

demonstrating that this process is slow, completing Phase I in more than 24 h of moisturizing.

The initial moisture content of the seeds was 12.3%. However, it was found that after pre-wetting, Phase II of the three-phase model of hydration reached moisture between 69.9 and 77.3% for 33 and 153 hours of hydration, respectively (Figure 1). In this hydration period, the seeds completed Phase II with radicle emission, thereby being observed that this hydration period is sufficient to reactivate the embryo metabolism.

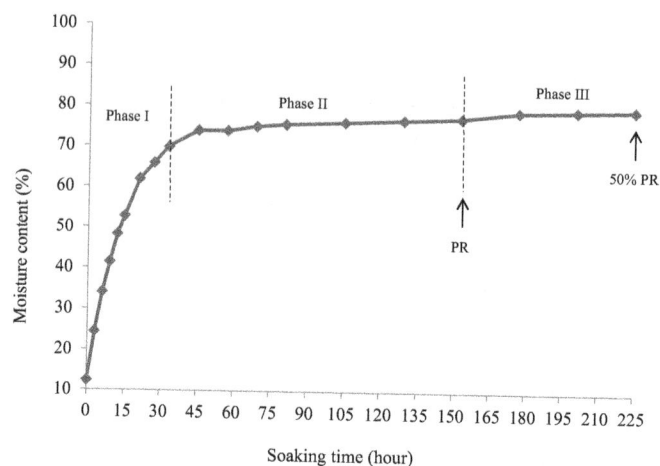

Figure 1. Soaking curve of seeds of *Simira gardneriana* M.R. Barbosa & Peixoto at a temperature of 30 °C. PR = Protrusion of the radicle; 50% PR = Protrusion of the radicle in 50% of seeds.

It was also observed that Phase I showed rapid water uptake during the first 33 hours and Phase III began at 153 hours after initiation of hydration. Thus, it was possible to perform the soaking curve of the *S. gardneriana* seeds by establishing the premoistening period (144 hours), i.e., the time required for the seeds to reach Phase II intermediate, avoiding that they reached Phase III (153 hours). Pre-wetting corresponds to the seeds preparation phase to facilitate penetration of the tetrazolium solution and development of a uniform coloration (Brasil, 2009). Besides, according to Marcos-Filho et al. (1987) and AOSA (2009), pre-hydration favors not only tissues softening, integument removal and longitudinal section of the embryos, but also provides a clearly evident coloration, improved test quality and makes assessment easier.

According to Zonta et al. (2009), what is recommended for the *Coffea arabica* L. seeds is soaking for 24 hours at 30 °C. Although the species of this study is from the same botanical family, the time recommended is insufficient to soften the *S. gardneriana* seeds, which complicates the extraction of embryos. Thus, in this study the pre-wetting period recommended may be 144 hours of soaking for the occurrence of sufficient activation of the enzyme system, which improves and speeds up the coloration process.

After determining the pre-wetting period (144 hours), embryos were exposed to concentrations of tetrazolium solution and coloration periods at each temperature (30 and 35 °C). There was no difference for the concentration factor of the tetrazolium solution at 30 °C, but at 35 °C this factor showed significance of 1% in the F-test. There was a significant effect of 1% probability of the coloration period factor and the interaction between the solution concentration and coloration period factors for experiments conducted at temperatures of 30 and 35 °C. In addition, there was a significant difference between the estimates of seed viability obtained by the tetrazolium test and the results of the germination test, as can be seen by the significance of the control (germination percentage) versus factorial contrast (Table 1).

Table 1. Summary of the analysis of variance seen in the tetrazolium test of seeds *Simira gardneriana* M.R. Barbosa & Peixoto compared to the results of the germination test (control).

Sources of variation	Mean square	
	Temperature (30 °C)	Temperature (35 °C)
Treatments	1359.692^{**}	4251.026^{**}
Control vs Factorial	11996.308^{**}	5123.308^{**}
Concentration of tetrazolium solution (C)	127.111^{ns}	1582.556^{**}
Coloration period (P)	1164.000^{**}	14763.000^{**}
C x P	268.444^{**}	1935.889^{**}
Error	49.231	55.077
Average	25.385	43.615
C.V. (%)	27.64	17.02
LSD (least significant difference)	14.41	15.24

** = significant at 1% probability by the F-test; ^{ns} = nonsignificant at 5% probability by the F-test.

As for the results for the percentage of viable seeds by the tetrazolium test at different concentrations and exposure times and the germination test (control) for experiments conducted at temperatures of 30 and 35 °C, they are presented in Table 2. For the temperature of 30 °C, the means of seeds viability were higher as the seed coloration period increased, although without difference between periods of 2 and 4 hours. It was found also that the means for seed viability differed from the results of the germination test (control – 78%) for all concentrations (0.075, 0.1, 0.5 and 1.0%) and time (2, 4 and 6 hours) tested within this temperature.

Under the temperature of 35 °C, the six-hour period of coloration gave the highest estimate of viability, regardless of the concentration of the tetrazolium solution, except at 1.0%, whose estimate did not differ from the periods of 2 and 4 hours (Table 2), and higher values for coloration periods were observed at concentrations of 0.1 and 0.5%. The six-hour period at concentrations of 0.075% (85% of viable seeds), 0.1% (93% of viable seeds) and 0.5% (90% of viable seeds) was favorable for the assessment of the seeds, providing estimates similar to the results from the germination test (78%). As for the other combinations of periods and concentrations, there was a difference between these and the control in view of the inadequate coloration, which underestimated seeds viability.

Among the tested temperatures, 35 °C were which provided the highest viability values.

Table 2. Viable seeds of *Simira gardneriana* M.R. Barbosa & Peixoto obtained by tetrazolium test at different concentrations (0.075; 0.1; 0.5 and 1.0%) and periods of coloration (2, 4 and 6 horas) at 30 and 35 °C compared to the results of the germination test (control).

Periods (hours)	Concentration of tetrazolium solution (%)			
	0.075	0.1	0.5	1.0
...............................Temperature (30 °C).............................				
2	12 Aby	15 Aby	13 Aby	16 Aay
4	16 Aby	18 Aby	20 Aaby	20 Aay
6	43 Aay	37 ABay	28 Bay	14 Cay
...............................Temperature (35 °C).............................				
2	13 Aby	13 Aby	24 Acy	27 Aay
4	8 Cby	13 Cby	57 Aby	32 Bay
6	85 Aaz	93 Aaz	90 Aaz	33 Bay
Germination = 78 z				

Means followed by the same uppercase letters (A, B, C) in the line and lowercase (a, b, c) in the column do not differ significantly at 5% probability by the Tukey's test.
Means followed by the same letter (z, y) between germination (control – germination test) and viability obtained in the tetrazolium test do not differ significantly at 5% probability by Dunnett's test.

These results support the claims observed by AOSA (2009), in which seeds coloration in the tetrazolium test settles faster at higher temperatures. For this reason, Marcos-Filho et al. (1987) recommend that seeds immersed in tetrazolium solution be placed in a chamber set at temperatures of 30 to 40 °C. In this study, the best combinations for assessing the viability of *S. gardneriana* seeds were recorded at temperature 35 °C whereas at 30 °C all combinations of concentrations and periods statistically differed from the germination test.

Work involving the standardization of the tetrazolium test methodology on other species of Rubiaceae can be found, focusing on reducing the time to detect the quality of seed lots. However, there has been different results according to the species studied. Fava and Albuquerque (2013), for example, have shown that the best procedure to estimate the viability of *Palicourea rigida* Kunth seeds was by using the tetrazolium solution at the concentration of 0.5% at 40 °C for 3 hours. On the other hand, Zonta et al. (2009), with *Coffea arabica* L. seeds have found, as the best methodology to assess the seeds viability of this species, soaking for 24 hours at 30 °C, removal of the embryo and exposure to tetrazolium solution at 0.1%, during 16 hours at 35 °C.

Coloration patterns observed in embryos ranged from light pink in viable seeds to white in dead seed (Figure 2). Viable embryos exhibited uniform light pink or bright red colors along their entire length, demonstrating that tissues were alive and vigorous. On the other hand, unviable embryos, when exposed to tetrazolium solution, exhibited an intense red coloration (tissue deterioration) or milky white (dead tissue) in its full extent, or an intense red color only at the end of the radicle.

The choice of the appropriate method for the tetrazolium test must be based on ease of identification of viable and unviable tissues and the ability to differentiate seed vigor (Azerêdo et al., 2011). The color differences observed in the seeds after coloration in tetrazolium solution are the main features that should be considered when interpreting the test results. The color intensity of the seeds in the tetrazolium test varies among species. For example, the pink color seen in viable seeds of Leucaena (*Leucaena leucocephala* (Lam.) de Wit) is lighter (Costa and Santos, 2010) than that seen in seeds of *Brachiaria brizantha* (Hochst. ex A. Rich) Stapf (Dias and Alves, 2008). In this species, the color which indicates a viable tissue is red or deep pink. In Leucaena seeds, that deep pink color means tissue deterioration. This is because the terminology used to determine the colors observed in seeds in the tetrazolium test is usually established by the authors. Therefore they may vary among studies.

Exposure of embryos in tetrazolium solution at 30 °C was less efficient in viable seeds classification, for it presented problems for the coloration of the embryo, many did not color at the lowest concentration (0.075%), and some only colored the ends at higher concentrations (0.5 and 1.0%), thus complicating the interpretation of results (Figures 2F, 2G and 2H).

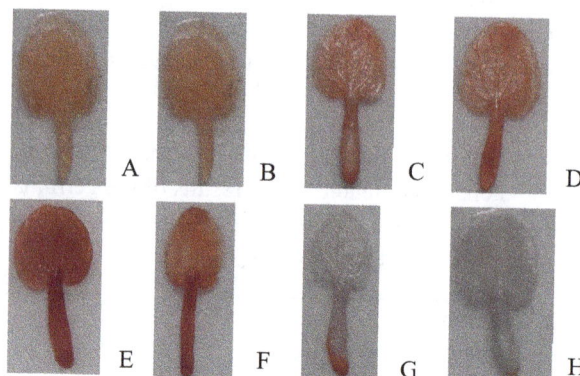

Figure 2. Viable seeds of *Simira gardneriana* M.R. Barbosa & Peixoto: an embryo with light pink (A, B) and bright red (C) colorations; hypocotyl-radicle axis with an intense red coloration in the cortex, but without reaching the central cylinder (D). Unviable seeds of pereiro-vermelho (*Simira gardneriana* M.R. Barbosa & Peixoto): an embryo with an intense red coloration in all its extension (E) or hypocotyl-radicle axis with an intense red coloration reaching the central cylinder (F); an embryo with discolored/milky white and/or intense red areas reaching the central cylinder (G, H).

At 30 °C, viability values at different concentrations and coloration periods did not represent the values observed in the germination test, probably due to the fact that the tetrazolium solution did not spread uniformly, thus hindering coloration and evaluation. In this case, embryos that did not develop optimal coloration may have been considered unviable when the absence of coloration could have been caused by the incubation temperature, combined with different concentrations and times. It was also observed that in many embryos exposed at the 0.5 and 1.0% concentrations, an intense red coloration developed at the ends of the plumule or bottom of the hypocotyl-radicle axis (Figure 2F), probably because these areas were more exposed to tetrazolium solution than the others.

In the evaluation of the tetrazolium test of *S. gardneriana* seeds at a temperature of 35 °C, variations were observed in the seeds coloration according to the treatments used, and light shades were observed at the lowest concentration of the tetrazolium solution (0.075%). The coloration periods of 2 and 4 hours were insufficient to estimate the viability of *S. gardneriana* seeds at all tested concentrations for they did not promote an adequate coloration that would enable to distinguish living from dead or damaged tissues.

On the other hand, the 6-hour coloration period was effective to assess the seeds viability of this species at concentrations of 0.075%, 0.1% and 0.5%, for they provided a clear coloration of the embryos, facilitating the analysis and interpretation of results (Figures 2A, 2B, 2C and 2D). Furthermore, it allowed the identification of lesions in areas that are critical for the embryo, such as the hypocotyl-radicle axis. It should be noted, however, that the 6-hour coloration period, associated with a concentration of 1.0% tetrazolium solution (Figure 2E) provided more than 60% of embryos with an intense red color (deteriorated seeds – unviable), different from the results obtained in the germination test (Table 2).

Tetrazolium test using concentrations of 0.075; 0.1 and 0.5% during 6 hours of coloration at 35 °C was efficient to estimate the viability of *S. gardneriana* seeds. In other periods, regardless of the temperature, tetrazolium salt concentrations did not promote an adequate coloration that would enable to distinguish living from dead or damaged tissues. However, lower concentrations are more indicated by their lower salt cost and due to allowing a better visualization of the coloration disorders and identification of different types of injuries (AOSA, 2009). Thus, tetrazolium test using a concentration of 0.075% can be used as a supplement to the germination test in assessing the viability of *S. gardneriana* seeds.

Conclusions

The pre-wetting of *S. gardneriana* seeds for 144 hours in the preparation phase for performing the tetrazolium test facilitates the extraction of embryos and does not affect the test results.

Tetrazolium salt concentration at 0.075% for six hours at 35 °C is an effective combination to assess the viability of *S. gardneriana* seeds.

Acknowledgments

To the Centro de Referência para Recuperação de Áreas Degradadas da Caatinga (Reference Center for Recovery of Degraded Areas of the Caatinga) at Federal University of São Francisco Valley (UNIVASF, Universidade Federal do Vale do São Francisco) (CRAD/UNIVASF) represented by Dr. José Alves de Siqueira Filho, for granting the seeds used in the experiment.

References

ABBADE, L.C.; TAKAKI, M. Teste de tetrazólio para avaliação da qualidade de sementes de *Tabebuia roseoalba* (Ridl.) Sandwith – Bignoniaceae, submetidas ao armazenamento. *Revista Árvore*, v.38, n.2, p.233-240, 2014. http://www.scielo.br/pdf/rarv/v38n2/03.pdf

AOSA. Association of Official Seed Analysts. *Seed vigor testing handbook*. Ithaca, 2009. 340p.

AZERÊDO, G.A.; PAULA, R.C.; VALERI, S.V. Viabilidade de sementes de *Piptadenia moniliformis* Benth. pelo teste de tetrazólio. *Revista Brasileira de Sementes*, v.33, n.1, p.61-68, 2011. http://www.scielo.br/pdf/rbs/v33n1/07.pdf

BARBOSA, M.R.V.; PEIXOTO, A.L. A new species of *Simira* (Rubiaceae, Rondeletieae) from Northeastern Brazil. *Novon*, v.10, n.2, p.110-112, 2000. https://www.jstor.org/stable/3393006?seq=1#page_scan_tab_contents

BRASIL. Ministério da Agricultura, Pecuária e Abastecimento. *Regras para análise de sementes*. Ministério da Agricultura, Pecuária e Abastecimento. Secretaria de Defesa Agropecuária. Brasília: MAPA-ACS, 2009. 395p. http://www.agricultura.gov.br/arq_editor/file/2946_regras_analise__sementes.pdf

COSTA, C.J.; SANTOS, C.P. Teste de tetrazólio em sementes de leucena. *Revista Brasileira de Sementes*, v.32, n.2, p.66-72, 2010. http://www.scielo.br/pdf/rbs/v32n2/v32n2a08.pdf

CRIPA, F.B.; FREITAS, L.C.N.; GRINGS, A.C.; BORTOLINI, M.F. Tetrazolium test for viability estimation of *Eugenia involucrate* DC. and *Eugenia pyriformis* Cambess. seeds. *Journal of Seed Science*, v.36, n.3, p.305-311, 2014. http://www.scielo.br/pdf/jss/v36n3/aop0514.pdf

DIAS, M.C.L.L.; ALVES, S.J. Avaliação da viabilidade de sementes de *Brachiaria brizantha* (Hochst. ex A. Rich) Stapf pelo teste de tetrazólio. *Revista Brasileira de Sementes*, v.30, n.1, p.145-151, 2008. http://www.scielo.br/pdf/rbs/v30n1/a26v30n1.pdf

FAVA, C.L.F.; ALBUQUERQUE, M.C.F. Viabilidade e emergência de plântulas de *Palicourea rigida* Kunth em função de diferentes métodos para superação de dormência. *Enciclopédia Biosfera*, v.9, n.17, p.2620-2629, 2013. http://www.conhecer.org.br/enciclop/2013b/CIENCIAS%20AGRARIAS/VIABILIDADE%20E%20EMERGENCIA.pdf

GASPAR-OLIVEIRA, C.M.; MARTINS, C.T.; NAKAGAWA, J. Método de preparo das sementes de mamoneira (*Ricinus communis* L.) para o teste de tetrazólio. *Revista Brasileira de Sementes*, v.31, n.1, p.160-167, 2009. http://www.scielo.br/pdf/rbs/v31n1/a18v31n1.pdf

KAISER, D.K.; FREITAS, L.C.N.; BIRON, R.P.; SIMONATO, S.C.; BORTOLINI, M.F. Adjustment of the methodology of the tetrazolium test for estimating viability of *Eugenia uniflora* L. seeds during storage. *Journal of Seed Science*, v.36, n.3, p.344-351, 2014. http://www.scielo.br/pdf/jss/v36n3/10.pdf

LAMARCA, E.V.; BARBEDO, C.J. Methodology of the tetrazolium test for assessing the viability of seeds of *Eugenia brasiliensis* Lam., *Eugenia uniflora* L. and *Eugenia pyriformis* Cambess. *Journal of Seed Science*, v.36, n.4, p.427-434, 2014. http://www.scielo.br/pdf/jss/v36n4/aop1114.pdf

LAZAROTTO, M.; PIVETA, G.; MUNIZ, M.F.B.; REINIGER, L.R.S. Adequação do teste de tetrazólio para avaliação da qualidade de sementes de *Ceiba speciosa*. *Semina: Ciências Agrárias*, v.32, n.4, p.1243-1250, 2011. http://www.uel.br/portal/frm/frmOpcao.php?opcao=http://www.uel.br/revistas/uel/index.php/semagrarias

MARCOS-FILHO, J.; CÍCERO, S.M.; SILVA, W.R. *Avaliação da qualidade das sementes*. Piracicaba: FEALQ, 1987. 230p.

NOGUEIRA, N.W.; TORRES, S.B.; FREITAS, R.M.O. Teste de tetrazólio em sementes de timbaúba. *Semina: Ciências Agrárias*, v.35, n.6, p.2967-2976, 2014. http://www.uel.br/revistas/uel/index.php/semagrarias/article/view/14704/pdf_521

OLIVEIRA, L.M.; CARVALHO, M.L.M.; DAVIDE, A.C. Teste de tetrazólio para avaliação da qualidade de sementes de *Peltophorum dubium* (Sprengel) Taubert – Leguminosae Caesalpinioideae. *Cerne*, v.11, n.2, p.159-166, 2005a.

OLIVEIRA, L.M.; CARVALHO, M.L.M.; NERY, M.C. Teste de tetrazólio em sementes de *Tabebuia serratifolia* Vahl Nich. e *T. impetiginosa* (Martius ex A. P. de Candolle) Standley – Bignoniaceae. *Revista Ciência Agronômica*, v.36, n.2, p.169-174, 2005b. http://www.ccarevista.ufc.br/seer/index.php/ccarevista/article/view/264/259

SILVA, F.A.S.E.; AZEVEDO, C.A.V. Versão do programa computacional Assistat para o sistema operacional Windows. *Revista Brasileira de Produtos Agroindustriais*, v.4, n.1, p.71-78, 2002. http://www.deag.ufcg.edu.br/rbpa/rev41/Art410.pdf

ZONTA, J.B.; SOUZA, L.T.; DIAS, D.C.F.S.; ALVARENGA, E.M. Comparação de metodologias do teste de tetrazólio para sementes de cafeeiro. *Idesia*, v.27, n.2, p.17-24, 2009. http://www.scielo.cl/pdf/idesia/v27n2/art02.pdf

11

Physiological quality of soybean seeds produced under artificial rain in the pre-harvesting period

Elisa de Melo Castro[1*], João Almir Oliveira[1], Amador Eduardo de Lima[1],
Heloísa Oliveira dos Santos[1], José Igor Lopes Barbosa[1]

ABSTRACT – The objective of this experiment was to evaluate responses of soybean cultivars seeds (with different levels of lignin) considering harvest postponement under the incidence of water and the effect of storage. The experiment was conducted in Iraí de Minas, Brazil using a randomized block design with three replications under a 5 x 3 x 3 factorial arrangement, with five soybean cultivars (NK 7059 RR, SYN 1163 RR, SYN 9070 RR, AS 7307 RR and SYN 1283 RR), three harvesting periods (R8, R8 + one rain simulation in the pre-harvesting period and R8 + two simulations in the pre-harvesting period) under three storage times (0, 90 and 180 days). Seeds were evaluated for their chemical composition (lignin contents), the percentage of moisture damage using the tetrazolium test and physiological quality (germination, accelerated aging, conductivity test, seedling emergence and emergence index). Cultivar AS 7307 RR had the highest lignin content in the integument, the lowest percentage of damage by moisture and the highest physiological quality. Cultivars NK 7059 RR and SYN 1163 RR had the lowest lignin contents in the integument and the highest moisture damage. Electrical conductivity increased after storing all cultivars.

Index terms: *Glycine max*, lignin, storage.

Qualidade fisiológica de sementes de soja submetidas à chuva artificial na pré-colheita

RESUMO – Objetivou-se com este trabalho avaliar o desempenho de sementes de cultivares de soja com diferentes teores de lignina considerando o retardamento da colheita sob incidência artificial de chuva, e ao longo do armazenamento. O ensaio foi conduzido em Iraí de Minas, MG, em blocos casualizados com três repetições e esquema fatorial 5 x 3 x 3, sendo cinco cultivares de soja (NK 7059 RR, SYN 1163 RR, SYN 9070 RR, AS 7307 RR e SYN 1283 RR), três épocas de colheita (R8, R8 + uma simulação de chuva na pré-colheita e R8 + duas simulações de chuva na pré-colheita), e três épocas de armazenamento (0, 90 e 180 dias). As sementes foram avaliadas quanto à composição química (teor de lignina), porcentagem de danos por umidade usando o teste de tetrazólio, e a qualidade fisiológica (germinação, teste de envelhecimento, condutividade elétrica, emergência, índice de emergência). A cultivar AS 7307 RR apresentou maior teor de lignina no tegumento, menor porcentagem de danos por umidade e melhor qualidade fisiológica. As cultivares NK 7059 RR e SYN 1163 RR apresentam menores teores de lignina no tegumento e maiores danos por umidade. Os valores de condutividade elétrica aumentam com o armazenamento para todas as cultivares.

Termos para indexação: *Glycine max*, lignina, armazenamento.

Introduction

Soybeans (*Glycine max* L. Merrill) are one of the most cultivated plants in the world, having great importance in the Brazilian economy. The development of soybean crops is associated with new technologies, especially those related to the production of high quality seeds, free of pathogens and with potential to develop high vigor seedlings (Pelúzio et al., 2008).

Genetic characteristics and environmental effects during the development, harvesting, processing and storage stages are key factors in the seed viability period, which is extremely variable. In seeds submitted to unfavorable conditions in any of these stages there may be physiological damage that can hinder the seeds quality and the intensity of such damage is variable with genetic factors and intrinsic to each cultivar (Gris et al., 2010).

[1]Departamento de Agricultura, UFLA, Caixa Postal 3037, 37200-000 – Lavras, MG, Brasil.
*Corresponding author < elisaagro@yahoo.com.br>

Soybean seeds reach their physiological ripening at the R7 stage, when they have maximum physiological quality and should be harvested. Due to this high water content in seeds at this stage, the recommended harvest point is stage R8, when 95% of the pods have the typical coloration of ripe pods. However, due to several factors, it is not always possible to harvest the seeds at the right time and delayed harvest often occurs (Diniz et al., 2013). Delayed harvest can lead to losses in physiological and sanitary quality of seeds (Henning et al., 2011).

In the period between physiological ripening and harvest, there may be damage by "moisture" in the seeds, which results from their exposure in alternating cycles of wet and dry environmental conditions in the post-ripening phase. Such damage can be even more aggressive in tropical regions, predominantly due to hot and humid weather, which can further accelerate the seeds deterioration process. Forti et al. (2013) have found that the seeds physiological potential is reduced by moisture damage and this may be one of the most detrimental factors to the quality of soybean seeds.

In some research it has been observed that the lignin content of the integument can influence the seeds physiological quality (Panobianco et al., 1999). The possibility of seeds with a certain degree of impermeability that have greater tolerance to deterioration from moisture has motivated researchers. Thus, the aim of this study was to evaluate soybean cultivars seeds performance with different levels of lignin in terms of harvest postponement with artificial incidence of rain during storage.

Material and Methods

The experiment was conducted in a seed production field deployed in October 2013 in an area irrigated by center-pivot in the Brazilian city of Iraí de Minas, in the state of Minas Gerais. The city is located in the (area in the west of the state of Minas Gerais, Brazil) "Triângulo Mineiro" region under geographical coordinates 18° 59' 23" LS and 47° 28' 33' LW' and altitude of 1029 meters.

The experiment was conducted in a randomized block design with three replications and a 5 x 3 factorial arrangement where five cultivars were evaluated: NK 7059 RR, SYN 1163 RR, SYN 9070 RR, AS 7307 RR, SYN 1283 RR and three harvest times: at the R8 stage, in R8 more rainfall simulation in the pre-harvest, and in R8 two more rain simulations in pre-harvest. Rain simulations were performed by means of a center-pivot irrigation, using an intensity of approximately 30 mm of water until the pods were soaked and reached 18% of water content. The plots consisted of six rows of 12 meters with spacing of 0.5 m between rows and 20 seeds/m. And the two rows of the central region were used as floor area of the plot. In

March 2014 there was a manual harvesting of plants and drying in the sun until the seeds reached 13% of water content. Seeds were processed with a plant threshing with beater cylinder. Then they were packed in paper bags and sent to the Brazilian city of Lavras, MG, where they were stored in the Seeds Processing Unit at Federal University of Lavras. Testing were performed on seeds retained in a circular screen 13 sieve.

To characterize the moisture damage caused by the incidence of artificial rain in the pre-harvest, tetrazolium test was performed, which was conducted with four subsamples of 25 seeds per plot, totaling 300 seeds per treatment. Seeds were preconditioned between paper towel moistened with distilled water for 16 hours at 25 °C and then they were immersed in a 2,3,5-triphenyl-2H-tetrazolium chloride solution) at 0.075% under temperature of 40 °C for three hours in the dark. The evaluation was conducted according to the methodology proposed by França-Neto et al. (1998) considering the percentage of seeds with moisture damage. To determine the lignin content in the integument, six replications of 50 seeds were used for each cultivar, following the methodology proposed by Capeleti et al. (2005).

Laboratory testing to evaluate the physiological quality of the seeds were performed with two subsamples of 50 seeds per plot totaling 300 seeds per treatment, performed at zero, 90 and 180 days of storage by the following tests: *Germination*: according to Brasil (2009); *Accelerated aging*: gerbox-type transparent acrylic boxes (11.0 x 11.0 x 3.5 cm) were used, adapted with a hanging aluminum screen, containing 40 mL of distilled water and a single layer of seeds on the hanging screen. Subsequently, these boxes were kept in a B.O.D. (Biochemical Oxygen Demand) at 42 °C for 72 hours (Dutra and Vieira, 2004). After this period, the seeds were submitted to the germination test. *Electrical conductivity*: performed with the aid of a digital conductivity meter (Digimed CD-21). Results were expressed in $\mu S\ cm^{-1}.g^{-1}$ according to the method described by Vieira and Krzyzanowski (1999). *Seedling emergence*: seeds were sown in plastic trays containing soil and sand as a substrate, in the 2:1 ratio, moistened to 60% of the moisture holding capacity. The trays were kept in a plant growth chamber at a temperature of 25 °C and a photoperiod of 12 hours, with daily assessments of normal seedlings emergence and a final score at 14 days after sowing. The final percentage of emergence and emergence speed index (ESI) were considered (Maguire,1962).

For statistical analysis of germination, accelerated aging, electrical conductivity, seedling emergence and emergence speed index tests, the treatments were arranged in a 5 x 3 x 3 factorial design involving five cultivars, three harvest times and three storage times. A complete randomized block design (RBD) with three replications was used. Data were submitted to analysis of variance using the statistical program Sisvar (Ferreira, 2011). Means were compared by the Scott-Knott test at 5% probability.

For the tetrazolium test, statistical analysis was performed separately by considering only the non-stored seeds, and moisture damage values were previously transformed into $(x + 1)^{0.5}$ using the statistical program Sisvar in RBD, 5 x 3 factorial arrangement, being five cultivars and three harvest times.

Lignin content was determined for the five cultivars studied and six replicates were analyzed. Statistical analysis was performed separately in a completely randomized design (CRD) using the statistical program Sisvar at 5% probability.

Results and Discussion

Lignin content

There were differences for the five cultivars studied. Cultivar AS 7307 RR showed higher content (0.4433 g%); cultivars SYN 1283 RR (0.3017 g%), SYN 9070 RR (0.2967 g%) showed intermediate contents and did not differ from each other. Cultivars SYN 1163 RR (0.2233 g%) and NK 7059 RR (0.1933 g%) showed the lowest contents of lignin.

The cultivar with the highest content of lignin, AS 7307 RR (0.4033 g%), throughout the study showed, in general, better physiological quality (Tables 1, 2, 3, 4, 5). The high physiological quality can be associated with the lignin content.

Menezes et al. (2009) have observed that the highest lignin content is associated with a higher percentage of normal seedlings in the accelerated aging test. However, there was a negative correlation to the germination rate. Santos et al. (2007) have found that soybean seeds with brown integument have a higher lignin content, ranging between 4 and 6% of lignin compared to the total weight of the integument, and higher vigor due to the slower rate of soaking.

Carvalho et al. (2014) have noticed that soybean cultivar seeds with high lignin content in the integument do not necessarily have better physiological quality. Cultivar Baliza RR, with lignin content of 0.4171 g%, an intermediate content between the cultivars studied, showed higher physiological quality before and after storage. Thus, seed quality can also be related to other factors inherent in the genotype.

Tetrazolium

The sources of variation cultivar and harvest season, as well as the interaction between these two factors, significantly differed in the evaluation of the tetrazolium test (damage caused by moisture 1-8).

Comparing cultivars at harvest season R8, it was observed that cultivar AS 7307 RR showed lower percentage of moisture

damage (10.67%), followed by cultivars SYN 9070 RR (22.00%) and SYN 1283 RR (24.33%). Cultivars SYN 1163 RR and NK 7059 RR showed higher percentage of damage (45.33% and 50.33%, respectively). AS 7307 RR and SYN 9070 RR were present in the group least affected by the effect of moisture in harvest times R8 + 1 and R8 + 2 (Table 1).

Table 1. Percentage of seeds with moisture damage for different soybean cultivars with a delayed harvest.

Cultivar	Harvest season		
	R8	R8 + 1	R8 + 2
NK 7059 RR	50.33 Ac	51.20 Ab	63.14 Ab
SYN 1163 RR	45.33 Ac	52.00 Ab	50.67 Ab
SYN 9070 RR	22.00 Ab	35.00 Ba	21.67 Aa
AS 7307 RR	10.67 Aa	39.33 Ba	28.33 Ba
SYN 1283 RR	24.33 Ab	26.67 Aa	41.67 Bb

Means followed by the same letter, lowercase in the column and uppercase in the row, do not differ at 5% probability by the Scott-Knott test. R8: harvest stage; R8 + 1: harvest stage R8 plus one rain simulation in the pre-harvest and R8 + 2: harvest stage R8 plus two rain simulations in the pre-harvest. The original means were presented, but data were compared according to the data transformed (Transformation in $(x + 1)^{0.5}$).

For cultivars AS 7307 RR and SYN 1283 RR, it was observed that the stress caused by the delayed harvest, coupled with humidity, caused an increase in the percentage of moisture damage. The simulation of one and two rains in the pre-harvest did not cause an increase in the percentage of moisture damage for NK 7059 RR and SYN 1163 RR. However, there was a higher frequency of damage in three harvest times for these cultivars. Genotype AS 7307 RR had a higher lignin content and a lower percentage of moisture damage in the three harvest times (Table 1).

Forti et al. (2013), studying moisture damage in soybean seeds, related tetrazolium test results with image analysis by X-ray and confirmed that moisture damage interferes with the seeds physiological potential, depending on their extent and location. Giurizatto et al. (2003) have observed that the vigor estimated by the tetrazolium test was influenced by the harvest season and the genotypes. Higher percentages of vigor were obtained when the harvest was held at the R8 stage and delayed harvest caused a reduction in vigor. In this study, it is noted that the increase in the percentage of moisture damage with a delayed harvest occurred only to cultivars AS 7307 RR and SYN 1283 RR, enhancing the effect of genotype for each harvest season.

Physiological analyses

For the results of the physiological tests, the sources of variation cultivar and harvest season significantly influenced all response variables. Regarding storage, there was no statistical difference for electrical conductivity, accelerated aging, emergence speed index and percentage of emergence at 14 days in the emergence test. For interactions, there was a significant effect cultivar*harvest season in all tests, cultivar*storage for electrical conductivity, emergence speed index and germination percentage at 14 days in the emergence test. For the interaction harvest season*storage time, only in the electrical conductivity test there was a significant effect. There was no significant effect for triple interaction (cultivar*harvest season*storage time).

Germination

Regarding the interaction cultivar*harvest season, it is noted that in the R8 harvest season the five cultivars studied were statistically different. For harvest season R8 + 1, cultivar AS 7303 RR showed germination percentage statistically equal to cultivar SYN 1283 RR, which were higher than the others. Cultivar SYN 1163 RR showed lower performance. For harvest season R8 + 2, all cultivars showed high germination percentages, and cultivars AS 7307 RR, SYN 9070 RR, SYN 1163 RR and NK 7059 RR did not differ from each other and showed higher germination percentage than cultivar SYN 1283 RR (Table 2).

Table 2. Seed germination, electrical conductivity, accelerated aging in different soybean cultivars with a delayed harvest.

Cultivars	Germination (%)			Electrical conductivity (uS cm^{-1}.g^{-1})			Accelerated aging (%)		
	R8	R8 + 1	R8 + 2	R8	R8 + 1	R8 + 2	R8	R8 + 1	R8 + 2
NK 7059 RR	83 Bc	81 Bb	90 Aa	62.94 Ab	72.24 Ba	68.58 Ba	48 Bb	51 Bc	69 Ab
SYN 1163 RR	66 Be	64 Bd	91 Aa	82.47 Bc	88.45 Bc	74.11 Ab	32 Bc	35 Bd	73 Ab
SYN 9070 RR	78 Bd	72 Cc	93 Aa	76,58 Bc	81,07 Bb	66,76 Aa	53 Bb	52 Bc	89 Aa
AS 7307 RR	98 Aa	92 Ba	94 Ba	51.55 Aa	65.87 Ba	63.80 Ba	94 Aa	84 Aa	88 Aa
SYN 1283 RR	89 Ab	88 Aa	86 Ab	63.67 Ab	67.83 Aa	74.34 Bb	86 Aa	72 Bb	72 Bb

Means followed by the same letter, lowercase in the column and uppercase in the row, do not differ at 5% probability by the Scott-Knott test. R8: harvest stage; R8 + 1: harvest stage R8 plus one rain simulation in the pre-harvest and R8 + 2: harvest stage R8 plus two rain simulations in the pre-harvest.

Considering the breakdown of harvest times in the variation source cultivar, it may be noted that with the exception of cultivars AS 7307 RR and SYN 1283 RR, the others showed higher germination percentage in season R8 + 2 (Table 2). Cultivar AS 7307 RR, with higher lignin content, had a higher germination percentage at harvest season R8, and the germination percentages in times R8 + 1 and R8 + 2 did not differ statistically. Cultivar SYN 1283 RR had germination percentage statistically equal for the three harvest times.

Giurizatto et al. (2003), studying nine genotypes harvested at the R8 stage and at 14 days after this stage, found that delayed harvest reduced germination, and the occurrence of rainfall close in time to the harvest season was possibly one of the factors contributing to seeds deterioration. These findings do not corroborate those of the present study. Probably the artificial incidence of rain coupled with the delayed harvest did not occur in sufficient quantities to cause deterioration in seeds.

Lima et al. (2007) have realized that the germination potential tends to decrease as the seeds remain in the field after physiological ripening, and this fact has not occurred in this study. However, these authors allowed a longer time of exposure to the elements, since the seeds remained in the field until stage R8, R8 + 15 days and R8 + 30 days. Minuzzi et al. (2010) indicate that obtaining higher quality of seeds occurs when the crop is carried out seven days after stage R7.

Electrical conductivity

As for the interaction cultivar*harvest season, it is possible to see that cultivar AS 7307 RR was placed in the group of lower electrical conductivity in the three harvest times. Cultivar SYN 1163 RR was placed among cultivars with higher values of electrical conductivity in the three harvest times, behaving physiologically lower than the others (Table 2).

Analyzing the behavior of cultivars in the three harvest times, it can be seen that cultivars AS 7307 RR, NK 7059 RR and SYN 1283 RR had a tendency to increase the electrical conductivity with delayed harvest, showing quality loss (Table 2). For cultivar AS 7307 RR, a positive correlation with the germination test is observed, where there was a decrease of germination vigor with a delayed harvest (Table 2).

Considering the conductivity results of the seeds non-submitted to storage, cultivar SYN 1163 RR showed higher electrical conductivity than the others studied. When analyzing the values of electrical conductivity at 90 and 180 days of storage, there was a stratification of cultivars in three lots of vigor, and cultivar AS 7307 RR was superior individually, indicating high physiological potential. Cultivars SYN 1283 RR and NK 7059 RR are shown to be intermediate and cultivars SYN 1163 RR and SYN 9070 RR were considered as having less vigor due to showing higher values of electrical conductivity (Table 3).

Analyzing each cultivar during storage time, a correlation between the increase in the amount of leached ones in the soaking solution, i.e., greater release of exudates that occurs over time, was found. For cultivars SYN 1163 RR and SYN 9070 RR, a gradual and significant increase in the electrical conductivity values for the storage times was seen. As for cultivars NK 7059 RR, AS 7307 RR and SYN 1283 RR, there was an increase of the electrical conductivity value of the first season for the 90 days of storage. However, this behavior was stabilized in this period, with no increased electrical conductivity at 180 days of storage (Table 3).

Changes in electrical conductivity among cultivars were noticed by Oliveira et al. (2012) when evaluating soybeans seeds physical, physiological and sanitary quality of two regions in the Brazilian state of Mato Grosso. Panobianco et al. (1999) have noticed the existence of close links between lignin content of soybean cultivars and the results of the electrical conductivity test, which was also observed in this study, since cultivar AS 7307 RR showed higher lignin content and lower electrical conductivity values in the three harvest times.

Generally it is observed that for non-stored seeds harvest season did not influence electrical conductivity values. Lower electrical conductivity with 90 days of storage for seeds of all crops when harvested in R8 was observed. At 180 days, the highest electrical conductivity values were found in the seeds of all cultivars when harvested in the R8 stage with one rain simulation. Whereas in each harvest season it is possible to identify the increase of electrical conductivity with increasing storage period, indicating increased permeability and loss of membrane integrity with storage time evolution (Table 4).

Table 3. Electrical conductivity of the seeds, emergence and emergence speed index in different soybean cultivars during storage (0, 90, 180 days).

Cultivar	Conductivity (uS cm^{-1}.g^{-1})			Emergence (%)			ESI		
	0	90	180	0	90	180	0	90	180
NK 7059 RR	53.35 Aa	71.44 Bb	78.36 Bb	87 Ab	77 Bc	91 Aa	13.51 Ab	8.79 Cb	10.98 Bb
SYN 1163 RR	63.27 Ab	84.02 Bc	97.74 Cc	78 Ac	66 Bd	79 Ac	11.81 Ac	7.60 Cb	9.50 Bc
SYN 9070 RR	53.60 Aa	78.87 Bc	91.94 Cc	87 Ab	81 Ab	85 Ab	13.78 Ab	10.15 Ba	10.42 Bc
AS 7307 RR	48.07 Aa	63.31 Ba	69.86 Ba	97 Aa	88 Ba	95 Aa	15.55 Aa	10.20 Ca	12.09 Ba
SYN 1283 RR	54.40 Aa	72.75 Bb	78.68 Bb	93 Aa	75 Bc	94 Aa	13.68 Ab	8.23 Cb	11.31 Bb

Means followed by the same letter, lowercase in the column and uppercase in the row, do not differ at 5% probability by the Scott-Knott test.

Table 4. Electrical conductivity (uS cm^{-1}.g^{-1}) of soybean seeds submitted to a delayed harvest and stored.

Harvest season	Storage (days)		
	0	90	180
R8	51.49 Aa	63.38 Ba	81.45 Ca
R8 + 1	55.32 Aa	78.82 Bb	91.43 Cb
R8 + 2	56.75 Aa	74.18 Bb	77.53 Ba

Means followed by the same letter, lowercase in the column and uppercase in the row, do not differ at 5% probability by the Scott-Knott test. R8: harvest stage; R8 + 1: harvest stage R8 plus one rain simulation in the pre-harvest and R8 + 2: harvest stage R8 plus two rain simulations in the pre-harvest.

Accelerated aging

For soybean seeds vigor, assessed by the accelerated aging test (Table 2), there was a higher germination percentage after aging for cultivars AS 7307 RR and SYN 1283 RR when harvested in R8, AS 7307 RR when harvested in R8 + 1, AS 7307 RR and SYN 9070 RR when harvested in R8 + 2.

Analyzing the effect of the harvest season for each cultivar, it is possible to see that the result was similar to that observed in the germination test (Table 2). From cultivars with the highest levels of vigor on the sample taken in stage R8, cultivar AS 7307 RR maintained physiological quality when the harvest was delayed, while SYN 1283 RR showed decline in the germination percentage after accelerated aging with a delayed harvest (Table 2).

Gris et al. (2010) have observed that the largest decreases in vigor for soybean seeds, evaluated by the accelerated aging test, occurred for cultivars Jataí and Silvânia RR when delaying the harvest. When searching for the response from 15 soybean genotypes with delayed harvest, Braccini et al. (2003) have also observed a significant reduction in the germination percentage and seed vigor when they were submitted to harvest 30 days after the R8 stage of development.

Emergence

Cultivar AS 7307 RR had the highest emergence percentage of seedlings in the three harvest times, being statistically equal to cultivar SYN 1283 RR when harvested in R8 + 1 and SYN 9070 RR when harvested in R8 + 2. SYN 1163 RR was among the cultivars with lower performance in all harvest times (Table 5).

There was an uneven behavior for the cultivars studied when analyzing the effect of delayed harvest. Seeds of cultivars

NK 7059 RR and AS 7307 RR showed high physiological quality when harvested in R8 and R8 + 2, and seeds of cultivars SYN 1163 RR and SYN 9070 RR in R8 + 2. Cultivar SYN 1283 RR did not differ statistically for harvest times (Table 5).

Table 5.   Seeds emergence and emergence speed index in different soybean cultivars with a delayed harvest.

Cultivars	Emergence (%)			ESI		
	R8	R8 + 1	R8 + 2	R8	R8 + 1	R8 + 2
NK 7059 RR	85 Ab	78 Bb	90 Ab	10.98 Ab	10.64 Ab	11.51 Ab
SYN 1163 RR	71 Bc	63 Cc	89 Ab	9.23 Bc	7.91 Cc	11.76 Ab
SYN 9070 RR	82 Bb	77 Bb	93 Aa	11.61 Bb	9.90 Cb	12.84 Aa
AS 7307 RR	96 Aa	89 Ba	95 Aa	13.07 Aa	11.99 Aa	12.79 Aa
SYN 1283 RR	88 Ab	88 Aa	85 Ab	11.13 Ab	11.06 Ab	11.03 Ab

Means followed by the same letter, lowercase in the column and uppercase in the row, do not differ at 5% probability by the Scott-Knott test. R8: harvest stage; R8 + 1: harvest stage R8 plus one rain simulation in the pre-harvest and R8 + 2: harvest stage R8 plus two rain simulations in the pre-harvest.

Diniz et al. (2013), when studying the relationship between the seeds physiological quality of eight soybean cultivars submitted to three harvest times, found that seedling emergence in the field of seeds harvested 30 days after the R8 stage showed lower seedling emergence than those harvested in stages R8 and R8 + 15 days. This behavior was not found in this study. Possibly, delayed harvest combined with artificial incidence of rainfall did not occur in sufficient time and/or quantity to cause this physiological wear on seeds.

There are variations during the storage in the behavior of cultivars (Table 3). Except for cultivar SYN 9070 RR, which did not differ statistically for storage periods, there are larger emergence percentages of normal seedlings for stored and non-stored seeds for 180 days. At 90 days of storage there was a decrease in the emergence percentage.

Schuab et al. (2006) point out that the interpretation of the laboratory and field emergence tests results should not consider only the correlation analysis, as this can lead to incorrect interpretations due to insufficient data.

Emergence speed index

It is noted that the ESI varied among cultivars, with higher values seen for seedlings of cultivar AS 7307 RR, which presented indices of 13.07, 11.99 and 12.79 when the seeds were collected in R8, R8 + 1 and R8 + 2, respectively. In stage R8 + 2 it did not differ statistically from cultivar SYN 9070 RR (Table 5).

SYN 1163 RR was among the cultivars that had the lowest levels of vigor in every harvest season (Table 5). Low vigor seeds take time for the restoration of organelles and damaged tissues before commencing the embryonic axis growth (Villiers, 1973).

Observing the effect of delayed harvest (Table 5), it can be seen that seeds from cultivars SYN 1163 RR and SYN 9070 RR had higher values of emergence speed index when harvested in R8 + 2, followed by the seeds that received treatment R8 and later by those which were collected in R8 + 1. These results not corroborate those found by Gris et al. (2010), who observed lower emergence speed in soybean seeds collected in stage R8 + 20 in relation to that observed in seeds collected in stages R7 and R8.

For non-stored seeds (Table 3), there was a predominance of higher values of emergence speed index, and in this period cultivar AS 7307 RR stood out regarding the others, showing higher ESI (15.55), followed by cultivars SYN 9070 RR, SYN 1283 RR and NK 7059 RR, which did not differ from each other, and cultivar SYN 1163 RR showed lower ESI. Seeds with higher ESI have better performance and consequently, higher emergence rate in the farming field, resisting better to stresses that may occur during emergence (Dan et al., 2010).

At 90 days of storage, cultivars AS 7307 RR and SYN 9070 RR showed higher vigor. With 180 days of storage it is noted that the higher index was seen for cultivar AS 7307 RR, followed by NK 7059 RR and SYN 1283 RR and the lowest vigor levels occurred for SYN 1163 RR and SYN 9070 RR (Table 3).

Observing each of the cultivars during storage (Table 3), it is seen that in general the seeds of the cultivars had higher vigor when not stored, lower throughput at 90 days of storage and an intermediate performance after 180 days of storage.

The highest emergence speed index at 180 days, when compared at 90 days, may be due to the fact that, even under controlled conditions, the temperature in the chamber of 25 °C in August (the time when the assessment at 90 days was carried out) may have been influenced at night by the outside temperature when the photoperiod of 12 hours of dark occurred, because as the lamps were off, the temperature may have been reduced because there was no heating system, which probably did not influence the other storage times, which were held in early May and in November, high temperature periods in the city of Lavras. This result corroborates that obtained in the percentage of normal seedlings at 14 days in the emergence test in trays.

Given the results achieved in the implementation of the physiological tests, it was found that the damage caused by the artificial incidence of water in the pre-harvest was small due to the probable low water absorption by seeds. In the period before the harvests, rainfall indices were low, which shows the effect of the treatment applied. Although the amount of water applied by the irrigation system was relatively high (30 mm), it occurred in a concentrated form, which probably favored the low water uptake by the seeds and the rapid loss of moisture by pods, differently from what occurs when there is a natural rainfall with atmospheric conditions of high relative humidity, which provides higher absorption of water in the seeds and consequently higher moisture damage.

Conclusions

Among the cultivars studied, AS 7307 RR has a higher lignin content in the seeds tegument and a smaller percentage of moisture damage evaluated by the tetrazolium test in all harvest times.

Cultivars NK 7059 RR and SYN 1163 RR have the lowest lignin content in the integument and the highest percentages of moisture damage evaluated by the tetrazolium test in all harvest times.

Cultivar AS 7307 RR has the highest physiological performance and cultivar SYN 1163 RR the lowest, evaluated by germination and vigor tests.

Electrical conductivity values increase with seed storage for all cultivars.

The percentage of normal seedlings emergence and the emergence speed index vary with storage.

Acknowledgments

The authors thank FAPEMIG (Research Support Foundation of the Brazilian State of Minas Gerais), CNPq (National Counsel of Technological and Scientific Development)] and CAPES (Coordination of Improvement of Higher Education Personnel) for the financial support during the conduction of this study.

References

BRACCINI, A.L.; ALBRECHT, L.P.; ÁVILA, M.R.; SCAPIM, C.A.; BIO, F.E.I.; SCHUAB, S.R.P. Qualidade fisiológica e sanitária das sementes de quinze cultivares de soja (Glycine max (L.) Merrill) colhidas na época normal e após o retardamento de colheita. Acta Scientiarum Agronomy, v.25, n.2, p 449-457, 2003. http://dx.doi.org/10.4025/actasciagron.v25i2.2153

BRASIL. Ministério da Agricultura, Pecuária e Abastecimento. Regras para análise de sementes. Ministério da Agricultura, Pecuária e Abastecimento. Secretaria de Defesa Agropecuária. Brasília: MAPA/ACS, 2009. 395p. http://www.agricultura.gov.br/arq_editor/file/2946_regras_analise__sementes.pdf

CAPELETI, I.; FERRARESE, M.L.L.; KRZYZANOWSKI, F.C.; FERRARESE FILHO, O. A new procedure for quantification of lignin in soybean (Glycine max (L.)Merrill) seed coat and their relationship with the resistance to mechanical damage. Seed Science and Technology, v.33, p.511-515, 2005. http://www.scielo.br/scielo.php?script=sci_nlinks&ref=000084&pid=S1413-705420100002000150006&lng=pt

CARVALHO, E.R.; OLIVEIRA, J.A.; CALDEIRA, C.M. Qualidade fisiológica de sementes de soja convencional e transgênica RR produzidas sob aplicação foliar de manganês. Bragantia, v.73, n.3, p.219-228, 2014. http://www.scielo.br/pdf/brag/2014nahead/aop_brag_0096_pt.pdf

DAN, L.G.M; DAN, H.A.; BARROSO, A.L.L; BRACCINI, A.L. Qualidade fisiológica de sementes de soja tratadas com inseticidas sob efeito do armazenamento. Revista Brasileira de Sementes, v.32, n.2, p.131-139, 2010. http://www.scielo.br/pdf/rbs/v32n2/v32n2a16.pdf

DINIZ, F.O.; REIS, M.S.; DIAS, L.A.S.; ARAÚJO, E.F.; SEDIYAMA, T., SEDIYAMA, C.A. Physiological quality of soybean seeds of cultivars submitted to harvesting delay and its association with seedling emergence in the field. Journal of Seed Science, v.35, n.2, p.147-152, 2013. http://www.scielo.br/scielo.php?pid=S2317-15372013000200002&script=sci_arttext

DUTRA, A.S.; VIEIRA, R.D. Envelhecimento acelerado como teste de vigor para sementes de milho e soja. Ciência Rural, v.34, n.3, p.715-721, 2004. http://www.scielo.br/scielo.php?pid=S0103-84782004000300010&script=sci_arttext

FERREIRA, D.F. Sisvar: a computer statistical analysis system. Ciência e Agrotecnologia, v.35, n.6, p.1039-1042, 2011. http://www.scielo.br/scielo.php?pid=S1413-70542011000600001&script=sci_arttext

FORTI, V.A.; CARVALHO, C.; TANAKA, F.A.O; CICERO, S.M. Weathering damage in soybean seeds: assessment, seed anatomy and seed physiological potential. Seed Technology, v.35, n.2, p.213-224, 2013. http://www.researchgate.net/profile/Cristiane_De_Carvalho/publication/268385544_Weathering_Damage_in_Soybean_Seeds_Assessment_Seed_Anatomy_and_Seed_Physiological_Potential/links/5469cdbe0cf2f5eb18051c35.pdf

FRANÇA-NETO, J.B.; KRZYZANOWSKI, F.C.; COSTA, N.P. O teste de tetrazólio em sementes de soja. Londrina: Embrapa Soja, 1998. 72p. (Embrapa-CNPSo. Documentos, 116). http://www.scielo.br/scielo.php?script=sci_nlinks&ref=000079&pid=S0101-31222012000100002000009&lng=en

GIURIZATTO, M.I.K.; SOUZA, L.C.F.; ROBAINA, A.D.; GONÇALVES, M.C. Efeito da época de colheita e da espessura do tegumento sobre a viabilidade e o vigor de sementes de soja. Ciência e Agrotecnologia, v.27, n.4, p.771-779, 2003. http://www.scielo.br/scielo.php?script=sci_arttext&pid=S1413-70542003000400005

GRIS, C.F.; VON PINHO, E.V.R.; ANDRADE, T.; BADONI, A.; CARVALHO, M.L.M. Qualidade fisiológica e teor de lignina no tegumento de sementes de soja convencional e transgênica RR submetidas a diferentes épocas de colheita. Ciência e Agrotecnologia, v.34, n.2, p.374-381, 2010. http://www.scielo.br/pdf/cagro/v34n2/15.pdf

HENNING, F.A.; JACOB JUNIOR, E.A.; MERTZ, L.M.; PESKE, S.T. Qualidade sanitária de sementes de milho em diferentes estádios de maturação. Revista Brasileira de Sementes, v.33, n.2, p.316-321, 2011. http://www.scielo.br/scielo.php?pid=S0101-31222011000200014&script=sci_arttext

LIMA, W.A.A.; BORÉM, A.; DIAS, D.C.F.S.; MOREIRA, M.A.; DIAS, L.A.S.; PIOVESAN, N.D. Retardamento de colheita como método de diferenciação de genótipos de soja para qualidade de sementes. *Revista Brasileira de Sementes*, v.29, n.1, p.186-192, 2007. http://www.scielo.br/ scielo.php?script=sci_arttext&pid=S0101-31222007000100026

MAGUIRE, J.D. Speed of germination-aid in selection and evaluation for seedling emergence and vigor. *Crop Science*, v.2, n.1, p.176-177, 1962. http:// www.scielo.br/scielo.php?script=sci_nlinks&ref=000106&pid=S0100-67622003000500001000017&lng=pt

MENEZES, M.; VON PINHO, E.V.R.; JOSÉ, S.C.B.R.; BALDONI, A.; MENDES, F.F. Aspectos químicos e estruturais da qualidade fisiológica de sementes de soja. *Pesquisa Agropecuária Brasileira*, v.44, n.12, p.1716-1723, 2009. http://www.scielo.br/scielo.php?pid=S0100-204X2009001200022&script=sci_arttext

MINUZZI, A.; BRACCINI, A.L.; RANGEL, M.A.S.; SCAPIM, C.A.; BARBOSA, M.C.; ALBRECHT, L.P. Qualidade de sementes de quatro cultivares de soja, colhidas em dois locais no estado do Mato Grosso do Sul. *Revista Brasileira de Sementes*, v.32, n.1, p.176-185, 2010. http://www. scielo.br/scielo.php?pid=S0101-31222010000100020&script=sci_arttext

OLIVEIRA, G.P.; ARAÚJO, D.V.; ALBUQUERQUE, M.C.F.; MAGNANI, E.B.Z.; MAINARDI, J.T. Avaliação física, fisiológica e sanitária de sementes de soja de duas regiões de Mato Grosso. *Revista Agrarian*, v.5, n.16, p.106-114, 2012. http://www.periodicos.ufgd.edu.br/index.php/agrarian/article/ viewArticle/1039

PANOBIANCO, M.; VIEIRA, R.D.; KRZYZANOWSKI, F.C.; FRANÇA-NETO, J.B. Electrical conductivity of soybean seed and correlation with seed coat lignin content. *Seed Science and Technology*, v.27, n.3, p.945-949, 1999. http://www.scielo.br/scielo.php?script=sci_nlinks&ref=000103&pid=S0006-87052010000100026000019&lng=pt

PELÚZIO, J.M.; FIDELIS, R.R.; JÚNIOR, D.A.; SANTOS, G.R.; DIDONET, J. Comportamento de cultivares de soja sob condições de várzea irrigada no sul do estado do Tocantins, entressafra 2005. *Bioscience Journal*, v.24, n.1, p.75-80, 2008. http://www.seer.ufu.br/index.php/biosciencejournal/ article/view/6734

SANTOS, E.L.; PÓLA, J.N.; BARROS, A.S.R.; PRETE, C.E.C. Qualidade fisiológica e composição química das sementes de soja com variação na cor do tegumento. *Revista Brasileira de Sementes*, v.29, n.1, p.20-26, 2007. http://www. scielo.br/scielo.php?pid=S0101-31222007000100003&script=sci_arttext

SCHUAB, S.R.P.; BRACCINI, A.L.; FRANÇA-NETO, J.B.; SCAPIM, C.A.; MESCHEDE, D.K. Potencial fisiológico de sementes de soja e sua relação com a emergência das plântulas em campo. *Acta Scientiarum Agronomy*, v.28, n.4, p.553-561, 2006. http://www.scielo.br/scielo.php?script=sci_ nlinks&ref=000113&pid=S1807-86212011000200016000020&lng=en

VIEIRA, R.D.; KRZYZANOWSKI, F.C. Teste de condutividade elétrica. In: KRZYZANOWSKI, F.C.; VIEIRA, R.D.; FRANÇA-NETO, J.B. (Ed.). *Vigor de sementes*: conceitos e testes. Londrina: ABRATES, 1999. p. 1-26.

VILLIERS, T.A. Ageing and longevity of seeds in field conditions. In: HEYDECKER, W. (Ed.). *Seed Ecology*, London: The Pennsylvania State University, 1973. p.265-288.

12

Desiccation tolerance and longevity of germinated *Sesbania virgata* (Cav.) Pers. seeds

Maria Cecília Dias Costa[1*], José Marcio Rocha Faria[2], Anderson Cleiton José[2], Wilco Ligterink[1], Henk W.M. Hilhorst[1]

ABSTRACT- Seed desiccation tolerance (DT) and longevity are necessary for better dissemination of plant species and establishment of soil seed bank. They are acquired by orthodox seeds during the maturation phase of development and lost upon germination. DT can be re-induced in germinated seeds by an osmotic and/or abscisic acid treatment. However, there is no information on how these treatments affect seed longevity. Germinated *Sesbania virgata* seeds were used as a model system to investigate the effects of an osmotic treatment to re-establish DT on seed longevity. Longevity of germinated *S. virgata* seeds treated and non-treated by an osmoticum was analysed after storage or artificial ageing. The radicle is the most sensitive organ, the cotyledons are the most resistant, and the ability to produce lateral roots is the key for whole seed survival. Germinated *S. virgata* seeds with 1mm protruded radicle tolerate desiccation and storage for up to three months without significant losses in viability. An osmotic treatment can improve DT in these seeds, but not longevity. Germinated *S. virgata* seeds are a good model to study DT uncoupled from longevity. Further studies are necessary to unveil the molecular mechanisms involved in both DT and longevity.

Index terms: desiccation tolerance, storage, germination, osmotic stress.

Tolerância à dessecação e longevidade de sementes germinadas de *Sesbania virgata* (Cav.) Pers.

RESUMO - Em sementes, tolerância à dessecação (TD) e longevidade são necessárias para a melhor dispersão da espécie e para o estabelecimento de um banco de sementes no solo. Ambas são adquiridas durante fase de maturação do desenvolvimento das sementes e perdidas durante a germinação. A TD pode ser re-induzida em sementes germinadas por um tratamento osmótico e/ ou com ácido abscísico. Entretanto, não há informações sobre como esses tratamentos afetam a longevidade das sementes. No presente estudo, utilizou-se sementes germinadas de *Sesbania virgata* como modelo experimental para investigar os efeitos na longevidade de um tratamento osmótico para re-induzir TD. A longevidade de sementes germinadas de *S. virgata* submetidas ou não a um tratamento osmótico foi analisada após armazenamento ou envelhecimento acelerado. A radícula é o órgão mais sensível e os cotilédones são o órgão mais resistente. A habilidade de produzir raízes laterais é imprescindível para a sobrevivência das sementes. Sementes germinadas de *S. virgata* com radícula de 1mm de comprimento toleram dessecação e armazenamento por até três meses sem redução significativa na longevidade. O tratamento osmótico melhora a TD nessas sementes, mas não a longevidade. Sementes germinadas de *S. virgata* são um bom modelo para estudar a TD desacoplada da longevidade.

Termos para indexação: tolerância à dessecação, armazenamento, germinação, estresse osmótico.

Introduction

Species from the genus *Sesbania* (Fabaceae) are distributed mainly in the African and American continents. Due to their fast growth, easy propagation, high biomass production, and potential to form symbiosis with nitrogen-fixing bacteria, they are used on a large scale in agroforestry and for ecological restorations (Florentino et al., 2009; Kwesiga et al., 1999; Ståhl et al., 2005; Zanandrea et al., 2009). One species of this genus commonly used in agroforestry in Brazil is *Sesbania virgata*, a fast-growing pioneer species that tolerates long periods of flooding and has a highly branched root system that protects the soil against erosion (Florentino et al., 2009; Zanandrea et al., 2009).

Seeds of *S. virgata* are orthodox, meaning that they are able to tolerate desiccation and survive in the dehydrated state

[1]Laboratory of Plant Physiology, Wageningen University, Droevendaalsesteeg 1, 6708 PB Wageningen, Netherlands.

[2]Departamento de Ciências Florestais, Universidade Federal de Lavras, Caixa Postal 3037, 37200-000- Lavras, MG, Brasil.
*Corresponding author <maria.diascosta@wur.nl>

for long periods of time (Pammenter and Berjak, 1999). In orthodox seeds, desiccation tolerance (DT) and longevity are considered necessary for the completion of their life cycle, permitting the plant to store seeds, and ensure better dissemination of the species (Ramanjulu and Bartels, 2002).

During the maturation phase, orthodox seeds acquire DT and longevity, enter a dormant or quiescent state and can remain apparently inactive for very long periods (Ooms et al., 1993; Toldi et al., 2009). During germination, upon imbibition, DT remains for some time but then starts to be lost when DNA synthesis and (somewhat later) cell division resume (Faria et al., 2005). After the loss of DT, the existence of a small developmental window during which DT can be re-established by treatment with an osmoticum and/or the plant hormone abscisic acid (ABA) was demonstrated in a number of species, including *Cucumis sativus*, *Impatiens walleriana* (Bruggink and Van der Toorn, 1995), *Medicago truncatula* (Buitink et al., 2003; 2006) and *Arabidopsis thaliana* (Maia et al., 2011; 2014). When DT is fully rescued, seeds seem to be in a stage resembling the developmental stage that they were in prior to germination (Buitink et al., 2006; Maia et al., 2011).

The ability of germinated seeds to re-acquire DT is thought to optimize successful seedling establishment under unpredictable environmental conditions (Dekkers et al., 2015). This ability, in conjunction with longevity, could represent an ecologically important stress tolerance mechanism that allows germinating/germinated seeds to remain viable in the dry state for a certain time. In the last few years, several studies have been carried out focusing on the acquisition, loss and re-induction of DT in seeds of model species (Dekkers et al., 2015). However, some diversity in stress-tolerance mechanisms is expected, raising the expectations that valuable information can be generated using non-model species that have to cope with such stress in their natural environments. In their natural habitat, *S. virgata* seedlings are subjected to irregular precipitation patterns at the start of the wet season. Despite the numerous studies on the re-induction of DT in germinated seeds, no information is available concerning the effects of treatments to re-induce DT on longevity. In the present study, the longevity of germinated *S. virgata* seeds was investigated. The results show that germinated *S. virgata* seeds with 1 mm protruded radicle can be dried back and be stored, and that an osmotic treatment can improve DT, but not longevity.

Material and Methods

Mature seeds of *S. virgata* were collected from 12 trees at Lavras (21°22"S, 45°1"W, Minas Gerais, Brazil) and stored in a cold room at 4 °C. Prior to germination tests, seeds were immersed in concentrated sulphuric acid (H_2SO_4) for 30 min

and washed with abundant running water to remove physical dormancy. Germination assays were carried out in moist rolled paper, at 30 °C, under constant light (30 W m^{-2}) for 36 h.

To determine the moment of loss of DT, three replicates of 20 seeds were selected according to their protruded radicle length (1 mm, 3 mm and 5 mm) and dried for three days. Throughout the study, drying treatments were performed by placing the seeds for three days at 40% relative humidity (RH) at 22 °C, resulting in a water content as low as 0.14 g H_2O g^{-1} dry weight (Figure 1). Water content was assessed gravimetrically for triplicate samples of 10 seeds, by determination of the fresh weight and subsequent dry weight, after 18 h in an oven at 105 °C. Water content is expressed on a dry weight basis. After drying, seeds were pre-humidified in air of 100% RH for 24 h at 22 °C in the dark to prevent possible imbibitional damage and subsequently rehydrated in water on a Copenhagen Table under a 12/12 h dark/light regime at 22 °C. Germinated seeds were evaluated according to the survival of their primary root, presence of green cotyledons (cotyledon survival) and growth resumption with both green cotyledons and development of a (secondary) root system (seedling survival) (Maia et al., 2011).

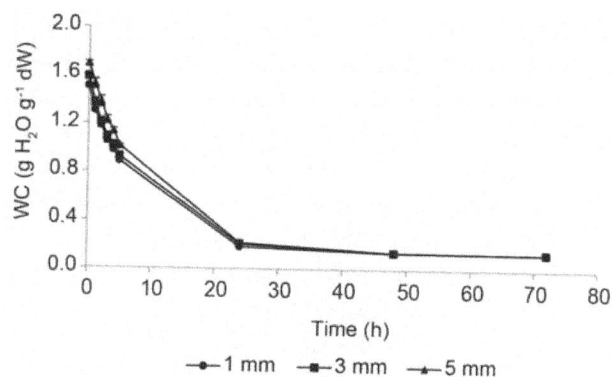

Figure 1. Changes in water content upon dehydration of germinated *S. virgata* seeds at 40% RH with 1 mm, 3 mm, and 5 mm protruded radicles. Each data point is the average of three replicates of 20 seeds. Bars represent standard error.

The re-establishment of DT in sensitive seeds was evaluated in three replicates of 20 germinated seeds selected according to their protruded radicle length (1 mm and 3 mm). These germinated seeds were dried (control) or treated with an osmoticum (incubation in 20 mL PEG solution with an osmotic potential ranging from -1.5 to -3.0 MPa at 4 °C or 22 °C (Villela and Beckert, 2001) in the dark, for 72 h) and dried. Possible hypoxic conditions were avoided by using an amount of PEG solution enough to cover the radicles, but not entire seeds. After incubation in PEG, seeds were rinsed thoroughly in distilled water, dried, pre-humidified, rehydrated and evaluated as described above.

Longevity was evaluated in triplicates of 15 germinated seeds with 1mm protruded radicle and dried as described above. These seeds were stored in sealed plastic bags for three to eight months at 4 °C. Additionally, longevity was estimated based on survival after an accelerated aging assay consisting of storage for two to seven days at 80% RH and 40 °C in the dark. Survival was evaluated as described above and by a tetrazolium test. For the tetrazolium test, three replicates of six seeds were moistened on filter paper in Petri dishes for 24 h at 22 °C. After the removal of the seed coat, they were soaked in 0.5% (w/v) 2,3,5-triphenyltetrazolium chloride solution for 2 h in the dark at 30 °C and scored using location of staining as criteria (Camargos et al., 2008).

In order to evaluate the influence of a treatment to re-establish DT on longevity, germinated seeds with 1 mm protruded radicle incubated in -2.5 MPa PEG at 4 °C and dried were subjected to the same longevity tests as described for non-treated seeds.

Data were statistically analysed with SPSS 22.0 for Windows (IBM Corporation, Somer, NY, USA) using one-way ANOVA followed by Duncan post-hoc test ($P \leq 0.05$).

Results and Discussion

Seeds of more than 90% of angiosperm species for which data are available are orthodox (Royal Botanic Gardens Kew, 2008). During germination of these seeds, water is taken up and metabolic processes are resumed, leading to the progressive loss of DT (reviewed by Dekkers et al., 2015). The point during germination when DT is lost varies among species. For example, within legume species, *Copaifera langsdorfii* and *Pisum sativum* lose DT before radicle protrusion (Pereira et al., 2014; Reisdorph and Koster, 1999), while *Glycine max*, *Medicago truncatula* and *Peltophorum dubium* lose DT after radicle protrusion (Buitink et al., 2003; Guimarães et al., 2011; Senaratna and McKersie, 1983). Here it is shown that *S. virgata* (also a legume) seeds lose DT progressively after radicle protrusion (Figure 2), being almost completely sensitive when the protruded radicle length reaches 5 mm.

More than 50% of germinated *S. virgata* seeds with 1 mm protruded radicle are able to survive desiccation without any previous treatment. As the radicle grows to 3 mm, there is a considerable drop in seedling formation to around 10%. Germinated seeds with 5 mm protruded radicle are sensitive to desiccation. DT is first lost by the radicle, the most sensitive organ in seeds of *S. virgata* and in other species (Buitink et

al., 2003; Maia et al., 2011). The cotyledons are the most resistant organs, but the crucial point to determine if the germinated seed will survive is the ability to produce lateral roots (Bruggink and Van der Toorn, 1995; Maia et al., 2011). Germinated seeds are able to produce lateral roots when there are viable tissues that can differentiate to root primordia, as shown by the tetrazolium staining. When the non-stained areas included the hypocotyl, the germinated seeds did not produce root primordia and survival of the whole seed was compromised. On the other hand, when the hypocotyl was stained, lateral roots were produced to replace the main root and to enable the seedling to establish.

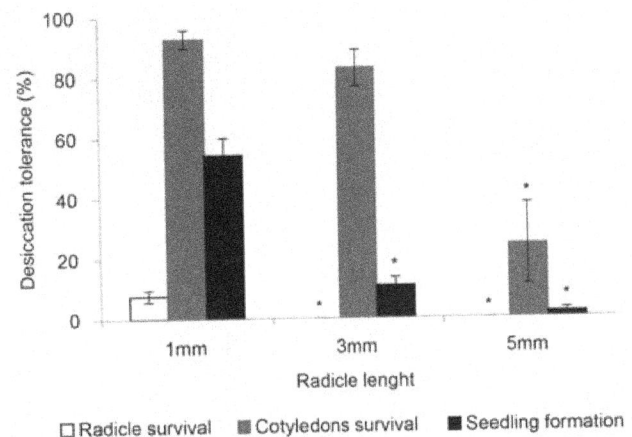

Figure 2. Desiccation tolerance in germinated *S. virgata* seeds with different protruded radicle lengths. Each data point is the average of three replicates of 20 seeds. Vertical bars represent standard errors. * indicate significant differences ($P \leq 0.05$) comparing to germinated seeds with 1 mm protruded radicle.

Previous studies have shown that after the loss of DT, a small developmental window is opened during which DT can be re-established by treatment with an osmoticum and/or ABA (Bruggink and Van der Toorn, 1995; Buitink et al., 2003; Maia et al., 2014). Attempts to re-induce DT in germinated *S. virgata* seeds by incubation in ABA failed, as ABA was not effective in arresting growth (data not shown). Figure 3 shows the percentage of DT obtained at different water potential/temperature combinations for germinated seeds with 1 mm and 3 mm protruded radicle. The osmotic treatment might inhibit radicle growth until enough ABA has accumulated and operates (Buitink et al., 2003). The water potential had considerably more impact on the re-establishment of DT than the temperature. For germinated seeds with 1 mm protruded radicle length, the water potential of -2.5 MPa led to the highest percentage of radicle survival. Still, even under these conditions, DT could not be re-established in all radicles. At the lowest water potential tested (-3.0 MPa), DT was observed in the lowest percentages of seeds. For germinated seeds with 3 mm protruded radicle length, some of the osmotic treatments significantly improved DT compared to non-treated seeds with the same radicle length, but the maximal percentage of seedling formation was lower than 40% and radicle survival was not improved. DT is a complex trait and it is possible that in *S. virgata*, as the radicle grows, the mechanisms needed for DT are progressively lost and cannot be fully re-activated anymore. Based on these results, the best radicle length and treatment to re-establish DT in germinated *S. virgata* seeds is 1 mm radicle length and incubation in an osmoticum of -2.5 MPa at 4 °C.

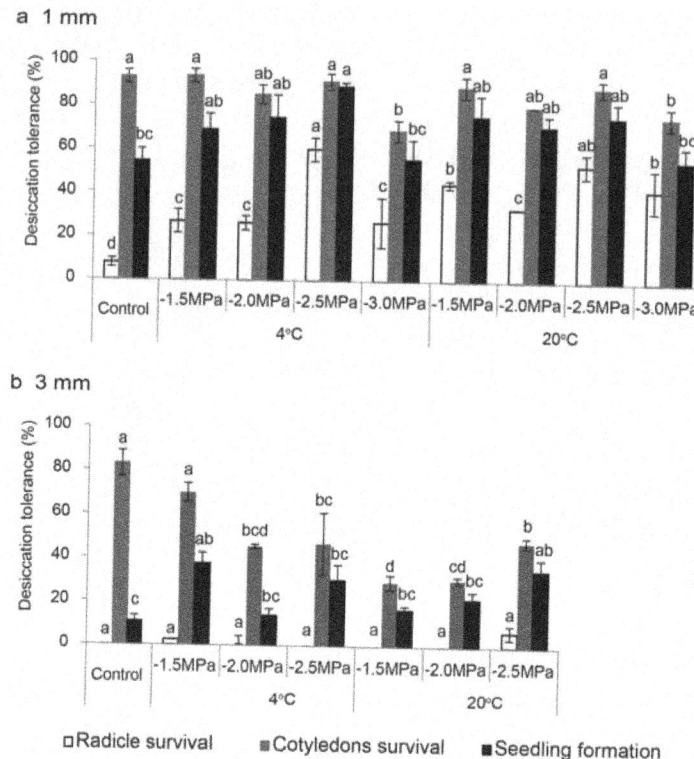

Figure 3. Osmotic potential and temperature effects on the re-establishment of desiccation tolerance of germinated *S. virgata* seeds. (a) Germinated seeds with 1mm protruded radicle. (b) Germinated seeds with 3 mm protruded radicle. Each data point is the average of three replicates of 20 seeds. Bars represent standard errors. Control = germinated seeds dried directly at 40% RH without previous osmotic treatment. Different letters above bars indicate significant differences ($P \leq 0.05$).

Seed DT and longevity are crucial for long-term survival of orthodox seeds after dispersal (Waterworth et al., 2015). As a result, both longevity and DT share several mechanisms. For example, mechanisms responsible for protection of cellular macromolecules and cellular structures in the dry state, damage repair during germination and the minimization of oxidative stress damage are well documented in relation to both DT and longevity (Rajjou and Debeaujon, 2008; Waterworth et al., 2015). Seed longevity can be analysed by storage and artificial aging. One of the methods to test artificial aging is the accelerated aging test, during which seeds are incubated at both high humidity and temperature. Cold storage (storage at 4 °C) and an accelerated aging test were used to determine longevity of germinated seeds with 1 mm protruded radicle and of these seeds incubated in an osmoticum at -2.5 MPa at 4 °C (Figure 4). The radicle was the most sensitive organ with respect to aging and the osmotic treatment did not improve its survival. The cotyledons were the most tolerant organs. The osmotic treatment significantly improved cotyledon survival and seedling formation after accelerated aging for four days, but this positive effect was not observed in any of the other treatments. Overall, the osmotic treatment had a positive effect on seedling formation before aging, but was not able to improve longevity of the seeds.

The tetrazolium test (Figure S1) corroborates these results. In the treatments with high seedling survival, the staining was homogeneous from the radicle tip to the cotyledons in most of the seeds. Low seedling survival was correlated with a bigger extent of non-stained areas starting from the radicle tip and reaching the hypocotyl.

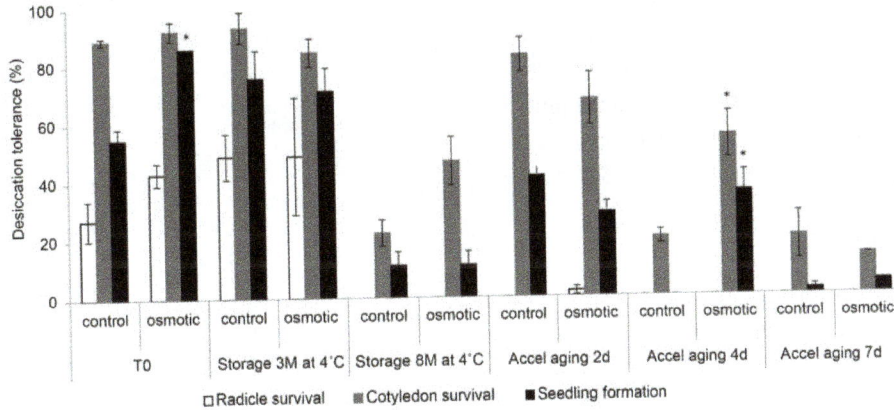

Figure 4. Desiccation tolerance after storage or accelerated aging test (Accel aging) of germinated *S. virgata* seeds with 1 mm protruded radicle. Storage was performed at 4 °C for three and eight months (M). Accelerated aging was performed for 2, 4 and 7 days (d). Each data point is the average of three replicates of 15 seeds. Bars represent standard error. T0 indicates germinated seeds dried directly at 40% RH without previous osmotic treatment. * indicate significant differences at $P \leq 0.05$ comparing control and seeds treated with an osmoticum (-2.5 MPa at 4 °C) within the same aging treatment.

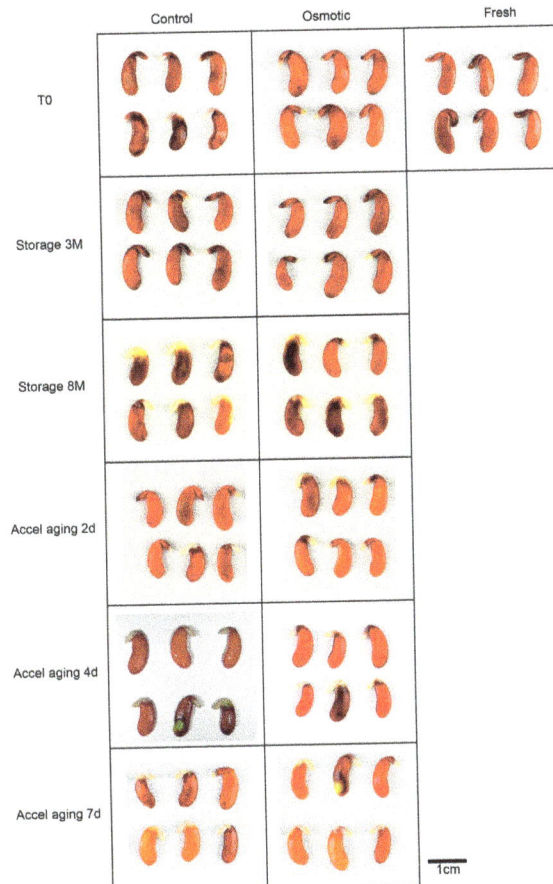

Figure S1. Tetrazolium staining of germinated *S. virgata* seeds with 1mm protruded radicle after storage or accelerated aging test (Accel aging). Storage was performed at 4 °C for three and eight months (M). Accelerated aging was performed for 2, 4 and 7 days (d). Fresh seeds were imbibed, decoated and stained with tetrazolium. Two seeds per replicate are shown.

Conclusions

Germinated seeds could be stored for three months at 4 °C without significant losses in viability. Overall, longevity was not improved by the osmotic treatment.

Once DT is lost upon germination, an osmotic treatment can re-establish it up to a certain point during development. The same may not hold true for longevity. More studies are necessary to unveil the molecular similarities and differences between DT and longevity.

Acknowledgments

The authors received financial support from the "Conselho Nacional de Desenvolvimento Científico e Tecnológico" (CNPq).

References

BRUGGINK, T.; VAN DER TOORN, P. Induction of desiccation tolerance in germinated seeds. *Seed Science Research*, v.5, n.1, p.1–4, 1995. http://www.journals.cambridge.org/abstract_S096025850000252X

BUITINK, J.; LEGER, J. J.; GUISLE, I.; VU, B. L.; WUILLÈME, S.; LAMIRAULT, G.; LE BARS, A.; LE MEUR, N.; BECKER, A.; KÜSTER, H.; LEPRINCE, O. Transcriptome profiling uncovers metabolic and regulatory processes occurring during the transition from desiccation-sensitive to desiccation-tolerant stages in *Medicago truncatula* seeds. *The Plant Journal*, v.47, n.5, p.735–750, 2006. http://www.ncbi.nlm.nih.gov/pubmed/16923015

BUITINK, J.; VU, B. L.; SATOUR, P.; LEPRINCE, O. The re-establishment of desiccation tolerance in germinated radicles of *Medicago truncatula* Gaertn. seeds. *Seed Science Research*, v.13, n.4, p.273–286, 2003. http://journals.cambridge.org/abstract_S0960258503000278

CAMARGOS, V. N.; CARVALHO, M. L. M.; ARAÚJO, D. V.; MAGALHÃES, F. H. L. Superação da dormência e avaliação da qualidade fisiológica de sementes de *Sesbania virgata*. *Ciência Agrotecnologia*, v.32, n.8, p.1858–1865, 2008. http://www.scielo.br/pdf/cagro/v32n6/v32n6a26.pdf

DEKKERS, B. J. W.; COSTA, M. C. D.; MAIA, J.; BENTSINK, L.; LIGTERINK, W.; HILHORST, H. W. M. Acquisition and loss of desiccation tolerance in seeds: from experimental model to biological relevance. *Planta*, v.241, n.3, p.563–577, 2015. http://www.ncbi.nlm.nih.gov/pubmed/25567203

FARIA, J. M. R.; BUITINK, J.; VAN LAMMEREN, A. A. M.; HILHORST, H. W. M. Changes in DNA and microtubules during loss and re-establishment of desiccation tolerance in germinating *Medicago truncatula* seeds. *Journal of Experimental Botany*, v.56, n.418, p.2119–2130, 2005. http://www.ncbi.nlm.nih.gov/pubmed/15967778

FLORENTINO, L. A.; GUIMARÃES, A. P.; RUFINI, M.; SILVA, K.; MOREIRA, F. M. D. S.; MOREIRA, M. D. S. *Sesbania virgata* stimulates the occurence of its microsymbiont in soils but does not inhibit microsymbionts of other species. *Scientia Agricola*, v.66, n.5, p.667–676, 2009. http://www.scielo.br/scielo.php?pid=S0103-90162009000500012&script=sci_arttext

GUIMARÃES, C. C.; FARIA, J. M. R.; OLIVEIRA, J. M.; SILVA, E. A. A. Avaliação da perda da tolerancia à dessecação e da quantidade de DNA nuclear em sementes de *Peltophorum dubium* (Spreng.) Taubert durante e após a germinação. *Revista Brasileira de Sementes*, v.33, n.2, p.207–215, 2011. http://www.scielo.br/pdf/rbs/v33n2/02.pdf

KWESIGA, F. R.; FRANZEL, S.; PLACE, F.; PHIRI, D.; SIMWANZA, C. P. *Sesbania sesban* improved fallows in eastern Zambia: their inception, development and farmer enthusiasm. *Agroforestry Systems*, v.47, p.49–66, 1999. http://link.springer.com/article/10.1023/A:1006256323647

MAIA, J.; DEKKERS, B. J. W.; DOLLE, M.; LIGTERINK, W.; HILHORST, H. W. M. Abscisic Acid (ABA) sensitivity regulates desiccation tolerance in germinated *Arabidopsis* seeds. *New Phytologist*, v.203, n.1, p.81–93, 2014. http://onlinelibrary.wiley.com/doi/10.1111/nph.12785/full

MAIA, J.; DEKKERS, B. J. W.; PROVART, N. J.; LIGTERINK, W.; HILHORST, H. W. M. The re-establishment of desiccation tolerance in germinated *Arabidopsis thaliana* seeds and its associated transcriptome. *PloS One*, v.6, n.12, p.e29123, 2011. http://www.pubmedcentral.nih.gov/articlerender.fcgi?artid=3237594&tool=pmcentrez&rendertype=abstract

OOMS, J. J. J.; LEON-KLOOSTERZIEL, K. M.; BARTELS, D.; KOORNNEEF, M.; KARSSEN, C. M. Acquisition of desiccation tolerance and longevity in seeds of *Arabidopsis thaliana*. *Plant Physiology*, v.102, n.4, p.1185–1191, 1993. http://www.plantphysiol.org/content/102/4/1185.short

PAMMENTER, N. W.; BERJAK, P. A review of recalcitrant seed physiology in relation to desiccation-tolerance mechanisms. *Seed Science Research*, v.9, p.13–37, 1999. http://journals.cambridge.org/abstract_S0960258599000033

PEREIRA, W. V. S.; FARIA, J. M. R.; TONETTI, O. A. O.; SILVA, E. A. A. Loss of desiccation tolerance in *Copaifera langsdorffii* Desf . seeds during germination. *Brazilian Journal of Biology*, v.74, n.2, p.501–508, 2014. http://www.scielo.br/scielo.php?pid=S1519-69842014000200501&script=sci_arttext&tlng=es

RAJJOU, L.; DEBEAUJON, I. Seed longevity: survival and maintenance of high germination ability of dry seeds. *Comptes Rendus Biologies*, v.331, n.10, p.796–805, 2008. http://www.ncbi.nlm.nih.gov/pubmed/18926494

RAMANJULU, S.; BARTELS, D. Drought- and desiccation-induced modulation of gene. *Plant Cell and Environment*, v.25, p.141–151, 2002. http://www.scielo.br/scielo.php?script=sci_nlinks&ref=000160&pid=S0101-31222009000100030000034&lng=pt

REISDORPH, N. A.; KOSTER, K. L. Progressive loss of desiccation tolerance in germinating pea (*Pisum sativum*) seeds. *Physiologia Plantarum*, v.105, p.266–271, 1999. http://onlinelibrary.wiley.com/doi/10.1034/j.1399-3054.1999.105211.x/pdf

ROYAL BOTANIC GARDENS KEW, 2008. Seed Information Database (SID). http://data.kew.org/sid/ . Accessed on: Jul, 13th 2015.

SENARATNA, T.; MCKERSIE, B. Dehydration injury in germinating soybean (*Glycine max* L. Merr.) seeds. *Plant Physiology*, v.72, p.620–624, 1983. http://www.plantphysiol.org/content/72/3/620.short

STÅHL, L.; HÖGBERG, P.; SELLSTEDT, A.; BURESH, R. J. Measuring nitrogen fixation by *Sesbania sesban* planted fallows using 15N tracer technique in Kenya. *Agroforestry Systems*, v.65, n.1, p.67–79, 2005. http://link.springer.com/10.1007/s10457-004-6072-8

TOLDI, O.; TUBA, Z.; SCOTT, P. Vegetative desiccation tolerance: Is it a goldmine for bioengineering crops? *Plant Science*, v.176, n.2, p.187–199, 2009. http://linkinghub.elsevier.com/retrieve/pii/S0168945208002781

VILLELA, F. A.; BECKERT, O. P. Potencial osmótico de soluções aquosas de polietileno glicol 8000. *Revista Brasileira de Sementes,* v.23, n.1, p.267–275, 2001. http://www.scielo.br/scielo.php?script=sci_serial&pid=2317-1537&lng=en&nrm=iso

WATERWORTH, W. M.; BRAY, C. M.; WEST, C. E. The importance of safeguarding genome integrity in germination and seed longevity. *Journal of Experimental Botany*, v.66, n.12, p.3549–3558, 2015. http://jxb.oxfordjournals.org/lookup/doi/10.1093/jxb/erv080

ZANANDREA, I.; ALVES, J. D.; DEUNER, S.; GOULART, P. F. P.; HENRIQUE, P. C.; SILVEIRA, N. M. Tolerance of *Sesbania virgata* plants to flooding. *Journal of Botany*, v.57, n.8, p.661–669, 2009. http://www.publish.csiro.au/?paper=BT09144

13

Quality of soybean seeds treated with fungicides and insecticides before and after storage

Thaís Francielle Ferreira[1*], João Almir Oliveira[1], Rafaela Aparecida de Carvalho[1], Laís Sousa Resende[1], Cassiano Gabriel Moreira Lopes[1], Valquíria de Fátima Ferreira[1]

ABSTRACT - The timing of seed treatment application is important to keep soybean seeds quality. Therefore, the aim of this study was verify the effect of fungicides and insecticides treatment in soybean seeds quality before and after storage. Seeds of NS 7494, NS 8693 and NS 7338 IPRO were utilized and analyses separately, through a factorial scheme 3x6, with three application moments: treated and assessed; treated, stored and assessed; stored, treated and assessed; and six combination of fungicides and insecticides: Cropstar® + Derosal Plus®; Cropstar® + Maxim xl®; Cruiser® + Derosal Plus®; Cruiser®+ Maxim xl®; Standak Top® and the control group. Germination, seedling emergence, accelerate aging, cold and health tests were performed. It was determined that the combination with Cruiser® doesn't affect the physiological quality of soybean seeds treated and assessed, and treated after two month of storage. Cropstar® + Derosal plus® keeps physiological quality of soybean seeds stored and treated for two months, while Standak top® has negative effect. The combination with Cropstar® damages the physiological quality of soybean seeds treat after two months of storage. The fungicide Derosal plus® improves the health quality of soybean seeds regardless treatment moment.

Index terms: *Glycine max* L., seed treatment, health.

Qualidade de sementes de soja tratadas com inseticidas e fungicidas antes e após o armazenamento

RESUMO - O momento de aplicação do tratamento de sementes é importante para a manutenção da qualidade de sementes de soja. Assim, objetivou-se verificar o efeito do tratamento fungicida e inseticida na qualidade de sementes de soja antes e após o armazenamento. Utilizou-se sementes das cultivares NS 7494, NS 8693 e NS 7338 IPRO, que foram analisadas separadamente, em esquema fatorial de 3x6. Três momentos de aplicação em que as sementes foram: tratadas e avaliadas; tratadas, armazenadas e avaliadas; armazenadas, tratadas e avaliadas; e seis combinações de inseticidas e fungicidas: Cropstar® + Derosal Plus®; Cropstar® + Maxim xl®; Cruiser® + Derosal Plus®; Cruiser®+ Maxim xl®; Standak Top® e o controle. Realizou-se testes de germinação, emergência, envelhecimento acelerado, frio e sanidade. Concluiu-se que as misturas com Cruiser® não afetam a qualidade fisiológica de sementes de soja tratadas e avaliadas, e tratadas após dois meses de armazenamento. Cropstar®+ Derosal Plus®mantém a qualidade fisiológica de sementes de soja armazenadas tratadas por dois meses, enquanto que o Standak Top®tem efeito negativo. As misturas com Cropstar® prejudicam a qualidade fisiológica de sementes de soja tratadas após dois meses de armazenamento. O fungicida Derosal Plus® melhora a qualidade sanitária de sementes de soja independentemente do momento de tratamento.

Termos para indexação: *Glycine max* L., tratamento de sementes, sanidade.

Introduction

Seed treatment is a technology that, when associated with plant genetic improvement and biotechnology, allows high productivity of soybean culture and producer satisfaction in meeting the demands of the market. This market, in its turn, is supported by the use of quality seeds, which, as a vehicle of technology and innovations, is responsible for stimulate it and make it competitive, since the advance of Brazilian agribusiness is characterized largely by the development of the system of seed production.

This aspect brings in response the search for the conquest

[1]Departamento de Agricultura, Universidade Federal de Lavras, Caixa Postal 3037, 37200-000 - Lavras, MG, Brasil.
*Corresponding author <franthata@yahoo.com.br>

of new markets by large companies investing in values and very impactful strategies as the launching of large-scale agricultural inputs to the commercialization and deployment of innovative technologies as the industrial seed treatment.

Today, industrial seed treatment guarantees the maximum efficiency of the products used and the quality of the seeds, no risk for operators, while in the past decades, most of the seed treatment was carried out on the properties so that both the quality of the seeds as the health of the operator were committed (Zambom, 2013).

However, after the resolution of the operational aspects of the treatment of seeds, certain limitations are of concern, such as the possible effects of the active ingredients in the quality of the seeds during storage and in the field (Brzezinski et al., 2015).

In this way, the need arises for studies that assess the effect of seed treatment products used commercially and launched recently in the market and the right time of application on the quality of soybean seeds. Considering that, according to the MAPA (2016), until the present time, there are twenty fungicides and twenty-three insecticides intended for the chemical treatment of seeds.

Thus, the objective of this study was to check the effect of the fungicide and insecticide treatment on the quality of soybean seeds before and after storage.

Material and Methods

The experiments were conducted on University Federal of Lavras, in Lavras, MG, in the Departments of Agriculture and Plant Pathology, with physiological analyses carried out at the Central Laboratory for Seed Testing and phyto-sanitary analysis, at Laboratory of Seed Pathology.

The seed used for the experiment were provided by company Nidera Seeds, with the cultivars NS 7494, NS 8693 and NS 7338 IPRO.

A completely randomized factorial 3x6 was used, with three times of application of chemical treatment and six mixtures of fungicides and insecticides, with four replications The cultivars data were analyzed separately.

Seeds from each cultivar were subjected to three treatments which differed from each other by the time they received the chemical treatment: 1) treated and assessed (the seeds were assessed immediately after the chemical treatment); 2) treated, stored and assessed (the seeds were stored with chemical treatment for two months and then assessed); 3) stored, treated and assessed (the seeds were stored without chemical treatment for two months, and then were treated and immediately assessed).

The products used for the chemical treatment of the seeds

were the insecticides Cropstar®, which active ingredients are imidacloprid and tiocarbe; and Cruiser®, with active ingredient thiamethoxam; the fungicide Derosal Plus®, which has carbendazim and thiram in its composition; and Maxim xl®, chemical based on fludioxonil and metalaxyl-M; Standak Top®, that is a mixture containing the insecticide fipronil and ready the fungicides pyraclostrobin and thiophanate methyl.

For the treatment of seeds, the products have been combined as follows with the respective dosing: Cropstar® (5 mL.Kg^{-1}) + Derosal Plus® (2 mL.Kg^{-1}), Cropstar® (5 mL.Kg^{-1}) + Maxim xl® (1 mL.Kg^{-1}), Cruiser® (2.5 mL.Kg^{-1}) + Derosal Plus® (2 mL.Kg^{-1}), Cruiser® (2.5 mL.Kg^{-1}) + Maxim xl® (1 mL.Kg^{-1}), Standak Top® (2 mL.Kg^{-1}) and the Control group, which received only addition of water.

The amount of the solution of the mixture to the treatments was determined according to the dose recommended for each commercial product and amount of water until the maximum volume of 10 mL.Kg^{-1} of seeds.

The amount of 500 g of seeds for each of the mixtures was used. For the application of the products in the seeds plastic bags of 2 Kg capacity were used. The products were previously mixed in a Petri dish (+ fungicide insecticide + water) and placed in plastic bags, followed by the seeds. The set was shaken until it obtained a homogeneous mixture of seeds. After treatment, the seeds were laid out in the shade, at a temperature of approximately 25 °C for 20 minutes, for drying of the product on the surface of the seed.

After drying, the seeds were placed in multiwall paper packaging and held for two months in storage, in uncontrolled conditions, average temperature of 20.4 °C and average of relative humidity in 65.5% (Dantas et al., 2007).

For the evaluation of the seed quality germination, emergence, accelerated aging, cold test and seed health tests were carried out.

In the germination test four replications of 50 seeds per plot were used, sown on two sheets of towel paper and covered with a sheet, moistened with distilled water in quantity equivalent to 2.5 times the weight of the dry paper. The rolls were kept in a germinator at 25 °C, the evaluations were conducted at eight days after sowing and the results expressed as percentage of normal seedlings (Brasil, 2009).

In the accelerated aging test, 250 seed samples were placed on metallic screens in plastic boxes adapted containing 40 mL of distilled water. These boxes were kept in an incubator, at 42 °C for 48 hours, on incubator type BOD (Marcos-Filho, 1999). Then we proceeded to the germination test with four subsamples of 50 seeds. The evaluation took place five days after sowing. The results were expressed as percentage of mean normal seedlings for each treatment (Brasil, 2009).

In the emergence test, the sowing was carried out in plastic trays containing substrate, soil + sand in the ratio 2:1. Four replications of 50 seeds were sown. After sowing, the trays were kept in plant growth chamber at a temperature of 25 °C, in alternating light and dark regime (12 hours). The number of emerged seedlings was determined 14 days after sowing. The results were expressed as percentage of seedlings.

In the cold test, the same procedures of the emergence test were performed, however, after sowing, the trays were placed in cold chamber at 10 °C for seven days and then taken to germinator at 25 °C for seven more days. Then the number of emerged seedlings was determined. The results were expressed as percentage of seedlings.

For the health test, we used the method of incubation in filter paper (Neergaard, 1979), with eight subsamples of 25 seeds. The seeds were distributed in Petri dishes of 15 cm in diameter containing three previously sterilized filter paper sheets and moistened with water, and 2.4 D sterile agar. The dishes were kept in incubation at 20 °C and photoperiod of 12 hours, where they remained for seven days, and then assessed for the presence of pathogens (Brasil, 2009). For the identification of pathogens present in the seeds, a stereoscopic magnifying glass and an optical microscope were used. The incidence was assessed in percentage of fungi found.

Statistical analysis was performed using the statistical software Sisvar® (Ferreira, 2011). In the analysis, when significant effect of the treatments was verified, to test the significance of differences between the averages of the treatments, the medium test Scott-Knott was used, with 5% probability. Fungal incidence values were previously transformed into ($\sqrt{x+1}$).

Results and Discussion

The interaction between the factors times of application and mixtures of insecticides and fungicides was significant for the germination, accelerated aging, cold and emergency tests, to the level of 1% and 5% of probability (Table 1A).

The results of the test of germination, the seeds from cultivars NS 7494 and NS 7338 IPRO treated before testing obtained generally better performance when compared to the seeds stored with or without treatment, which proves that the pre-planting treatment favours the establishment of plants in the field and grain production in relation to the anticipated treatment (Table 1) (Brzezinski et al., 2015).

Table 1. Germination (%) of seeds from three soybean cultivars treated with different products at different times of application.

Cultivars	Products	Application Time		
		Treat	Treat + Stored	Stored + Treat
NS 7494 (VC=2.26%)	Cropstar® + Derosal Plus®	87 aC	83 bC	70 cD
	Cropstar® + Maxim xl®	89 aC	92 aB	71 bD
	Cruiser® + Derosal Plus®	97 aA	91 bB	98 aA
	Cruiser® + Maxim xl®	94 aB	85 bC	81 cC
	Standak Top®	93 bB	97 aA	80 cC
	Control	95 aB	85 cC	91 bB
NS 8693 (VC=3.50%)	Cropstar® + Derosal Plus®	90 aB	92 aA	57 bC
	Cropstar® + Maxim xl®	91 aB	91 aA	56 bC
	Cruiser® + Derosal Plus®	95 aA	87 bB	82 cA
	Cruiser® + Maxim xl®	81 bD	92 aA	79 bA
	Standak Top®	87 aC	85 aB	73 bB
	Control	95 aA	94 aA	70 bB
NS 7338 IPRO (VC=3.33%)	Cropstar® + Derosal Plus®	87 aB	74 bD	64 cD
	Cropstar® + Maxim xl®	91 aB	88 aB	73 bC
	Cruiser® + Derosal Plus®	96 aA	85 bC	84 bB
	Cruiser® + Maxim xl®	93 aA	90 aB	90 aA
	Standak Top®	92 aA	85 bC	86 bB
	Control	90 bB	98 aA	86 bB

The averages followed by the same low case letter in the lines and capital letter in the column do not differ between each other by the Scott-Knott test, at 5% probability.

The lower averages of germinated seeds, cultivar NS 7494, were obtained when the seed treatment was carried out after two months of storage in conventional warehouse conditions, with the exception of treatment Cruiser® + Derosal Plus®, since the mixture provided an increase in germination of these seeds when comparing to the control, both in post treatment storage and pre-treatment tests. Regarding the seeds treated and stored for two months, the Standak Top® product was the one which provided the largest percentage of normal seedlings, contrary to the results found by Souza et al. (2015), in which even without storing the treated seeds, concluded that the Standak Top® was unfavorable to the development of seedlings.

When treatment with Cruiser® + Derosal Plus® was held in seeds of cultivar NS 8693 before the quality tests, the percentage of germinated seeds remained the same when copared to the control, unlike other treatment with products that contributed to the reduction of seed germination. The seeds that were stored with this same treatment obtained less percentage of germination, unlike the seeds who received the treatment after two months of storage, which obtained an increase in germination. This shows that the product Cruiser® + Derosal Plus® is detrimental to seed during storage. However, when the quality tests are done before, regardless of whether the seeds were stored or not, the treatment helps to enhance germination. Dan et al. (2013) when studying the physiological potential of soybean seeds treated with thiametoxan, a.i. of Cruiser®, and submitted to storage, observed that there is a reduction in germination of seeds treated during the storage, however, when the treatment is performed pre sowing, the product acts like bio-activator and increments germination, emergence and agronomic characteristics of seedlings, which corroborates with the results described.

As well as mixing Cruiser® + Derosal Plus®, the Standak® Top product was also harmful to seed germination of NS 8693 cultivar when these were stored. Cunha et al. (2015), with the objective of evaluating the phyto-toxic effect of seed treatment products on the physiological quality of soybean seeds, have observed a lower performance of seeds treated with Standak Top® in the accelerated aging test, which allows to evaluate the potential for seed storage. This lower performance in accelerated aging test is related with the lowest germination of stored seeds treated with Standak Top®.

The combinations containing the insecticide Cruiser® in its composition have contributed to a better germination of the cultivar NS 8693, while other products have reduced germination when the treatment was held after two months of storage.

For the cultivar NS 7338 IPRO, the timing of the treatment of the seeds did not influence seed germination performance when treated with Cruiser® + Maxim xl®. And the seeds treated with Cruiser® + Maxim xl®, obtained storage after a larger percentage of germination when compared to other products treatment, which demonstrates that this

association is compatible since it reaches the expected result with treatment: maintaining or increasing the quality of seeds (Follmann et al., 2014).

Seed germination of cultivar NS 7338 IPRO treated before the quality tests was higher when the products used were Standak Top® and containing the insecticide Cruiser® combined with the fungicide when compared to the control, due to the low deleterious effect of these products the seeds (Zilli et al., 2010; Dan et al., 2013).

Regardless of the product used for the treatment and of cultivars, seed germination was reduced in the treated and stored and the smallest seedlings were obtained when the seeds quality after storage were assessed. This reduction in germination of seeds is due to many circumstances such as the environmental conditions during the production of seeds, insect attack, lipid and water content in seeds, presence of mechanical damage arising from transport and processing, storage conditions, and especially, in the case of seed with high quality, of the chemicals used in seed treatment (Sales et al., 2011). In this study, when the seeds were stored there was phytotoxic effect treated by the contact time of the products with the seeds. And when the treatment was done after there was storage phytotoxicity by the entry of products within seeds due to the damage caused by storage fungi (Table 5).

In the same way that occurred in the first count of germination, it is observed that when the treatment was done after two months of storage, there was lower percentage of germination, when comparing to other moments of seed treatment, especially when using the Cropstar® product in combination with the fungicide, regardless of the plant variety which received the treatment.

In the accelerated aging test, seed treatments containing in its formula the fungicide Derosal Plus® contributed to a greater percentage of normal seedlings for 'NS 7494', regardless of when that treatment was held (Table 2). This behavior is evidenced by the health check of seeds in which the seed treatment formulations containing the Derosal Plus® were able to control the fungus in almost its entirety for any time of treatment (Table 5).

The seeds of the cultivar NS 8693 treated with Cropstar® + Derosal Plus® obtained larger numbers of normal seedlings after accelerated aging seed regardless of how they were treated, which goes against the results obtained by Cunha et al. (2015) who observed the maintenance of quality of seeds treated with Cropstar® after accelerated aging of seeds. For this same cultivar, chemical treatment of the seed has contributed to an increase in the number of normal seedlings after accelerated aging of seeds that received treatment after storage.

Table 2. Accelerated ageing (%) of seeds from three soybean cultivars treated with different products at different times of application.

Cultivars	Products	Application Time		
		Treat	Treat + Stored	Stored + Treat
NS 7494 (VC=2.87%)	Cropstar® + Derosal Plus®	42 cA	85 aA	78 bA
	Cropstar® + Maxim xl®	26 cB	76 aD	71 bB
	Cruiser® + Derosal Plus®	42 cA	86 aA	80 bA
	Cruiser® + Maxim xl®	24 cB	76 aD	63 bC
	Standak Top®	21 cC	82 aB	70 bB
	Control	21 cC	80 aC	71 bB
NS 8693 (VC=3.72%)	Cropstar® + Derosal Plus®	22 cA	82 aA	63 bA
	Cropstar® + Maxim xl®	17 cB	80 aB	63 bA
	Cruiser® + Derosal Plus®	22 cA	70 aC	57 bB
	Cruiser® + Maxim xl®	3 cD	65 aD	52 bC
	Standak Top®	14 cC	62 aE	53 bC
	Control	6 cD	84 aA	48 bD
NS 7338 IPRO (VC=4.36%)	Cropstar® + Derosal Plus®	69 bA	73 aB	59 cC
	Cropstar® + Maxim xl®	48 cB	72 aB	62 bC
	Cruiser® + Derosal Plus®	34 cD	73 aB	69 bB
	Cruiser® + Maxim xl®	43 cC	63 aC	54 bD
	Standak Top®	11 bE	83 aA	86 aA
	Control	7 cF	25 bD	61 aC

The averages followed by the same low case letter in the lines and capital letter in the column do not differ between each other by the Scott-Knott test, at 5% probability.

For NS 7338 IPRO cultivar, mixing products the Cropstar® + Derosal Plus® contributed to an increase in the number of normal seedlings in treated seeds before the quality tests, which shows that these products do not have phytotoxic effect on the seeds over a period of two months in contact with the seeds and offer high protection as the results found by Conceição et al. (2014). As well as on a study by Cunha et al. (2015), the Standak Top® product contributed to a poor performance in the accelerated aging when comparing seeds treated before the tests against the seeds stored and treated seeds after storage.

In general, regardless of the cultivars, there was low germination after accelerated aging of seeds which served as control. This is due to the presence of storage and field fungi (Table 5). The seeds that have received treatment without storing had the smallest average germination obtained, as well as of the product concentration, time of contact with the seed treatment was lower and this implies a smaller action of active ingredients on the fungi that came directly from the field. However, when the seeds were stored, there has been a natural reduction of fungi that came from the field, as is noted in the control and in the seeds who received treatment after the storage (Table 5). In this way, the seeds stored handled in addition to being protected against fungi had storage fungi of field eliminated by the action of chemical treatment. The seed treatment with fungicides, in addition to controlling important pathogens transmitted via seeds, is important to ensure

adequate populations of plants when the soil and climate conditions are unfavorable (Zorato and Henning, 2001).

The application of products held prior to storage of seed has contributed to a better expression of quality in the accelerated aging test. In the accelerated aging test, products tend to focus, as there are high relative humidity and high temperature, which can cause damage to seed due to penetration of products in the membrane. Thus, seed germination is reduced by the fact that the toxic potential of treatment products is intensified in stress conditions (Pereira et al., 2007).

The chemical treatment of seeds has contributed to increase the virgor expression by the stress caused by the accelerated aging test. Regarding field levels, this result was found by Brzezinski et al. (2015) who observed in their study that the beneficial effect of treatment products to control pathogens and insects was higher when coincided with adverse conditions of temperature and rainfall distribution during the establishment of culture.

In the cold test, to cultivar seeds NS 7494 treated before the quality tests, there was improved performance in vigor when the treatment was with Cropstar® + Plus Derosal® and it remained with the Standak Top® treatment (Table 3). The same result was found by Dan et al. (2010) and Castro et al. (2008) that, when testing the Cropstar® insecticide on soybean seeds, observed higher vigor from seeds treated with these insecticide even after 45 days of storage. The other products contributed

Quality of soybean seeds treated with fungicides and insecticides before and after storage

Table 3. Cold test (%) of seeds from three soybean cultivars treated with different products at different times of application.

Cultivars	Products	Application Time		
		Treat	Treat + Stored	Stored + Treat
NS 7494 (VC=1.76%)	Cropstar® + Derosal Plus®	77 cA	93 aB	85 bB
	Cropstar® + Maxim xl®	60 cD	98 aA	89 bA
	Cruiser® + Derosal Plus®	61 cD	96 aA	84 bB
	Cruiser® + Maxim xl®	64 cC	92 aB	76 bC
	Standak Top®	69 cB	86 aC	77 bC
	Control	71 cB	93 aB	90 bA
NS 8693 (VC=1.96%)	Cropstar® + Derosal Plus®	75 bD	93 aC	93 aB
	Cropstar® + Maxim xl®	63 cE	98 aA	93 bB
	Cruiser® + Derosal Plus®	83 bB	100 aA	98 aA
	Cruiser® + Maxim xl®	92 bA	95 aB	96 aA
	Standak Top®	80 cC	98 aA	94 bB
	Control	78 bC	91 aC	92 aB
NS 7338 IPRO (VC=2.32%)	Cropstar® + Derosal Plus®	79 cA	95 aA	83 bC
	Cropstar® + Maxim xl®	65 cC	91 aB	82 bC
	Cruiser® + Derosal Plus®	63 cC	90 aB	87 bB
	Cruiser® + Maxim xl®	55 cD	94 aA	84 bC
	Standak Top®	72 cB	84 aC	77 bD
	Control	70 cB	94 aA	90 bA

The averages followed by the same low case letter in the lines and capital letter in the column do not differ between each other by the Scott-Knott test, at 5% probability.

to a reduction in the vigor. Based on the principle of cold test, we can see that when the seeds were sown and left for seven days at a temperature of 10 °C under 70% of the moisture retention capacity, there was the sudden entry of the seed treatment products causing an underwhelming performance for that specific cultivar, probably due to phytotoxicity.

The seeds of this same variety which were stored treated with Cropstar® + Derosal Plus®, Cruiser® + Maxim xl®, Cropstar® + Maxim xl® e com Cruiser® + Derosal Plus® showed better performance regarding vigor. The percentage of normal seedlings emerged after the cold period remained when the seeds of the plant variety NS 7494 were treated after storage the products Cropstar® + Maxim xl®.

As well as cultivar NS 7494, the seeds of the 'NS 8693' that were treated with the Standak Top® product kept vigor after the cold test when the treatment was done pre test, however, the expression of quality was accentuated when the treatment was done with Cruiser® + Maxim xl®. Treatment prior to storage has contributed to better expression of vigor regardless of the products used as treatment. The quality of the seeds that were stored and later treated remained for most products. Mixtures containing the insecticide Cruiser® in its composition have contributed to better performance of these seeds vigor. Different from the result found by Tonim et al. (2014), researching the effect of treatment products on the quality of corn seed, a reduction in vigor were observed in storage for seeds treated with thiametoxan, i. a. of Cruiser®,

and with imidacloprido and tiocarbe, i. a. of Cropstar®, and attributed the decrease of the vigor to the negative effects of active products to development of seedlings under adverse conditions after a short storage period.

The seeds of the cultivar NS 7338 IPRO, when treated preoperatively with products Cropstar® + Derosal Plus® obtained greater expression of the vigor regarding the control. With the exception of the Standak Top® product, which kept the quality of seed, the other products have contributed to the reduction of the vigor by the cold stress. For the seeds stored and treated after storage for two months, the worst treatment was with the Standak Top® product that has contributed to a lower expression of vigor of the seeds. This result is indicative that when in contact with the seeds for longer periods, the Top Standak® treatment product can have phytotoxic effects on seeds.

Likewise in the accelerated aging test, any one of the cultivars manifested high effect for treated seed and stored and low vigor to the treated seeds before the quality tests, which proves that the contact time of the fungicides with the seeds is important for the complete elimination of fungi since pathogens present in the soil or on the seeds reduce the soybean plants stand (Costamilan et al., 2012). Besides fungicide treatment not reducing physiological quality of seeds, it is extremely efficient in the control of pathogens, about 96% in the control of *Aspergillus* and *Fusarium* and 100% in control of *Penicillium* and *Cercospora*, such results can be confirmed in Table 5 with the health test results of this research (Pereira et al., 2007).

The percentage of seedling emergence from cultivar NS 7494 was greater when the treatment was performed after storage, except when the seeds were treated with the mixture Cropstar® + Maxim xl®, where it was more efficient when the treatment was performed before the quality tests (Table 4).

The product of treatment Cruiser® + Derosal Plus® caused a better performance of the stored seeds treated with highest number of seedlings emerged, and kept the quality when the seeds were stored before treatment, opposed to the results found by Dan et al. (2013) who observed a reduction in emergence of seedlings along the storage with seeds treated with the insecticide Cruiser®.

Table 4. Emergence (%) of seeds from three soybean cultivars treated with different products at different times of application.

Cultivars	Products	Application Time		
		Treat	Treat + Stored	Stored + Treat
NS 7494 (VC=2.14%)	Cropstar® + Derosal Plus®	87 bC	92 aB	94 aB
	Cropstar® + Maxim xl®	97 aA	79 cD	91 bC
	Cruiser® + Derosal Plus®	91 cB	96 bA	100 aA
	Cruiser® + Maxim xl®	85 bC	93 aB	93 aB
	Standak Top®	90 aB	83 bC	89 aC
	Control	94 bA	92 bB	98 aA
NS 8693 (VC=2.07%)	Cropstar® + Derosal Plus®	95 aA	89 bC	91 bB
	Cropstar® + Maxim xl®	93 bB	95 aB	96 aA
	Cruiser® + Derosal Plus®	96 bA	99 aA	96 aA
	Cruiser® + Maxim xl®	92 bB	96 aB	94 aA
	Standak Top®	95 aA	96 aB	85 bC
	Control	97 aA	88 bC	97 aA
NS 7338 IPRO (VC=2.21%)	Cropstar® + Derosal Plus®	85 bB	78 cE	92 aB
	Cropstar® + Maxim xl®	88 cB	97 aA	92 bB
	Cruiser® + Derosal Plus®	91 bA	97 aA	94 aA
	Cruiser® + Maxim xl®	90 aA	87 bD	90 aB
	Standak Top®	91 bA	90 bC	94 aA
	Control	91 aA	94 aB	80 bC

The averages followed by the same low case letter in the lines and capital letter in the column do not differ between each other by the Scott-Knott test, at 5% probability.

Either for cultivar NS 8693 or for cultivar NS 7338 IPRO, the best averages of seedlings were when seeds were treated with Cruiser® + Derosal Plus®, regardless the management of treatment used. The active ingredient thiametoxan Cruiser® of insecticide is a bioatctivator that brings increments in germination, emergence and agronomic characteristics of seedlings through unknown mechanisms of morphological changes and metabolism (Dan et al., 2013). In addition to the advantages of this insecticide, for the protective effect of the fungicide Derosal Plus®, which guarantees an ideal stand of seedlings in the field.

When the treatment was performed after storage of the seeds from cultivar NS 7338 IPRO, there has been greater seedling number, as well as on cultivar NS 7494, except for treatment with product Cropstar® + Maxim xl®. Storage for two months caused, naturally, a reduction of fungi that came from the field, the example of Fusarium in the percentage before the storage for the seeds to cultivar NS 7494 was 30% and after storage was 0%, and for cultivar NS 7338 IPRO

was of 15% and reduced to 0%, as shown in the results for the sanity of the seed (Table 5). Ludwig et al. (2011) also observed a reduction of Fusarium over soy seeds storage.

In assessing the sanitary quality of seeds, there was no significant effect, on all three cultivars, from the main fungi present in soy culture, Penicillium, Fusarium and Aspergillus (Table 2A).

In general, the products were efficient in the control of all cultivars seeds. It is observed that this efficiency is directly connected to the contact time of the products with the seeds (Table 5). The Maxim® fungicide and i. a. fungica action Standak Top® product require a contact period of two months with the seeds to act on reduction of fungi. On the other hand, the product Derosal Plus® was efficient in fungus control regardless of the moment in which the product application and grow.

However, despite the efficiency of treatment and storage in the control of fungi, reduction in germination and vigor of seeds more infested is observed, as has been shown in the results of physiological quality (Juhász et al., 2013).

Table 5. Incidence (%) of *Penicillium* sp. in seeds from three soybean cultivars treated with different products at different times of application.

Cultivars	Products	Penicillium			Fusarium			Aspergillus		
		Treat	Treat + Stored	Stored + Treat	Treat	Treat + Stored	Stored + Treat	Treat	Treat + Stored	Stored + Treat
NS 7494 (VC=26.52%)	Cropstar® + Derosal Plus®	0 aA	0 aA	0 aA	0 aA	0 aA	0 aA	0 aA	0 aA	0 aA
	Cropstar® + Maxim xl®	3 aB	4 aB	2 aA	9 cC	0 aA	3 bB	0 aA	4 bB	2 bA
	Cruiser® + Derosal Plus®	3 bB	0 aA	0 aA	2 aB	0 aA	0 aA	0 aA	0 aA	0 aA
	Cruiser® + Maxim xl®	13 bC	0 aA	13 bB	12 bC	0 aA	0 aA	9 bC	0 aA	10 bB
	Standak Top®	13 bC	0 aA	0 aA	4 bB	0 aA	2 bB	0 aA	0 aA	0 aA
	Control	54 bD	6 aB	9 aB	30 cD	7 bB	0 aA	3 aB	7 bC	15 cC
NS 8693 (VC=18.57%)	Cropstar® + Derosal Plus®	0 aA	0 aA	0 aA	2 aA	0 aA	2 aA	0 aA	0 aA	0 aA
	Cropstar® + Maxim xl®	3 bB	0 aA	3 bB	15 cB	9 bB	2 aA	3 aA	4 aB	12 bB
	Cruiser® + Derosal Plus®	0 aA	0 aA	0 aA	3 bA	0 aA	0 aA	0 aA	0 aA	0 aA
	Cruiser® + Maxim xl®	20 bC	0 aA	0 aA	58 cD	7 aB	12 bB	0 aA	0 aA	0 aA
	Standak Top®	40 bD	0 aA	0 aA	26 bC	0 aA	0 aA	0 aA	0 aA	0 aA
	Control	35 cD	8 aB	17 bC	85 bE	59 aC	66 aC	0 aA	5 bB	46 cC
NS 7338 IPRO (VC=17.06%)	Cropstar® + Derosal Plus®	0 aA	0 aA	0 aA	1 aA	0 aA	0 aA	0 aA	0 aA	0 aA
	Cropstar® + Maxim xl®	23 bB	15 aB	25 bB	13 cB	0 aA	6 bB	28 bB	10 aB	8 aB
	Cruiser® + Derosal Plus®	0 aA	0 aA	0 aA	3 bA	0 aA	5 bB	0 aA	0 aA	0 aA
	Cruiser® + Maxim xl®	67 bD	0 aA	0 aA	11 bB	0 aA	0 aA	53 cC	0 aA	6 bB
	Standak Top®	46 bC	0 aA	0 aA	11 bB	0 aA	0 aA	0 aA	0 aA	0 aA
	Control	81 bD	40 aC	70 bC	15 bB	0 aA	0 aA	50 cC	16 aB	36 bC

The averages followed by the same low case letter in the lines and capital letter in the column, of each fungi, do not differ between each other by the Scott-Knott test, at 5% probability. The original averages were presented but the data was compared with the transformed data (Transformation in $(x+1)^{0.5}$).

In general, the treatment before the seed quality tests was not detrimental to the quality, in ideal conditions, however, under conditions of stress, due to the concentration of the product, the performance of the seeds has been compromised, except with the use of mixtures containing the insecticide Cruiser®. In addition, the contact time of the products with the seeds was not enough to reduce the incidence of fungi *Fusarium*, *Aspergillus* and *Penicillium*, except by the use of products based on Derosal Plus®, which contributed to reducing all fungi regardless if the seeds were not stored or treated.

With the use of products Cropstar® + Derosal Plus®, there were better expression of the physiological quality of seeds stored treated, whereas Cropstar® insecticide did not cause negative effects on the seeds and the fungicide Derosal Plus® was effective in reducing the three genera of fungi presented regardless of the time of application of the products. On the other hand, although the Standak Top® product has provided an initial start-up, established by germination test subsequently contributed to reducing the vigor of seeds stored treated as seedlings showed signs of phytotoxicity.

The quality of the seeds which have received treatment after the reduced storage with the use of products containing the insecticide Cropstar® in its composition, contrary to the Cruiser® insecticide which contributed to better performance of vigor of seeds under these conditions.

Conclusions

The chemical treatment with mixtures containing Cruiser® (thiamethoxam) in its composition does not affect the physiological quality of soybean seeds when treated and assessed, and when application occurs after two months of storage.

Cropstar® products (imidacloprido and tiocarbe) + Derosal Plus® (carbendazim and tiram) maintains the physiological quality of soybean seeds stored treated for a period of two months.

The product Standak Top® (fipronil, pyraclostrobin and thiophanate methyl) has negative effect on the quality of soybean seeds assessed and stored for two months.

Mixtures containing the insecticide Cropstar® in its composition reduces the physiological quality of soybean seeds when application occurs after two months of storage.

The fungicide Derosal Plus® improves the sanitary quality of soybean seeds regardless of time of treatment.

References

BRASIL. Ministério da Agricultura, Pecuária e Abastecimento. *Regras para análise de sementes*. Ministério da Agricultura, Pecuária e Abastecimento. Secretaria de Defesa Agropecuária. Brasília: MAPA/ACS, 2009. 395p. http://www.agricultura.gov.br/arq_editor/file/2946_regras_analise__sementes.pdf

BRZEZINSKI, C. C.; HENNING, A. A.; ABATI, J.; HENNING, F. A.; FRANÇA-NETO, J. B.; KRZYANOWSKI, F. C.; ZUCARELI, C. Seeds treatment times in the establishment and yield performance of soybean crops. *Journal of Seed Science*, v.37, n.2, p.147-153, 2015. http://www.scielo.br/pdf/jss/v37n2/2317-1537-jss-37-02-00147.pdf

CASTRO, G. S. A.; BOGIANI, J. C.; SILVA, M. G.; GAZOLA, E.;ROSOLEM, C. A. Tratamento de sementes de soja com inseticidas e um bioestimulante. *Pesquisa Agropecuária Brasileira*, v.43, n.10, p.1311-1318, 2008. http://www.scielo.br/scielo.php?pid=S0100-204X2008001000008&script=sci_arttext&tlng=e

CONCEIÇÃO, G.M.; BARBIERI, A.P.P.; LÚCIO, A.D.; MARTIN, T.N.; MERTZ, L.M.; MATTIONI, N.M.; LORENTZ, L.H. Desempenho de plântulas e produtividade de soja submetida a diferentes tratamentos químicos nas sementes. *Bioscience Journal*, v.30, n.6, p.1711-1720, 2014. http://ainfo.cnptia.embrapa.br/digital/bitstream/item/113098/1/Desempenho-deplantulas-e-produtividade-de-soja-submetida-a-diferentes-tratamentosquimicos-nas-sementes.pdf

COSTAMILAN, L. M.; HENNING, A. A.; ALMEIDA, A. M. R.; GODOY, C. V.; SEIXAS, C. D. S.; DIAS, W. P. La Niña e os possíveis efeitos sobre a ocorrência de doenças de soja na safra 2010/2011. Londrina: Embrapa, 2012.

CUNHA, R.P.; CORRÊA, M.F.; SCHUCH, L.O.B.; OLIVEIRA, R.C.; ABREU JÚNIOR, J.S.; SILVA, J.D.G.; ALMEIDA, T.L. Diferentes tratamentos de sementes sobre o desenvolvimento de plantas de soja. *Ciência Rural*, v.45, n.10, 2015. http://www.scielo.br/pdf/cr/2015nahead/0103-8478-cr-cr20140742.pdf

DAN, L. G. M.; DAN, H. A.; BARROSO, A. L. L.; BRACCINI, A. L. Qualidade fisiológica de sementes de soja tratadas com inseticidas sob efeito do armazenamento, *Revista Brasileira de Sementes* v. 32, n.2, p. 131-139, 2010 .http://www.scielo.br/pdf/rbs/v32n2/v32n2a16.pdf

DAN, L.G.M.; BRACCINI, A.L.; BARROSO, A.L.L.; DAN, H.A.; PICCININ, G.G.; VORONIAK, J.M. Physiological potential of soybean seeds treated with thiamethoxam and submitted to storage. *Agricultural Sciences*, v.4, n.11, p.19-25, 2013. http://www.scirp.org/journal/PaperInformation.aspx?PaperID=40173

DANTAS, A. A. A.; CARVALHO, L. G.; FERREIRA, E. Classificação e tendências climáticas em Lavras, MG. *Ciência e Agrotecnologia*, v.31, n.6, p.1862-1866, 2007.

FERREIRA, D.F. Sisvar: A computer statistical analysis system. *Ciência e Agrotecnologia*, v.35, n.6, p.1039-1042, 2011. http://www.scielo.br/pdf/cagro/v35n6/a01v35n6.pdf

FOLLMANN, D. N.; SOUZA, V. Q.; NARDINO, M.; CARVALHO, I. R.; DEMARI, G. H. Diferentes associações para aditivos em pré-semeadura na cultura da soja e seus efeitos sobre a qualidade das sementes produzidas. *Enciclopédia Biosfera*, v.10, n.18. p.1284-1292, 2014. http://www.conhecer.org.br/enciclop/2014a/AGRARIAS/diferentes%20associacoes.pdf

JUHÁSZ, A. C. P.; PÁDUA, G.P.; WRUCH, D.S.M.; FAVORETO, L.; RIBEIRO,N.R. Desafios fitossanitários para a produção de soja. *Informe Agropecuário*, v.34, n. 276, p. 66-75, 2013. https://www.alice.cnptia.embrapa.br/alice/bitstream/doc/978383/1/cpamtwruck010033642013.pdf

LUDWIG, M.P.; LUCCA FILHO, O.A.; BAUDET, L.; DUTRA, L.M.C.; AVELAR, S.A.G.; CRIZEL, R.L. Qualidade de sementes de soja armazenadas após recobrimento com aminoácido, polímero, fungicida e inseticida. *Revista Brasileira de Sementes*, v.33, n.3, p.395-406, 2011. http://www.scielo.br/scielo.php?pid=S0101-31222011000300002&script=sci_arttext

MAPA- Ministério da Agricultura Pecuária e Abastecimento. http:\\www.agricultura.gov.br. Accessed on: Jan, 30th, 2016.

MARCOS-FILHO, J. Teste de envelhecimento acelerado. In: KRZYZANOWSKI, F.C.; VIEIRA, R.D.; FRANÇA-NETO, J.B. (Ed.). *Vigor de sementes*: conceitos e testes. Londrina: ABRATES, 1999. p.3.1-3.21.

NEERGAARD, P. *Seed Pathology*. London: Macmillan Press, 1979. 839p.

PEREIRA, C. E.; OLIVEIRA, J. A.; EVANGELISTA, J. R.E.; BOTELHO, F. J. E.; OLIVEIRA, G. E.; TRENTINI, P. Desempenho de sementes de soja tratadas com fungicidas e peliculizadas durante o armazenamento. *Ciência e Agrotecnologia*, v.31, n.3, p.656-665, 2007. http://www.scielo.br/pdf/cagro/v31n3/a09v31n3.pdf

SALES, J. F.; PINTO, J.E.B.P.; OLIVEIRA, J.A.; BOTREL, P.P.; SILVA, F.G.; CORRÊA, R.M. The germination of bush mint (*Hyptis marrubioides* EPL) seeds as a function of haverst stage, light, temperature and duration of storage. *Acta Scientiarum Agronomy*, v.33, n. 4, p.709-713, 2011. http://www.scielo.br/pdf/asagr/v33n4/22.pdf

SOUZA, V.Q.S.; FOLLMANM, D.N.; NARDINO, M.; BARETTA, D.; CARVALHO, I.R.; CARON, B.O.; SCHMIDT, D.; DEMARI, G.H.. Produção de sementes de soja e vigor das sementes produzidas com diferentes tratamentos de sementes. *Global Science Technology*, v.8, n.1, p.157-166, 2015. http://rv.ifgoiano.edu.br/periodicos/index.php/gst/article/view/703/454

TONIM, R.F.B.; LUCCA FILHO, O.A.; LABBE, L.M.B.; ROSSETO, M. Potencial fisiológico de sementes de milho híbrido tratadas com inseticidas e armazenadas em duas condições ambiente. *Scientia Agropecuária*, v.5, n.1, p. 7-16, 2014. http://www.scielo.org.pe/pdf/agro/v5n1/a01v5n1.pdf

ZAMBOM, S. Aspectos importantes do tratamento de sementes. *Anuário Abrasem*, Brasília, p. 24-25, 2013. http://www.abrasem. com.br/anuario-2013/

ZILLI, J. E.; CAMPO, R. J.; HUNGRIA, M. Eficácia da inoculação de *Bradyrhizobium* em pré-semeadura da soja. *Pesquisa Agropecuária Brasileira*, v.45, n.3, p.335-338, 2010. http://www.scielo.br/pdf/pab/v45n3/v45n3a15.pdf

ZORATO, M.; HENNING, A. Influência de tratamentos fungicidas antecipados, aplicados em diferentes épocas de armazenamento, sobre a qualidade de sementes de soja. *Revista Brasileira de Sementes*, v.23, n.2, p.236-244, 2001. http://www.abrates.org.br/revista/artigos/2001/v23n2/artigo33.pdf

14

Treatment with fungicides and insecticides on the physiological quality and health of wheat seeds

Julia Abati[1]*, Claudemir Zucareli[1], José Salvador Simoneti Foloni[1],
Fernando Augusto Henning[2], Cristian Rafael Brzezinski[1], Ademir Assis Henning[2]

ABSTRACT – Seed treatment with insecticides and fungicides has become an important practice for ensuring initial plant stand in establishing crops. In this context, the aim of this study was to evaluate the influence of chemical seed treatment with insecticides and fungicides on the physiological quality and health of the seeds of wheat cultivars. Seeds of the wheat cultivars BRS Pardela and BRS Gaivota were used, subjected to the following chemical treatments: 1- control, 2- carboxin + thiram + imidacloprid + thiodicarb, 3- carbendazim + thiram + imidacloprid + thiodicarb, 4- fipronil + thiophanate-methyl + pyraclostrobin, 5- triadimenol + imidacloprid + thiodicarb, 6- fipronil, and 7- imidacloprid + thiodicarb. Physiological quality was evaluated by tests of germination, accelerated aging, the length and dry weight of shoots and roots, and seedling emergence in the field. Seed health quality was evaluated by the blotter test method. The seeds of the wheat cultivars tested respond differently to the chemical treatments in regard to effects on germination and vigor. The treatment with triadimenol + imidacloprid + thiodicarb is harmful to seedling development. For the BRS Gaivota cultivar, the seed treatment with carboxin + thiram + imidacloprid + thiodicarb; and carbendazim + thiram + imidacloprid + thiodicarb improved seedling establishment in the field compared to the control.

Index terms: *Triticum aestivum* L., seed treatment, germination, vigor.

Tratamento com fungicidas e inseticidas na qualidade fisiológica e sanitária de sementes de trigo

RESUMO - O tratamento de sementes com inseticidas e fungicidas tem se tornado uma prática importante para assegurar o estande inicial de plantas na implantação das lavouras. Diante disso, o objetivo foi avaliar a influência do tratamento químico de sementes com inseticidas e fungicidas sobre a qualidade fisiológica e sanitária de sementes de cultivares de trigo. Foram utilizadas sementes das cultivares de trigo BRS Pardela e BRS Gaivota, submetidas aos seguintes tratamentos químicos: 1- testemunha, 2- carboxin + thiram + imidacloprido + tiodicarbe, 3- carbendazim + thiram + imidacloprido + tiodicarbe, 4- fipronil + tiofanato metílico + piraclostrobina, 5- triadimenol + imidacloprido + tiodicarbe, 6- fipronil e 7- imidacloprido + tiodicarbe. A qualidade fisiológica foi avaliada por meio dos testes de germinação, envelhecimento acelerado, comprimento e massa seca de parte aérea e raiz, emergência de plântulas em campo e a qualidade sanitária pelo método do *blotter test*. As sementes das cultivares de trigo testadas respondem de forma diferenciada aos tratamentos químicos quanto aos efeitos na germinação e vigor. O tratamento com triadimenol + imidacloprido + tiodicarbe é prejudicial ao desenvolvimento de plântulas. Para a cultivar BRS Gaivota o tratamento de sementes com carboxin + thiram + imidacloprido + tiodicarbe e carbendazim + thiram + imidacloprido + tiodicarbe favorece o estabelecimento das plântulas em campo, em relação a testemunha.

Termos para indexação: *Triticum aestivum* L., tratamento de sementes, germinação, vigor.

Introduction

The wheat (*Triticum aestivum* L.) crop holds an important position in the world and Brazilian domestic grain market. In Brazil, annual production of the cereal crop oscillates between five and six million tons, with 90% of this production centered in the southern region of the country (Embrapa, 2014). Nevertheless, around 50% of the wheat consumed within Brazil is imported (CONAB, 2014). Thus, there is great socioeconomic interest in the country in increasing its production so as to meet domestic demand and because of the benefits generated from growing it (Barbieri et al., 2013).

[1]Departamento de Agronomia, UEL, Caixa Postal 6001, 86051-990 - Londrina, PR, Brasil.

[2]Embrapa Soja, Caixa Postal 231, 86001-970 - Londrina, PR, Brasil.
*Corresponding author <juliaabati@yahoo.com.br>

Together with the use of quality seeds, the application of fungicides and insecticides through seed treatment is of utmost importance in obtaining high yields. This is because fungicides and insecticides act in protection against seed pathogens, whether from storage or present in the soil, and also act against the initial attack of pests specific to the soil (Antonello et al., 2009; Menten and Moraes, 2010; Pereira et al., 2010).

Chemical seed treatment is considered one of the most effective methods of control of fungi and insects. Nevertheless, Kashypa et al. (1994), in studies on wheat seeds, and Ludwig et al. (2011) on soybean seeds, found that some chemical products, when applied on the seeds, may lead to reduction in germination and hinder seedling survival. According to Antonello et al. (2009) and Deuner et al. (2014), who worked with maize seeds, the phytotoxicity effect depends on the product used and the time that the seeds were stored. According to Goulart (1988), the fungicide triadimenol, although it was effective for seedling protection (providing for total absence of disease lesions on wheat), proved to be phytotoxic, leading to the rise of abnormal seedlings in the germination process.

In contrast, Matos et al. (2013) observed that maize seeds treated with different fungicides exhibited germination and vigor superior to the untreated control in the cold test. This difference may be explained as a result of the chemical products controlling the pathogens present in the seeds since these seeds exhibited low quality of health.

Associated with chemical treatment, some insecticides, in addition to providing pest control in the crop, may act physiologically, assisting both initial growth and plant development (Dan et al., 2012). Corroborating this affirmation, Hossen et al. (2014) verified a greater percentage and speed of germination in wheat seeds treated with the insecticide thiamethoxam in relation to untreated seeds. Furthermore,

Barros et al. (2005) found similar results in common bean seeds treated with the insecticide fipronil.

The practice of seed treatment is widely adopted in various crops; however, in the wheat crop, the results are still in the initial stages, especially with the advent of new formulations and chemical products that are being introduced on the market, as well as new cultivars. It is noteworthy that together with the use of these new products and formulations, there is a trend of recommending the use of a smaller quantity of seeds per area, especially for these new cultivars. Thus, the correct recommendation of products for seed treatment becomes essential in assuring uniform germination and seedling emergence, especially considering the possibility of occurrence of stress conditions, seed colonization by fungi, and pest attack. It is also important to recommend the use of seeds of high physiological and genetic quality associated with physical purity and health to obtain crops with adequate stand and vigorous seedlings (Scheeren et al., 2010; Dan et al., 2012).

In light of the above, the aim of this study was to evaluate the influence of chemical treatment with insecticides and fungicides on the physiological quality and health of seeds of wheat cultivars.

Material and Methods

The experiment was conducted at the Seed and Grain Technological Center and in the experimental field of the Empresa Brasileira de Pesquisa Agropecuária (Embrapa), at the National Soybean Research Center (Embrapa Soja) in Londrina, Parana, Brazil.

The wheat cultivars used were BRS Pardela and BRS Gaivota, evaluated separately, with seven chemical seed treatments, involving fungicides and insecticides (Table 1).

Table 1. Active ingredients, commercial names, and application rates used for seed treatment of the wheat cultivars BRS Pardela and BRS Gaivota.

Treatments	Active ingredient (a.i.)	Commercial name	Commercial product application rate[1]
T1	Control	-	-
T2	carboxin[2] + thiram[2] + imidacloprid[3] + thiodicarb[3]	Vitavax-Thiram® +Cropstar®	250 + 100
T3	carbendazim[2] + thiram[2] + imidacloprid[3] + thiodicarb[3]	Derosal Plus® +Cropstar®	200 +100
T4	fipronil[3] + thiophanate methyl[2] + pyraclostrobin[2]	Standak Top®	100
T5	triadimenol[2] + imidacloprid[3] + thiodicarb[3]	Baytan® + Cropstar®	250 +100
T6	fipronil[3]	Fipronil®	100
T7	imidacloprid[3] + thiodicarb[3]	Cropstar®	250

[1]Commercial product application rate: mL.100 kg[-1] of seeds.
[2]Active ingredients of the fungicide class.
[3]Active ingredients of the insecticide class.

Chemical treatment of the seeds was carried out in plastic bags in which the products were added over the seeds, followed by shaking until the seeds were completely covered, with volume of the mixture of 600 mL.100 kg^{-1} of seeds (product + water).

After the seed treatment, the physiological quality and health of the seeds was evaluated by the following tests:

Germination: carried out on four subsamples of 50 seeds per replication, for a total of 600 seeds per treatment. The seeds were distributed over germitest paper with a water volume for soaking at the quantity of 2.5 times the dry weight of the substrate, in the form of rolls. The seeds were placed in a germinator at a temperature of 20 °C for eight days. After that, evaluations were carried out according to the recommendations of the Rules for Seed Testing [Regras para Análise de Sementes] (Brasil, 2009), and the results were expressed in percentage of normal and abnormal seedlings.

Accelerated aging: four subsamples of 50 seeds per replication were used, placed in gerbox type plastic boxes, with screen supports, containing 40 mL of distilled water at the bottom. The seeds were uniformly distributed on the screen in a single layer. After that, they were placed in a water-jacketed incubation chamber at 42 °C for 60 hours (Marcos- Filho, 1999). After this period, they were subjected to the germination test.

Length of total seedling, shoot, and root: four subsamples of 25 seeds per replication were used. The germination paper was moistened with distilled water at 2.5 times the dry weight of the substrate. The seeds were arranged on the upper third in the lengthwise direction of the paper. The rolls were placed in plastic bags which were arranged vertically in the germinator for seven days at 20 °C. After that, the length of the normal seedlings was checked (total length of the seedling, of the shoot, and of the root) with a ruler in millimeters. The results were expressed in centimeters per seedling.

Dry matter weight of the shoot and root: this was carried out on the normal seedlings obtained in the test of seedling length. After measuring seedling length, the remaining part of the seed was removed and the shoot and root were separated. These were then placed in paper bags and placed in an air circulation laboratory oven, where they remained for 24 hours at a temperature of 80 °C (Nakagawa, 1999). At the end of this period, dry weight was checked on a precision scale accurate to 0.0001 g, and the results were expressed in mg per seedling.

Seedling emergence in the field: seeds were sown in April 2013, using a density of 300 seeds per m^2. Plots consisted of six rows of six-meter length, spaced at 0.20 m, for a total area of 7.2 m^2 per plot. Only the four center rows were considered as useful area, leaving the 0.5 m at the beginning and end of the plot as a border. At 15 days after sowing, the total number of emerged seedlings was counted in a total area of 0.75 m^2, which was composed of three subsamples of 0.25 m^2, and the result was expressed in seedlings per m^2. The data on mean daily temperature and rainfall during the period of sowing and seedling emergence were obtained from the meteorological station of Embrapa Soja, located at approximately 2000 m from the experiment, in Londrina (Figure 1).

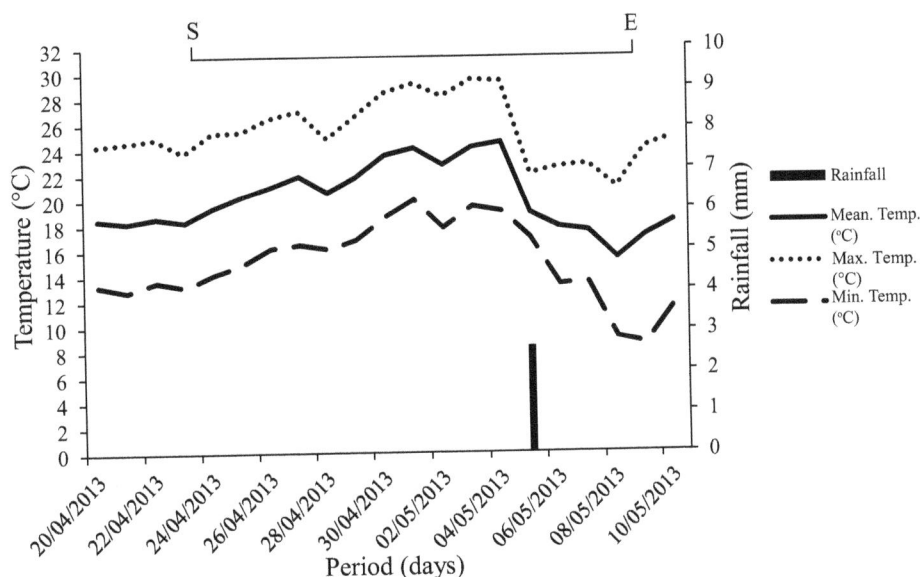

Figure 1. Maximum, minimum, and mean daily temperature (°C) and rainfall during the period of seedling emergence in the field. S: sowing; E: emergence.

Seed health analysis: the method used was the blotter test. The seeds were arranged in gerbox type plastic boxes, duly disinfected with sodium hypochlorite at 1.05%. The boxes contained three sheets of filter paper moistened with distilled water and autoclaved. Ten gerboxes with 20 seeds taken at random were used, with three replications, for a total of 600 seeds per treatment. After that, the seeds were placed in an incubation chamber where they remained for seven days at a temperature of 20 °C (±2 °C) under continuous fluorescent light (Henning, 2005). Fungi were identified with the aid of a binocular stereoscopic microscope and optical microscope. The results were expressed in percentage.

A completely randomized design was used for laboratory evaluations, with three replications and, for the field evaluation, a randomized block design was used, with four replications. The data obtained were subjected to analysis of variance in an independent way for each cultivar, and the mean values were compared by the Tukey test at 5% probability. Analyses were carried out by the computational program Sistema para Análise de Variância (System for Analysis of Variance) - SISVAR (Ferreira, 2011).

Results and Discussion

The presence of pathogenic fungi was not observed in seed health analysis; there was only the presence of fungi of the genus *Epicoccum* sp. and *Alternaria* sp. According to Celano et al. (2012), these fungi are considered saprophytes in the wheat crop and are not pathogenic agents.

The absence of pathogenic fungi, both in the treated seeds and in the control, may be associated with the use of commercial seed lots produced with high physiological quality and health. According to França-Neto et al. (2010), the use of healthy seeds, associated with physiological, genetic, and physical quality, is fundamental for obtaining an adequate stand of plants, and is thus a focus of seed industries.

The effects of chemical seed treatments on physiological quality in the cultivars BRS Pardela and BRS Gaivota are shown in the table of analysis of variance (Table 2).

Table 2. Summary of analysis of variance (mean squares) for the traits of physiological quality in seeds of wheat cultivars as a result of chemical treatment of seeds.

		Mean Squares				
		-------------- BRS Pardela --------------				
S.V.	DF	G	AS	AA	TSL	SL
Treatment	6	34.55**	21.96**	166.04**	57.31**	9.62**
Error	14	4.95	1.95	19.85	12.65	0.39
Mean		93.33	4.57	48.28	24.91	10.09
CV (%)		2.38	30.57	9.23	14.28	6.25
S.V.	DF	RL	SDW	RDW	DF	EF
Block[1]					3	607.23[ns]
Treatment	6	21.46 [ns]	0.54 [ns]	0.54[ns]	6	1071.23[ns]
Error	14	9.29	0.37	0.27	18	418.79
Mean		14.82	6.81	5.15		269.14
CV (%)		20.57	9.02	10.11		7.60
		-------------- BRS Gaivota --------------				
S.V.	DF	G	AS	AA	TSL	SL
Treatment	6	74.66**	31.63**	508.98**	43.27**	5.78**
Error	14	2.57	2.52	59.38	8.18	0.39
Mean		93.00	4.42	68.80	27.03	9.44
CV (%)		1.72	35.87	11.20	10.58	6.66
S.V.	DF	RL	SDW	RDW	DF	EF
Block[1]					3	2266.47*
Treatment	6	18.12 [ns]	0.12 [ns]	0.30 [ns]	6	2016.57*
Error	14	6.39	0.23	0.14	18	520.69
Mean		17.59	5.90	6.05		257.85
CV (%)		14.37	8.23	6.31		8.85

ns, not significant, ** and *, significant at 1% and 5% probability, respectively, by the F test.
[1] Randomized blocks.

G: germination test; AS: abnormal seedlings; AA: accelerated aging; TSL: total seedling length; SL: shoot length; RL: root length; SDW: shoot dry weight; RDW: root dry weight; EF: seedling emergence in the field.

In relation to the germination test in the cultivar BRS Pardela, the seeds treated with triadimenol + imidacloprid + thiodicarb (T5) showed inferior results; however, they were not significantly different from the other treatments. As for BRS Gaivota, seed treatment with carboxin + thiram + imidacloprid + thiodicarb (T2) brought about significant reduction in seed germination. Possibly, this reduction in germination was due to association of the fungicides carboxin + thiram and triadimenol with the insecticides imidacloprid + thiodicarb, because these insecticides separately (T7) did not show a negative effect (Table 3).

Still, all the values obtained for germination were above the standard established for commercialization of wheat seeds, which require minimum germination of 80% (MAPA, 2013).

The results of the germination test corroborate the evaluations of abnormal seedlings (Table 3). This indicates that some treatments may have led to phytotoxicity, which caused abnormality in seedlings. For soybean, França-Neto et al. (2000) reported that the phytotoxic effect of the chemical treatments may reduce seed quality. Fungicides of the triazole group, when used in seed treatment, may cause phytotoxicity to wheat seedlings, and may lead to reduction of the mesocotyl and cracks in the leaf tip (Picinini and Fernandes, 2003). Similar results were described by Goulart (1988) upon observing that the fungicide triadimenol applied on wheat seeds led to an increase in abnormal seedlings, as well as the appearance of seedlings with twisted, thick, and broadened leaves.

Table 3. Mean values of the properties of physiological quality and health of seeds of wheat cultivars as a function of the response to chemical seed treatment.

Treatment[1]	G (%)	AS (%)	AA (%)	TSL (cm)	SL (cm)	
		--------------- BRS Pardela ---------------				
T1	91 ab	7 c	52 ab	27.18 a	10.80 a	
T2	92 ab	7 c	50 ab	26.16 a	11.10 a	
T3	97 a	1 a	54 a	25.67 a	11.09 a	
T4	96 a	2 ab	52 ab	26.07 a	10.52 a	
T5	89 b	8 c	35 c	15.10 b	6.08 b	
T6	91 ab	6 bc	41 bc	27.10 a	10.64 a	
T7	97 a	2 ab	54 a	27.12 a	10.42 a	

Treatment[1]	G (%)	AS (%)	AA (%)	TSL (cm)	SL (cm)	EF (plants/m^2)
		--------------- BRS Gaivota ---------------				
T1	93 bc	5 ab	53 b	29.52 a	10.25 a	217 b
T2	83 d	10 c	65 ab	27.91 a	10.05 a	279 a
T3	96 abc	2 a	69 ab	28.80 a	10.15 a	285 a
T4	94 abc	4 ab	84 a	28.66 a	9.84 a	264 ab
T5	96 abc	2 a	85 a	18.71 b	6.33 b	247 ab
T6	92 c	7 bc	53 b	29.10 a	9.78 a	254 ab
T7	97 a	1 a	73 ab	26.56 a	9.72 a	259 ab

The mean values within each column followed by the same letter do not differ among themselves by the Tukey test (p≤0.05).
[1]Treatments: 1: control (without seed treatment); 2: carboxin + thiram + imidacloprid + thiodicarb; 3: carbendazim + thiram + imidacloprid + thiodicarb; 4: fipronil + thiophanate methyl + pyraclostrobin; 5: triadimenol + imidacloprid + thiodicarb; 6: fipronil; 7: imidacloprid + thiodicarb.
G: germination test; AS: abnormal seedlings; AA: accelerated aging; TSL: total seedling length; SL: shoot length; EF: emergence in the field.

In the accelerated aging test in the cultivar BRS Pardela, treatment 5 (triadimenol + imidacloprid + thiodicarb) led to reduction in vigor in comparison to the control. However, for the seeds of BRS Gaivota, this treatment did not result in the same behavior (Table 3). This treatment together with T4 (fipronil + imidacloprid + thiodicarb) led to smaller reductions in viability and greater seed vigor (Table 3).

For the data on total seedling length and shoot length, in both cultivars, the treatment with triadimenol + imidacloprid + thiodicarb (T5) differed from the others and from the

control, showing lower total length and shoot length of the seedlings (Table 3). Similar results were observed by Silva et al. (1993), where the fungicide triadimenol reduced the length of the coleoptile and of the mesocotyl in wheat and barley, and by Rampim et al. (2012) who, studying three wheat cultivars, observed that triadimenol led to lower values of shoot length, regardless of the cultivar tested. According to these authors, such results indicate that this product may cause a phytotoxic effect to wheat seedlings.

In regard to seedling emergence in the field, there was a

significant effect only in the cultivar BRS Gaivota, in which T2 (carboxin + thiram + imidacloprid + thiodicarb) and T3 (carbendazim + thiram + imidacloprid + thiodicarb) led to greater plant populations per m² in relation to the control (Table 3).

Furthermore, for this variable, the control (T1) led to lower values of emerged seedlings in the field, though not differing statistically from treatments 4, 5, 6, and 7 (Table 3). This result may be explained due to the absence of chemical treatment on the seeds for, although there was high initial seed health, there was a reduction in emergence because, when they are exposed to uncontrolled environmental conditions, such as occurs in the field (Figure 1), they are subject to the action of soil pathogens. This therefore indicates the importance of seed treatment to obtain an adequate plant stand.

In light of the above, it may be seen, in relation to germination and accelerated aging, that the cultivars respond in a differentiated manner to the different chemical products. Furthermore, some of the formulations used with fungicides and insecticides may lead to phytotoxicity in wheat seedlings.

Conclusions

The wheat cultivars tested respond in a differentiated manner to the chemical seed treatments in regard to the effects on germination and vigor.

The treatment with triadimenol + imidacloprid + thiodicarb is harmful to seedling development.

In the cultivar BRS Gaivota, the seed treatments with carboxin + thiram + imidacloprid + thiodicarb, and carbendazim + thiram + imidacloprid + thiodicarb improve seedling establishment in the field compared to the control.

References

ANTONELLO, L.M.; MUNIZ, M.B.; BRANDT, S.C.; GARCIA, D.; RIBEIRO, L.; SANTOS, V. Qualidade de sementes de milho armazenadas em diferentes embalagens. *Revista Ciência Rural*, v.39, n.7, p.2191-2194, 2009. http://www.scielo.br/scielo.php?script=sci_arttext&pid=S0103-84782009000700036

BARBIERI, A.P.P.; MARTIN, T.N.; MERTZ, L.M.; NUNES, U.G.; CONCEIÇÃO, G.M. Redução populacional de trigo no rendimento e na qualidade fisiológica das sementes. *Revista Ciência Agronômica*, v.44, n.4, p.724-731, 2013. http://www.scielo.br/scielo.php?script=sci_arttext&pid=S1806-66902013000400008

BARROS, R.G.; BARRIGOSSI, J.A.F.; COSTA, J.L.S. Efeito do armazenamento na compatibilidade de fungicidas e inseticidas, associados ou não a um polímero no tratamento de sementes de feijão. *Bragantia*, v.64, n.3, p.459-465, 2005. http://www.scielo.br/scielo.php?pid=S0006-87052005000300016&script=sci_arttext

BRASIL. Ministério da Agricultura, Pecuária e Abastecimento. *Regras para análise de sementes*. Ministério da Agricultura, Pecuária e Abastecimento. Secretaria de Defesa Agropecuária. Brasília: MAPA/ACS, 2009. 395p. http://www.agricultura.gov.br/arq_editor/file/2946_regras_analise__sementes.pdf

CELANO, M.M.; MACHADO, J.C.; JACCOUD FILHO, D.S.; GUIMARÃES, R.M. Avaliação do potencial de uso da restrição hídrica em teste de sanidade de sementes de trigo visando à detecção de fungos. *Revista Brasileira de Sementes*, v.34, n.4, p.613-618, 2012. http://www.scielo.br/scielo.php?pid=S0101-31222012000400012&script=sci_arttext

CONAB. Companhia Nacional de Abastecimento. *Acompanhamento da safra brasileira – Grãos*. Available at: http://www.conab.gov.br/OlalaCMS/uploads/arquivos/14_06_10_12_12_37_boletim_graos_junho_2014.pdf Accessed on: Jun. 28th, 2014.

DAN, L.G.M.; DAN, H.A.; PICCININ, G.; RICCI, T.T.; ORTIZ, A.H.T. Tratamento de sementes com inseticida e a qualidade fisiológica de sementes de soja. *Revista Caatinga*, v.25, n.1, p.45-51, 2012. http://periodicos.ufersa.edu.br/revistas/index.php/sistema/article/view/2073/pdf

DEUNER, C.; ROSA, K.C.; MENEGHELLO, G.E.; BORGES, C.T.; ALMEIDA, A.S.; BOHN, A. Physiological performance during storage of corn seed treated with insecticides and fungicide. *Journal of Seed Science*, v.36, n.2, p.204-212, 2014. http://www.scielo.br/scielo.php?pid=S2317-15372014000100005&script=sci_arttext

EMBRAPA. Empresa Brasileira de Pesquisa Agropecuária. *Cultivos – Trigo*. Available at: https://www.embrapa.br/trigo/cultivos Accessed on: Jun. 28th, 2014.

FERREIRA, D.F. Sisvar: A computer statistical analysis system. *Ciência e Agrotecnologia*, v.35, n.6, p.1039-1042, 2011. http://www.scielo.br/scielo.php?script=sci_arttext&pid=S1413-70542011000600001&lang=pt

FRANÇA-NETO, J.B.; HENNING, A.A.; YORINORI, J.T. *Caracterização dos problemas de fitotoxicidade de plântulas de soja devido ao tratamento de sementes com fungicida Rhodiauram 500 SC, na safra 2000/01*. Londrina: Embrapa Soja, 2000. 24p. (Circular Técnica, 27). http://ainfo.cnptia.embrapa.br/digital/bitstream/CNPSO/2596/1/circtec27.pdf

FRANÇA-NETO, J.B.; KRZYZANOWSKI, F.C.; HENNING, A.A. A importância do uso de sementes de soja de alta qualidade. *Informativo Abrates*, v.20, n.1, p.37-38, 2010. http://www.abrates.org.br/portal/images/stories/informativos/v20n12/artigo04.pdf

GOULART, A.C.P. Eficiência de três fungicidas no tratamento de sementes de trigo (*Triticum aestivum*) visando o controle do fungo *Helminthosporium sativum* P. K. & B., em condições de laboratório. *Revista Brasileira de Sementes*, v.10, n.1, p.55-61, 1988. http://www.abrates.org.br/revista/artigos/1988/v10n1/artigo05.pdf

HENNING, A.A. *Patologia e tratamento de sementes*: noções gerais. Londrina: Embrapa Soja, 2005. 52p.

HOSSEN, D.C.; CORREIA JÚNIOR, E.S.; GUIMARÃES, S.; NUNES, U.R.; GALON, L. Tratamento químico de sementes de trigo. *Pesquisa Agropecuária Tropical*, v.44, n.1, p.104-109, 2014. http://www.revistas.ufg.br/index.php/pat/article/viewFile/23117/16313

KASHYPA, R.K.; CHAUDHARY, O.P.; SHEORAN, I.S. Effects of insecticide seed treatments on seed viability and vigor in wheat cultivars. *Seed Science and Technology*, v.22, n.3, p.503-517, 1994. http://eurekamag.com/research/002/608/influences-insecticide-seed-treatments-seed-viability-vigour-wheat-varieties.php

LUDWIG, M.P.; LUCCA FILHO, O.A.; BAUDET, L.; DUTRA, L.M.C.; AVELAR, S.A.G.; CRIZEL, R.L. Qualidade de sementes de soja armazenadas após recobrimento com aminoácido, polímero, fungicida e inseticida. *Revista Brasileira de Sementes*, v.33, n.3, p.395-406, 2011. http://www.scielo.br/scielo.php?pid=S0101-31222011000300002&script=sci_arttext

MAPA. Ministério da Agricultura, Pecuária e Abastecimento. Padrões para produção e comercialização de sementes de trigo e de trigo duro. Brasília, DF, 17/09/2013, n° 45, Seção 1. http://apasem.com.br/site/wp-content/uploads/padroesim0452013.pdf. Accessed on: Apr. 04th, 2014.

MARCOS-FILHO, J. Teste de envelhecimento acelerado. In: KRZYZANOWSKI, F.C.; VIEIRA, R.D.; FRANÇA-NETO, J.B. (Ed.). *Vigor de sementes*: conceitos e testes. Londrina: ABRATES, 1999. p.3.1-3.21.

MATOS, C.S.M.; BARROCAS, E.N.; MACHADO, J.C.; ALVES, F.C. Health and physiological quality of corn seeds treated with fungicides and assessed during storage. *Journal of Seed Science*, v.35, n.1, p.10-16, 2013. http://www.scielo.br/pdf/jss/v35n1/01.pdf

MENTEN, J.O.; MORAES, H.M.D. Tratamento de sementes: histórico, tipos, características e benefícios. *Informativo Abrates*, v.20, n.3, p.52-53, 2010. http://www.abrates.org.br/portal/images/stories/informativos/v20n3/minicurso03.pdf

NAKAGAWA, J. Testes de vigor baseados no desempenho das plântulas. In: KRYZANOWSKI, F.C.; VIEIRA, R.D.; FRANÇA-NETO, J.B. (Ed.). *Vigor de sementes*: conceitos e testes. Londrina: ABRATES, 1999. p2.1-2.24.

PEREIRA, C.E.; OLIVEIRA, J.A.; COSTA NETO, J.; MOREIRA, F.M.S.; VIEIRA, A.R. Tratamentos inseticida, peliculização e inoculação de sementes de soja com rizóbio. *Revista Ceres*, v.57, n.5, p.653-658, 2010. http://www.scielo.br/scielo.php?script=sci_arttext&pid=S0034-737X2010000500014

RAMPIM, L.; RODRIGUES-COSTA, A.C.P.; NACKE, H.; KLEIN, J.; GUIMARÃES, V.F. Qualidade fisiológica de sementes de três cultivares de trigo submetidas à inoculação e diferentes tratamentos. *Revista Brasileira de Sementes*, v.34, n.4, p. 678-685, 2012. http://www.scielo.br/scielo.php?script=sci_arttext&pid=S0101-31222012000400020

SILVA, D.B.; CHARCHAR, M.J.D.; VIVALDI, L.J. Efeito do tratamento de sementes sobre a emergência de plântulas de trigo e de cevada em duas profundidades de semeadura. *Pesquisa Agropecuária Brasileira*, v.28, n.3, p.303-311, 1993. http://seer.sct.embrapa.br/index.php/pab/article/view/3879/1170

SCHEEREN, B.R.; PESKE, S.T.; SCHUCH, L.O.B.; BARROS, A.C.A. Qualidade fisiológica e produtividade de sementes de soja. *Revista Brasileira de Sementes*, v.32, n.3, p.35-41, 2010. http://www.scielo.br/pdf/rbs/v32n3/v32n3a04.pdf

PICININI, E.C.; FERNANDES, J.M.C. Efeito do tratamento de sementes com fungicida sobre o controle de doenças na parte aérea do trigo. *Fitopatologia Brasileira*, v.28, n.5, p. 515-520, 2003. http://www.scielo.br/scielo.php?pid=S0100-41582003000500008&script=sci_arttext

15

Biocontrol and seed transmission of *Bipolaris oryzae* and *Gerlachia oryzae* to rice seedlings

Andrea Bittencourt Moura[1*], Juliane Ludwig[2], Aline Garske Santos[1],
Jaqueline Tavares Schafer[1], Vanessa Nogueira Soares[3], Bianca Obes
Corrêa[1]

ABSTRACT - *Bipolaris oryzae* and *Gerlachia oryzae*, which cause rice brown spot and leaf scald, respectively, are mainly disseminated by seeds. The aim of this study was to evaluate the potential of seeds microbiolization to reduce transmission of these pathogens to seedlings by using the bacteria DFs185 (*Pseudomonas synxantha*), DFs223 (*P. fluorescens*), DFs306 (unidentified) and DFs418 (*Bacillus* sp.). Seeds naturally infested/infected with both pathogens were immersed in suspension of these bacteria ($A_{540} = 0.5$) individually or in saline solution (control treatment). After 30 minutes of agitation at 10 °C, 400 seeds were submitted to a sanity test through the blotter method and the isolate DFs223 was the best to reduce the incidence of *B. oryzae* and *G. oryzae* in both seed lots evaluated. Seeds treated like above were sowed in sterilized vermiculite. Seed transmission and growth promotion were recorded after 21 days of incubation in the same conditions. The isolates DFs185 and DFs306 reduced transmission of both pathogens, although the isolate DFs306 was the one wich gave the greatest growth increases. The evaluation of the *in vitro* antibiosis showed that isolates inhibited the mycelial growth of both pathogens, except DFs306. It is possible to affirm that these bacteria have the potential to be used as a seed treatment for seed-borne disease control.

Index terms: biological control, rice brown spot, rice leaf scald, seed treatment, *Oryza sativa* L.

Biocontrole e transmissão de *Bipolaris oryzae* e *Gerlachia oryzae* para plântulas de arroz

RESUMO - *Bipolaris oryzae* e *Gerlachia oryzae* causadores, respectivamente, da mancha parda e da escaldadura do arroz são principalmente disseminados por sementes. Objetivou-se com esse trabalho avaliar o potencial da microbiolização de sementes para a redução da transmissão destes patógenos das sementes para as plântulas usando as bactérias DFs185 (*Pseudomonas synxantha*), DFs223 (*P. fluorescens*), DFs306 (não identificado) e DFs418 (*Bacillus* sp.). Duas amostras de sementes portadoras de *B. oryzae* e *G. oryzae* foram imersas em suspensão dessas bactérias, sendo a testemunha imersa em solução salina ($A_{540} = 0,5$). Após agitação (30 min./10 °C), 400 sementes foram submetidas ao teste de sanidade pelo método do papel de filtro e o resultado indicou que o isolado DFs223 destacou-se na redução de *B. oryzae* e de *G. oryzae* nos dois lotes de sementes avaliados. Sementes tratadas conforme descrito foram dispostas em vermiculita autoclavada e a transmissão dos patógenos e promoção de crescimento avaliados após 21 dias de incubação em mesmas condições. Os isolados DFs185 e DFs306 reduziram a transmissão de ambos os patógenos, porém o isolado DFs306 foi o que proporcionou maiores incrementos de crescimento. Avaliação de antibiose *in vitro* mostrou que os isolados inibiram o crescimento micelial de ambos os patógenos, exceto DFs306. Pode se afirmar que estas bactérias apresentam potencial para tratar sementes visando ao controle de doenças transmitidas por estas.

Termos para indexação: controle biológico, mancha parda do arroz, escaldadura do arroz, tratamento de sementes, *Oryza sativa*.

Introduction

The transmission of pathogens through seeds is an efficient mechanism by which plant pathogens are spread over long distances, are introduced in new cultivation areas and are spread via plant population as random sources of primary inoculum (Malavolta et al., 2002). Therefore, necrotrophic pathogens use seeds as a vehicle for dissemination, as a shelter and as means of survival.

In Brazil, *Bipolaris oryzae* and *Pyricularia grisea* are

[1]Departamento de Fitossanidade, Universidade Federal de Pelotas, Caixa Postal, 54, 96010970 –Pelotas, RS, Brasil.
[2]Universidade Federal da Fronteira Sul, 97900000 - Cerro Largo, RS, Brasil.

[3]Departamento de Fitotecnia, Universidade Federal de Pelotas, Caixa Postal, 354, 96010970- Pelotas-RS, Brasil.
*Corresponding author < abmoura@ufpel.edu.br >

mentioned as major pathogens associated with rice seeds, followed by *Gerlachia oryzae, Cercospora oryzae, Phoma* spp., *Alternaria padwickii, Fusarium* spp., *Nigrospora oryzae* and *Tilletia barclayana* (Franco et al., 2001).

Damage caused by *B. oryzae* and *G. oryzae*, etiologic agents of brown spot and leaf scald, respectively, is due to the reduction in the number of seeds per panicle and their weight, reflecting on the quality of the grown seeds, causing a decrease in their germination. In addition, there is damage from the early epidemic in the field due to the high percentage of transmission of fungi from seeds to seedlings, which, in the case of *B. oryzae*, may reach 15.1% (Malavolta et al., 2002).

In the search for reducing damage caused by these pathogens, the use of resistant cultivars is recommended, although the cultivars available in the market do not always have the desirable levels of resistance and/or resistance to more than one pathogen (Nunes et al., 2004). Another aspect that limits the possibility of use of resistant cultivars is their reduced usable life time brought about by the emergence of new races of pathogens (Cornélio et al., 2003).

The brazilian market offers various compounds formulated and registered for the control of most diseases present in rice, but only four of them for seed treatment and, of these, only one active ingredient is recommended for the control of *B. oryzae* and *G. oryzae* (Agrofit, 2014). The chemical control is effective but its use increases production costs and its residues are accumulated in the environment besides increases selection pressure, allowing the emergence of pathogen populations resistant to these chemical compounds (Celmer et al., 2007). Faced with this scenario, biological control by microbiolization of seeds appears as a viable alternative. Reports on the potential of different isolates of *Pseudomonas* and *Bacillus* to control rice blast (*P. grisea*) (Krishhnamurthy and Gnanamanickam, 1998) and seedling blight (*Rhizoctonia solani*) (Commare et al., 2002; Wiwattanapatapee et al., 2004; Souza Júnior et al., 2010) have shown encouraging results. Studies on the biological control of brown spot and leaf scald in rice through the use of microorganisms are still scarce, but, Ludwig et al. (2009) reported the potential of bacteria for the control of *B. oryzae* and *G. oryzae* when these pathogens were inoculated on rice leaves. However, nothing is known about the effect of biocontrollers during seed germination and their impact on the transmission of seed borne pathogens.

Thus, the aim of this study was to evaluate the effect of selected bacteria for the biocontrol of *B. oryzae* and *G. oryzae* (Ludwig et al., 2009) on the incidence in seeds and transmission of these pathogens in seed lots naturally infested/infected with these pathogens.

Material and Methods

Effect of the microbiolization of seeds with biocontrollers on the incidence of pathogens

Seeds of cultivars Chumbinho (lot 376) and Formosa (lot 375) with incidence of *B. oryzae* and *G. oryzae* were microbiolized with isolates DFs185 *Pseudomonas synxantha* (Ehrenberg) Hollan; DFs223 *P. fluorescens* Migula; DFs418, *Bacillus* sp. Cohn; and DFs306 (unidentified). Seeds were immersed in a suspension of each bacterium with 24 hours of growth in medium 523 (10 g sucrose; 4.0 g yeast extract, 8.0 g casein hydrolysate; 0.3 g $MgSO_4$; 2.0 g K_2HPO_4; 15.0 g agar) prepared with saline solution (0.85% NaCl) and adjusted to A_{540}=0.5. The control was immersed in saline solution.

After a stirring period of 30 minutes at 10 °C, 400 seeds were placed in gerbox® boxes, according to the blotter test method (Brasil, 2009), incubated at 23 ± 2 °C under 12 hours light/12 hours dark conditions.

The incidence of pathogens was evaluated after seven days, individually examining the seeds in a stereoscopic microscope and an optical microscope to confirm the characteristics of conidia and conidiophores. The percentage of incidence was calculated compared to the control, considered 100%.

Effect of biocontrollers on the transmission of pathogens from seeds to seedlings

Seeds of the same lots from the previous trial were microbiolized as described in the previous section, and then planted in plastic cups with a capacity 50 mL containing sterilized vermiculite, placing one seed per cup. The assay was carried out inside transparent plastic boxes (41 cm x 30 cm x 30 cm) maintained in a moist chamber, where 25 cups were placed and incubated at 23 ± 2 °C, each box constituting one experimental plot. The statistical design was a randomized block with four replications of each treatment.

After 21 days of incubation in the above-mentioned conditions, we evaluated the number of seedlings (percentage of emerged seedlings). Subsequently, the severity of symptoms caused by the pathogens was evaluated. Each seedling received a score regarding the intensity of symptoms, where: 0 = seedling with no symptoms; 1 = slightly necrotic seedling; 2 = moderately necrotic seedling; 3 = highly necrotic seedling. We calculated the average severity of each plot relative to the total of emerged seedlings.

In addition, seeds that did not germinate (dead seeds) were removed from the substrate and sterilized in sodium hypochlorite at 1% for 1 minute. After that, they were washed in running water and subjected to a moist chamber aiming to determine the incidence of *B. oryzae* and *G. oryzae*, the average incidence of each pathogens was calculated regarding the total of non-germinated seeds in each plot.

Subsequently, of the developed seedlings, the root system was cuted off from the shoot at the neck, and the length of both parts was measured. To calculate the responses of each treatment, the value of the control for each variable was converted to 100%.

Antibiosis in vitro against B. oryzae and G. oryzae

The pathogens used in this test were isolated from the seeds infested with them, observed in the above test.

The four biocontrollers isolates (DFs185, DFs223, DFs306 and DFs418) were cultured in liquid medium 523 (10.0 g sucrose; 4.0 g yeast extract; 8.0 g casein hydrolysate; 0.3 g $MgSO_4$; K_2HPO_4) for a period of 72 h at 28 °C. Subsequently, 1 mL of each culture was centrifuged for 15 minutes at 9860 rpm (10,000 g). The supernatant was removed and subjected to ultrasound bath (Ultrasonic Cleaner 1440D) for a period of 20 min. for the disruption of bacterial cells still present in the cultures, thereby obtaining the metabolic liquid from each bacterial isolate. Separately, 10 mL of PDA medium (Acumedia Potato-dextrose-agar) were transferred to Petri dishes. After solidification of the medium, disks were removed using a punch of 5 mm diameter, forming four equidistant holes in the edges of the plates. The liquid formed by metabolites of each isolate was added to the holes (20 mL). Next, a 5 mm disk of G. oryzae or B. oryzae was placed in the center of each plate. As a control, a disk of mycelium was placed on a plate containing PDA without liquid metabolite. The plates were incubated at 23 ± 2 °C for up to 14 days. The evaluation occurred when the mycelial growth of the control reached the edge of the plates, observing the occurrence of a mycelial growth inhibition halo, considered an indicator of antibiosis. In this test, four replicates for each bacterial isolate were performed.

Statistical analyzes

The data obtained from the developed seedlings in the test of pathogen transmission through seeds (severity of symptoms, germination rate, length of root and shoot) were subjected to analysis of variance and means grouped by the Scott-Knott test at 5% probability.

Results and Discussion

The G. oryzae and B. oryzae incidence observed in blotter test method for untreated seeds (control) was respectively 36.75 and 11.25% for lot 375; and 24.25% and 4.75 for lot 376. All treatments, when assessed by blotter test, reduced the

incidence of pathogens (in media B. oryzae by 18.8% and G. oryzae by 20.5%), except in the isolate DFs418 in lot 375 for B. oryzae and DFs306 and DFs418 in lot 376 for G. oryzae (Figure 1). Larger decreases were observed in lot 376 for the fungus B. oryzae in the treatment with isolate DFs185 (Figure 1B - 78.9%). When considering the average of the two lots, the isolate DFs185 was the most effective for the control of B. oryzae and isolate DFs223 for G. oryzae, both providing control percentages of 40%. These isolates in the overall average (pathogens and lots) also showed a similar behavior, resulting in a reduction of 35 and 38% respectively.

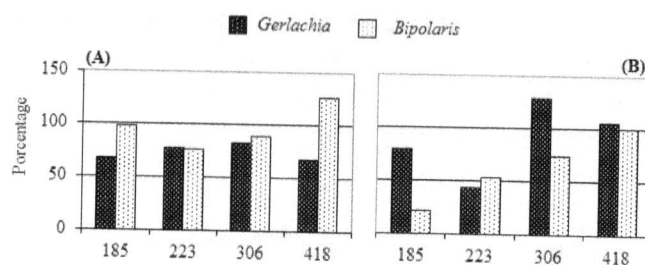

Figure 1. Relative percentage of *Gerlachia oryzae* and *Bipolaris oryzae* incidence in seed lots 375 (A) and 376 (B) naturally infested/infected and microbiolized with different biocontrollers bacteria isolates, as determined by the blotter test. Control considered as 100%.

The high incidence of pathogens resulted in lower emergence in sterile vermiculite, noting 44% of developed seedlings in the control of both lots evaluated. In non-germinated seeds in the control, a pathogen incidence of approximately 68 and 71% in lots 375 and 376, respectively, was observed, and G. oryzae represented 68% of these in both seed lots.

The effect of the biocontroller bacteria isolates, when evaluated in autoclaved vermiculite, was similar to that observed in blotter test. In general, all isolates reduced transmission of pathogens from seeds to seedlings, particularly in lot 376 (Figure 2). When considering the results for both seed lots, the most efficient biocontroller isolates in reducing this transmission, measured by the intensity of symptoms emerging from the seedlings, were DFs185 and DFs306, since these were placed in a distinct group from the control by Scott-Knott in both lots evaluated. On the other hand, the incidence of fungi on non-germinated seeds show that the biocontroller isolate DFs223 seemed most effective in reducing the incidence of B. oryzae and G. oryzae (24.3% and 27.5% respectively).

Growth promoting effects were observed for all bacterial treatments on the length of the roots, which, on average, resulted in an increase of 22 percentage points (Figure 3).

Again, the most intense effects were observed in lot 376. The most effective isolate as a whole, when considering all variables and both lots, was DFs306, resulting in an average increase of 16%.

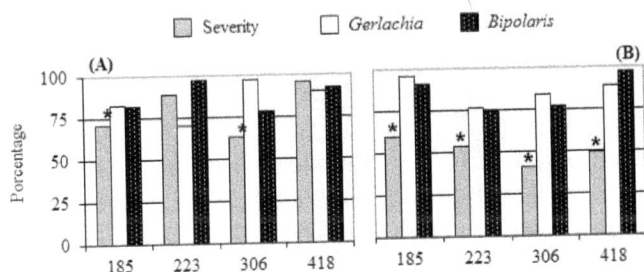

Figure 2. Relative percentage of transmission determined by the severity of symptons in plants and incidence of *Bipolaris oryzae* and *Gerlachia oryzae* in non-germinated seeds in seed lots 375 (A) and 376 (B) naturally infested/infected and microbiolized with biocontrollers bacteria isolates, determined in autoclaved vermiculite after 21 days of incubation at 23 ± 2 °C. Control considered as 100%.
*Significant values different from the control by the Scott-Knott test (α = 0.05).

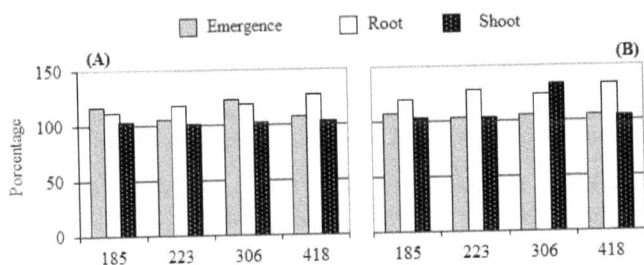

Figure 3. Growth promotion expressed as a percentage compared with the control: emergence rate, length of leaves and roots of rice seedlings originating from seeds of lots 375 (A) and 376 (B) naturally infested/infected by *Bipolaris oryzae* and *Gerlachia oryzae* and microbiolized with biocontrollers bacteria isolates, determined in autoclaved vermiculite after 21 days of incubation at 23 ± 2 °C. Control considered as 100%.

It could be verified that all the isolates have a potential to control pathogens by antibiosis, that is, they all produced at least one compound capable of inhibiting the pathogenic fungi tested, except for DFs306, which did not inhibit mycelial growth of both pathogens (Table 1).

The biocontrol of diseases transmitted by seeds to the seedlings as observed for *B. oryzae* and *G. oryzae* by the bacteria evaluated have already been reported. The microbiolization of seeds with biocontroller has also been

used to reduce transmission of pathogenic fungi such as *Fusarium oxysporum* f. sp. *ciceris* by treatment with *Bacillus subtilis* and *Trichoderma harzianum*, used individually or in combination (Herváz et al., 1998). Correa et al. (2008) showed the efficiency of this strategy when they microbiolized bean seeds with biocontrollers bacteria of *Xanthomonas axonopodis* pv. *phaseoli* to reduce the transmission of *Colletotrichum lindemuthianum* of naturally infested/infected seeds to seedlings. On the other hand, the transmission of pathogenic bacteria *Acidovorax avenae* subsp. *citruli* from the plant to the seeds, as well as from seeds to seedlings was strongly reduced by microbiolization of watermelon seeds with an *A. avenae* subsp. *avenae* or *P. fluorescens* isolate (Fessehaie and Walcott, 2005).

Table 1. Mycelial growth of pathogenic fungi provided by biocontrollers isolates determined by *in vitro* antibiosis using liquid metabolite obtained after 72 hours of cultivation at 28 °C of each bacterium individually.

	DFs185	DFs223	DFs306	DFs418
Bipolaris oryzae	+	+	-	+
Gerlachia oryzae	+	+	-	+

(+) presence of inhibition of mycelial growth; (-) absence of mycelial growth inhibition.

The growth promotion exhibited by the antagonists, especially as for the root length (Figure 3), is known in several plants species (Ahemad and Kibret, 2014), including rice (Lucas et al, 2014; Souza et al., 2013). The biostimulator effect in relation to seed physiological potential has been also reported. In this sense, the bacteria studied in this work, in the absence of pathogens and when used to microbiolize seed lots with low quality, showed positive effects, highlighting the isolate DFs185, which provided increased seed germination and seedling emergence (Soares et al., 2012).

The increment of the different variables, especially fast germination and intensive development of the root system, may result, in the first case, in an escape of main soil pathogens (Canteri et al., 1999), and, in the second case, in access to a greater volume of soil, providing better nutritional conditions and increasing tolerance to adverse weather conditions in the field (El-Abyad et al., 1993). Therefore, these bacteria are able to combine positive effects on the growth and health of seedlings, resulting in a highly interesting possibility of use.

In this study, isolates DFs185 and DFs306 stood out in the production of healthy rice seedlings (Figure 3), for both seeds lots evaluated *in vivo*, although DFs306 did not produce compounds capable of inhibiting the mycelial growth of the pathogens when confronted *in vitro*. In this sense, it is worth noting the possibility of DFs306 acting by resistance induction, since this bacterium was not able to inhibit the mycelial growth of *B. oryzae* and *G. oryzae*. On the other hand, the possibility of isolate DFs185 also acting by resistance induction is not ruled out, even knowing its *in vitro* antibiosis capacity. The possible occurrence of induction by both bacteria is based on the fact that when they were used in other studies, it was observed control of *Meloidogyne graminicola* (Ludwig et al., 2013) and *R. solani* (Ludwig and Moura, 2007), therefore exhibiting nonspecific control, besides, they caused increased activity of enzymes related to pathogenesis and induction of resistance (Ludwig and Moura, 2009).

On the other hand, the isolate DFs223 stood out as a whole, reducing both pathogens, both in blotter test and in autoclaved vermiculite (non-germinated seeds), although it did not provide the greatest reductions in all evaluations. Additionally, it provided *in vitro* inhibition halos. This bacterium, as well as DFs185, belongs to a well-studied group of biological control agents, which makes it possible to suggest that these isolates act by antibiosis for the control of these pathogens. This possibility is supported by the fact that these isolates, in addition to presenting antifungal activity against the fungi used in this study, also act by antibiosis against other rice pathogens, such as *Alternaria lunata*, *Curvularia lunata*, *R. solani* (Ludwig and Moura, 2009) and *M. graminicola* (Ludwig et al., 2013).

Another interesting aspect of the biocontrollers used in this study is that they have the ability to colonize the root system of plants of various cultivars of rice, thereby maintaining a high population level during the crop cycle (Ludwig and Moura, 2009). This feature allows the control to be effective, resulting in beneficial effects (Okubara et al., 2004) such as those presented in this study during germination and crop establishment, as well as throughout the crop development (Ludwig et al., 2009).

Finally, isolates DFs185, DFs223 and DFs306 present characteristics that allow their use in leaf scald and brown spot biological control programs. However, further studies are required in the search for the elucidation of the mechanisms of biological control/growth promotion, as well as studies of combinations of isolates with different mechanisms that allow a greater spectrum of activity and/or control amplitude.

Conclusions

All bacteria studied (DFs185, DFs223, DFs306 and DFs418) promote early seedling growth even in the presence of pathogens. Among these bacteria, DFs185 and DFs306

reduce transmission of *B. oryzae* and *G. oryzae* from seeds to seedlings.

Acknowledgments

The authors thank the brazilians National Council for Scientific and Technological Development (CNPq) for the first author's productivity scholarship; the Coordination for Improvement of Higher Education Personnel (CAPES) for the last author's postdoctoral scholarship and the remaining authors' master's scholarship; and the Foundation for Research Support of the State of Rio Grande do Sul (FAPERGS) and CAPES for funding the study.

References

AGROFIT. *Sistema de agrotóxicos fitossanitários.* http://extranet.agricultura.gov.br/agrofit_cons/principal_agrofit_cons. Access on July 7th, 2014.

AHEMAD, M.; KIBRET, M. Mechanisms and applications of plant growth promoting rhizobacteria: Current perspective. *Journal of King Saud University – Science*, v.26, p.1–20, 2014. http://dx.doi.org/10.1016/j.jksus.2013.05.001

BRASIL. Ministério da Agricultura, Pecuária e Abastecimento. *Regras para análise de sementes.* Ministério da Agricultura, Pecuária e Abastecimento. Secretaria de Defesa Agropecuária. Brasília: Mapa/ACS, 2009. 395p. http://www.agricultura.gov.br/arq_editor/file/12261_sementes_-web.pdf

CANTERI, M. G.; PRIA, M. D.; SILVA, O. C. *Principais doenças fúngicas para manejo econômico e ecológico.* Ponta Grossa: UEPG, 1999. 178p.

CELMER, A.; MADALOSSO, M.G.; DEBORTOLI, M.P.; NAVARINI L.; BALARDIN, R.S. Controle químico de doenças foliares na cultura do arroz irrigado. *Pesquisa Agropecuária Brasileira*, v.42, n.6, p.901-904, 2007. http://seer.sct.embrapa.br/index.php/pab/article/download/7645/4564

CORNÉLIO, V.M.O.; SOARES, A.A.; SOARES, P.C.; BUENO FILHO, J.S.S. Identificação de raças fisiológicas de *Pyricularia grisea* em arroz no estado de Minas Gerais. *Ciência e Agrotecnologia*, v.27, p.1016-1022, 2003. www.scielo.br/pdf/cagro/v27n5/a07v27n5.pdf

COMMARE, R.J.; NANDAKUMAR, R.; KANDAM, A.; SURESH, S.; BHARATHI, M.; RAGUCHANDER, T.; SAMIYAPPAN, R. *Pseudomonas fluorescens* based bio-formulation for the management of sheath blight disease and leaf folder insect in rice. *Crop Protection*, v.21, p.671-677, 2002. www.sciencedirect.com/science/journal/02612194/21/8.

CORREA, B.O.; MOURA, A.B.; DENARDIN, N.D.; SOARES, V.N.; SCHÄFER, J.T.; LUDWIG, J. Influência da microbiolização de sementes de feijão sobre a transmissão de *Colletotrichum lindemuthianum. Revista Brasileira de Sementes*, v.30, p.156-163, 2008. www.abrates.org.br/revista/artigos/2008/v30n2/artigo19.pdf

EL-ABYAD, M; EL-SAYED, M. A.; EL-SHANSHOURY, A. R.; EL-SABBAGH, S. H. Towards the biological control of fungal and bacterial diseases of tomato using antagonistic *Streptomyces* spp. *Plant and Soil*, v.149, p.185-195, 1993.

FESSEHAIE, A.; WALCOTT, R. R. Biological control protect watermelon blossoms and seed from infection by *Acidovorax avenae* subsp. *citrulli*. *Phytopathology*, v.95, n.4, p.413-419, 2005. www.seeds.iastate.edu/images/10Fessehaie_et_al%202005_10.pdf

FRANCO, D.F.; RIBEIRO, A.S.; NUNES, C.D.; FERREIRA, E. Fungos associados a sementes de arroz irrigado no Rio Grande do Sul. *Revista Brasileira de Agrociência*, v.7, n.3, p.235-236, 2001. www2.ufpel.edu.br/faem/agrociencia/v7n3/artigo16.pdf

HERVÁZ, A.; LANDA, B.; DATNOFF, L. E. R.; JIMÉNEZ-DÍAZ, M. Effects of commercial and indigenous microorganisms on *Fusarium* wilt development in chickpea. *Biological Control*, v.13, p.166-176, 1998. www.ias.csic.es/rmjimenez/docs/articulos/Hervas_et_al.BiolControl...

KRISHHNAMURTHY, K.; GNANMANICKAM, S.S. Biological control of rice blast by *Pseudomonas fluorescens* strain Pf7-14: evaluation of a marker gene and formulations. *Biological Control*, v.13, p.158–165,1998.

LUCAS, J.A.; GARCÍA-CRISTOBAL, J.; BONILLA, A.; RAMOS, B.; GUTIERREZ-MAÑERO, J. Beneficial rhizobacteria from rice rhizosphere confers high protection against biotic and abiotic stress inducing systemic resistance in rice seedlings. *Plant Physiology and Biochemistry*, v.82, p.44-53, 2014. http://dx.doi.org/10.1094/PHYTO-98-6-0666

LUDWIG, J.; MOURA, A.B. Controle biológico da queima-das-bainhas em arroz pela microbiolização de sementes com bactérias antagonistas. *Fitopatologia Brasileira*, v.32, n.5, p.381-386, 2007. http://www.scielo.br/pdf/fb/v32n5/v32n5a02.pdf

LUDWIG, J.; MOURA, A.B. Controle Biológico de *Bipolaris orizae* no arroz irrigado. In: BETTIOL, W.; MORANDI, M.A.B. (Eds.). Biocontrole de doenças de plantas: usos e perspectivas. Jaguariuna: Embrapa Meio Ambiente, 2009. p.317-330. https://www.embrapa.br/busca-de-publicacoes/-/publicacao/579954/biocontrole-de-doencas-de-plantas-uso-e-perspectivas

LUDWIG, J.; MOURA, A.B.; SANTOS, A.S.; RIBEIRO, A.S. Biocontrole da mancha parda e da escaldadura em arroz irrigado, pela microbiolização de sementes. *Tropical Plant Pathology*, v.34, n.5, p.322-328, 2009. http://www.scielo.br/pdf/fb/v32n5/v32n5a02.pdf

LUDWIG, J.; MOURA, J.; GOMES, C.B. Potencial da microbiolização de sementes de arroz com rizobactérias para o biocontrole do nematoide das galhas. *Tropical Plant Pathology*, v.38, n.3, p.264-268, 2013. http://dx.doi.org/10.1590/S1982-56762013005000007

MALAVOLTA, V.M.A.; PARISI, J.J.D.; TAKADA, H.M.; MARTINS, M.C. Efeito de diferentes níveis de infecção por *Bipolaris oryzae* em sementes de arroz sobre aspectos fisiológicos, transmissão do patógeno às plântulas e produtividade. *Summa Phytopathologica*, v.28, p.336-340, 2002.

NUNES, C.D.; RIBEIRO, A.S.; TERRES, A.L. Principais doenças do arroz irrigado e seu controle. In: GOMES, A.S.; MAGALHÃES JUNIOR, A.M. (Eds.) Arroz irrigado no sul do Brasil. Brasilia: Embrapa Informação Tecnológica, 2004. p.579-633.

OKUBARA, P. A.; KORNOELY, J. P.; LANDA, B.B. Rhizosphere colonization of hexaploid wheat by *Pseudomonas fluorescens* strains Q8r1-96 and Q2-87 is cultivar-variable and associated with changes in gross root morphology. *Biological Control*, v.30, p.392–403, 2004. naldc.nal.usda.gov/download/9824/PDF

SOARES, V.N.; TILLMANN, M.A.A.; MOURA, A.B.; ZANATTA, Z.G.C.N. Physiological potential of rice seeds treated with rhizobacteria or the insecticide thiamethoxam. *Revista Brasileira de Sementes*, v.34, n.4, p.563 – 572, 2012. www.scielo.br/pdf/rbs/v34n4/06.pdf

SOUZA, R.; BENEDUZI, A.; AMBROSINI, A.; BESCHOREN DA COSTA, P.; MEYER, J., VARGAS, L. K.; SCHOENFELD, R.; PASSAGLIA, L.M. P. The effect of plant growth-promoting rhizobacteria on the growth of rice (*Oryza sativa* L.) cropped in southern Brazilian fields. *Plant Soil*, v.366, p.585–603, 2013. http://link.springer.com/article/10.1007/s11104-012-1430-1.

SOUZA JUNIOR, I.T.; MOURA, A.B.; SCHAFER, J.T.; CORRÊA, B.O.; GOMES, C.B. Biocontrole da queima-das-bainhas e do nematoide-das-galhas e promoção de crescimento de plantas de arroz por rizobactérias. *Pesquisa Agropecuária Brasileira*, v.45, n.11, p. 1259-1267, 2010. http://dx.doi.org/10.1590/S0100-204X2010001100005

WIWATTANAPATAPEE, R.; PENGOO, A.; KANJANAMANEESATHIAN M.; MATCHAVANICH, W.; NILRATANA, L.; JANTHARANGSRI, A. Floating pellets containing bacterial antagonist for control sheath blight of rice: formulations, viability and bacterial release studies. *Journal of Controlled Release*, v.95, p.455-462, 2004. http://www.sciencedirect.com/science/journal/01683659/95/3.

16

Dormancy and enzymatic activity of rice cultivars seeds stored in different environments

Elizabeth Rosemeire Marques[1*], Roberto Fontes Araújo[2], Eduardo Fontes Araújo[3],
Sebastião Martins Filho[4], Plínio César Soares[2], Eduardo Gomes Mendonça[5]

ABSTRACT – The objective of this study was to assess the dormancy and the enzymatic activity of seeds of rice cultivars during storage in different environments. After harvesting, the seeds of two rice cultivars (BRS Ourominas and BRSMG Caravera) were dried in the sun, to reach a moisture content around 13%. Then, they were packed in paper and stored in four environments: 5 ± 2 °C / 70 ± 5% RH, 12 ± 2 °C / 70 ± 5% RH, 18 ± 2 °C / 65 ± 5% RH and in a natural condition. Germination and enzymatic activity were assessed at the beginning and at 3, 6, 9 and 12 months of storage. The experiment was conducted in split subplots in a completely randomized design with three replications. The dormancy of seeds stored in the natural environment was exceeded in a shorter time than the dormancy of seeds stored in cold. Catalase and ascorbate peroxidase activity increased during the storage period, the most obviously in storage in natural environment to cultivate BRSMG Caravera. The activity of the enzyme α-amylase decreased during the storage period.

Index terms: enzymes, germination, *Oryza sativa* L.

Dormência e atividade enzimática de sementes de cultivares de arroz armazenadas em diferentes ambientes

RESUMO – O objetivo deste trabalho foi avaliar a dormência e a atividade enzimática de sementes de cultivares de arroz durante o armazenamento em diferentes ambientes. Após a colheita, as sementes das cultivares de arroz BRS Ourominas (várzea) e BRSMG Caravera (terras altas) foram secadas ao sol, até atingirem teor de água em torno de 13%. Em seguida, foram acondicionadas em embalagem de papel e armazenadas em quatro ambientes: 5 ± 2 °C/70 ± 5% UR, 12 ± 2 °C/70 ± 5% UR, 18 ± 2 °C/65 ± 5% UR e em condição natural. A germinação e a atividade enzimática foram avaliadas no início e aos 3, 6, 9 e 12 meses de armazenamento. O experimento foi realizado no esquema de parcelas subdivididas no delineamento inteiramente casualizado com três repetições. A dormência das sementes armazenadas em ambiente natural foi superada em menor tempo do que as sementes armazenadas em câmara fria. Houve aumento da atividade das enzimas catalase e ascorbato peroxidase, com maior evidência no armazenamento em ambiente natural, para a cultivar BRSMG Caravera. A atividade da enzima α-amilase diminuiu durante o período de armazenamento.

Termos para indexação: enzimas, germinação, *Oryza sativa* L.

Introduction

Seed is one of the most important inputs in modern agriculture and, among the various stages through which it passes after harvest, storage is a compulsory step of a production program, assuming an important role, especially in Brazil, due to the tropical and subtropical climate conditions.

In the storage conditions, temperature and relative humidity are key factors in maintaining seed quality (Tonin and Perez, 2006), influencing the speed of the biochemical processes and indirectly interfering in water content of seeds.

The process of deterioration in seeds comprises a sequence of biochemical and physiological changes initiated immediately after physiological maturity, which lead to reduced vigor, culminating in the loss of germination capacity.

Changes in physiology of the seeds are indirectly related

[1]Departamento de Ciências Florestais, Universidade Federal de Lavras, Caixa Postal 3037, 37200-000 – Lavras, MG, Brasil.
[2]Empresa de Pesquisa Agropecuária de Minas Gerais, Caixa Postal 216, 36571-000 – Viçosa, MG, Brasil.
[3]Departamento de Fitotecnia, Universidade Federal de Viçosa, 36570-000 – Viçosa, MG, Brasil.

[4]Departamento de Estatística, Universidade Federal de Viçosa, 36570-000 – Viçosa, MG, Brasil.
[5]Departamento de Bioquímica Agrícola, Universidade Federal de Viçosa, 36570-000 – Viçosa, MG, Brasil.
*Corresponding author <bethagro@yahoo.com.br>

to the integrity of their cell membranes (Carvalho et al., 2009), which, in turn, depend on the nature of the enzymes and structural proteins of each species. The enzymes have been used in the assessment of physiological and biochemical changes in stored seeds (Santos et al., 2004).

Under normal physiological conditions, the oxidative stress produced is fought by a complex antioxidant defense system, related to increased production and activation of enzymes, that catalyzes the conversion of the hydrogen peroxide (H_2O_2) into water (H_2O), protecting cells from oxidative damage (Scandalios, 2005).

During the process of seed deterioration, there is a decrease in enzyme activity by its progressive inactivation or reduction and stoppage of its synthesis (Marcos-Filho 2005). The reduction of catalase activity in seeds reduces respiratory capacity, decreasing the supply of energy (ATP) and similar for the germination (Demirkaya et al., 2010).

In seeds of *Triticum aestivum*, reduction activity of catalase enzymes and superoxide dismutase was observed during artificial aging, resulting in loss of viability (Lehner et al., 2008). For *Medicago sativa* L., reduction of seed vigor during artificial aging was associated with decreased activity of catalase, superoxide dismutase and peroxidase (Cakmak et al., 2010).

In the quality control program of companies producing seeds, monitoring the physiological potential of dormant rice seeds is a fundamental aspect. The behavior is differentiated according to the cultivar and the cultivation system, and the dormancy is more intense in the floodplain system (Roberts, 1963).

Among certain inhibitor chemicals, which can be detected on the coat (glumes) of the seeds (Ketring, 1997), the phenolic compounds are highlighted, besides the presence of oxidizing agents, which promote oxidation, with the decrease in the concentration of O_2. Accordingly, the association of the phenolic compounds with a high respiratory activity in seed coat tissues limits the O_2 availability to the embryo (Bewley and Black, 1994). Therefore, during storage of seeds for a certain period, there is the possibility of a slow and gradual diffusion of O_2 into its interior, causing a gradual reduction in the amount of germination inhibitors, thereby favoring breaking dormancy and hence its germination (Olatoye and Hall, 1972).

In this context, the objective of this work was to assess the enzymatic activity and dormancy of seeds of rice cultivars of floodplains and uplands during storage under different conditions.

Material and Methods

The study was developed at the Department of Plant Science at Universidade Federal de Viçosa, Viçosa, MG, with seeds of the rice cultivars BRS Ourominas (floodplains)

and BRSMG Caravera (uplands), which were produced by Empresa de Pesquisa Agropecuária de Minas Gerais (EPAMIG), in Lambari, MG, in harvest 2009/2010.

Immediately after the harvest, the seeds were threshed and dried in the sun until they reached a water content of around 13%.

The seeds were stored in paper packing and stored in four environments: cold chamber 1 (5 ± 2 °C, $70 \pm 5\%$ of relative humidity), cold chamber 2 (12 ± 2 °C, $70 \pm 5\%$ of relative humidity), cold room (18 ± 2 °C, $65 \pm 5\%$ of relative humidity) and in a natural environment (without control of temperature and relative humidity). During storage, the monitoring of temperature and relative humidity of the natural environment was done (Figure 1).

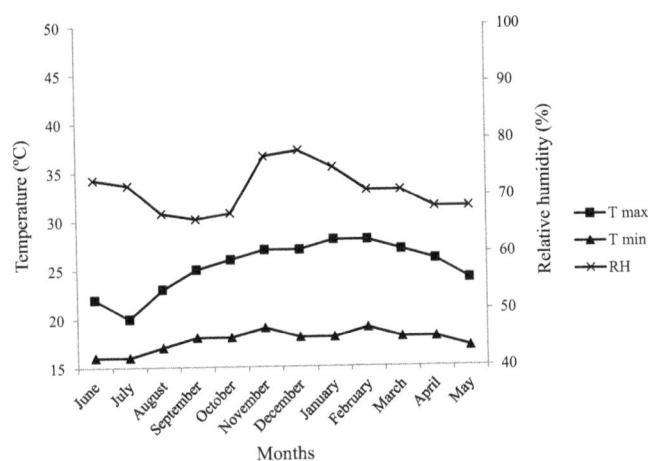

Figure 1. Monthly mean maximum (T max) and minimum temperatures (T min) and relative humidity (RH) in a natural environment, from June 2010 (beginning of the experiment) to May 2011 (end of the experiment).

Before storage and after 3, 6, 9 and 12 months the seeds were assessed according to the following methods:

Water content – it was determined by the oven method at 105 ± 3 °C for 24 hours, using two samples of each treatment according to the Rules for Seed Analysis (Brasil, 2009). The results were expressed as percentages (wet basis).

Germination –four samples of 50 seeds per replication were used, treated with the fungicide Captan at a dose of 2.4 g of product per kg of seed. Then they were sown in (three sheets of) germitest paper moistened with distilled water in quantities equivalent to 2.5 times the weight of the dry paper. The rolls made were kept in a germinator at at 25 °C. Assessments were performed at 5 and 14 days after sowing, and the results were expressed as percentage of normal seedlings (Brasil, 2009). To do so, at the end of the tests, the seeds that remained dormant had their viability assessed by the tetrazolium test, according to criteria established by the Rules for Seed Testing (Brasil, 2009), to prove their dormancy.

Enzyme assays: for the extraction of catalase and ascorbate peroxidase, were used 2 mL of the phosphate buffer 50 mM, pH 7.8 and 0,02 g of Polyvinylpolypyrrolidone (PVPP), and for α-amylase, were used 2 mL of the sodium phosphate buffer 0.02 M, pH 6.9 which were added to 0.2 g of crushed seed ice using mortar and pestle. Then the extract was centrifuged at 17,000 xg for 30 minutes at 4 °C. The supernatant was used for determination of enzyme activity and quantification of proteins. *Catalase activity* (CAT): enzyme activity was determined by assay containing 150 μL of crude enzyme extract and 1350 μL of a reaction medium consisting of: 850 μL of phosphate buffer 50 mM, pH 7.8 and 500 μL of H_2O_2 0,97 M. The decrease in absorbance at 240 nm at a temperature of 25 °C was measured for 2 minutes of reaction, and the CAT activity was determined using the slope of the line after the start of the reaction. *Ascorbate peroxidase activity* (APX): enzyme activity was determined by assay containing 100 μL of crude enzyme extract and 1400 μL of a reaction medium consisting of: 700 μL of phosphate buffer 50 mM, pH 7.6, 400 μL of ascorbic acid 0.25 mM containing EDTA 0.1 mM, and 300 μL of H_2O_2 0,3 mM. The decrease in absorbance at 290 nm at a temperature of 25 °C was measured during 2 minutes, and the APX activity was determined based in the inclination of the line after the reaction beginning.

Activity of α-amylase: enzyme activity was determined by the formation of reducing sugars from starch using the DNS reagent, by the test containing in glass tubes 50 μL of the enzyme extract and 450 μL of starch 1%; after a five-minute reaction 500 μL of the solution of the dinitrosalicylic acid were added – 3.5; then the tubes were placed to boil for five minutes. After cooling, 3 mL of deionized water were added to all. The reading was performed at a wavelength of 540 nm.

Statistical design – the experiment was conducted using a completely randomized design with three replications in a split subplot design. The environment factor was applied to the plots, cultivars in the subplots and storage time in the subsubplots. In the analysis of variance, regardless of significance, deploying the environment x cultivar x storage interaction was chosen. The levels of environment factors and cultivars were compared by Tukey test (5%) and the storage factor was subjected to a regression analysis. For data analysis, the Sistema de Análises Estatísticas e Genéticas (Statistical and Genetic Analysis System) – SAEG was used (SAEG, 2007).

Results and Discussion

In the characterization of germination and seed dormancy of rice cultivars recently harvested it was noted, by the tetrazolium test, a higher percentage of dormancy in the cultivar of floodplains BRS Ourominas (17%) and only 5% of dormant seeds in cultivar BRSMG Caravera (Table 1). Thus, it is possible to consider that after breaking dormancy, germination of the lot of both cultivars will be above 80%, which is a minimum standard for marketing rice seeds. The remaining seeds that did not germinate in the germination test were dead.

Table 1. Results of germination and seed dormancy of rice cultivars before storage.

Cultivars	Germination (%)	Dormancy (%)
BRS Ourominas	69 B	17 A
BRSMG Caravera	75 A	5 B

Means followed by the same letter do not differ significantly at 5% probability by Tukey test.

The contents of the initial water of the seeds were 13.6% (cultivar BRS Ourominas) and 13.3% (cultivar BRSMG Caravera), practically similar (Table 2). During storage, the variation in water content was less than 1%, suggesting that for different storage conditions the initial water content of the seeds of cultivars was close to the moisture equilibrium. The water content of the seeds was similar among cultivars, environment and storage periods. Consequently, only the effect of storage temperatures on physiological quality and enzyme activity of the seeds during storage was considered.

Analyzing the results of germination (Figure 2) on seeds stored at 5 °C, 12 °C, 18 °C and in a natural environment, it was found that the maximum germination for cultivar BRS Ourominas was 85% at 6.1 months, 83% at 6.4 months, 85% at 6.9 months and 80% at 3.5 months, respectively.

In general, the sharpest increase in seed germination of cultivar BRS Ourominas during the first six months of storage under different conditions is probably due to the fact that this cultivar has a higher percentage of dormant seeds (Table 1). The results suggest that the dormancy of cultivar BRS Ourominas has been overcome between 6 and 7 months of storage in seeds stored at controlled temperatures (5, 12 and 18 °C) (Figure 2). For the seeds of this cultivar stored under natural conditions, breaking dormancy has occurred early, at 3.5 months of storage (Figure 2); probably the alternating temperatures and the storage average maximum temperatures near 25 °C (Figure 1) have accelerated the process of breaking dormancy.

Dormancy and enzymatic activity of rice cultivars seeds stored in different environments

Table 2. Water content (%) of seeds of rice cultivars during storage under different environmental conditions.

Storage time (months)	Environments	Cultivars	
		BRS Ourominas	BRSMG Caravera
0	5 °C / 70% RU	13.6 aA[1]	13.3 aB
	12 °C / 70% RU	13.6 aA	13.3 aB
	18 °C / 65% RU	13.6 aA	13.0 bB
	Natural	13.6 aA	13.3 aB
3	5 °C / 70% RU	13.0 bA	12.8 cB
	12 °C / 70% RU	12.9 cA	12.9 cA
	18 °C / 65% RU	12.9 cB	13.1 bA
	Natural	13.5 aA	13.3 aB
6	5 °C / 70% RU	12.9 bB	13.6 aA
	12 °C / 70% RU	13.5 aB	13.7 aA
	18 °C / 65% RU	13.5 aA	12.6 bB
	Natural	13.5 aB	13.6 aA
9	5 °C / 70% RU	14.0 bB	14.6 aA
	12 °C / 70% RU	14.0 bB	14.2 cA
	18 °C / 65% RU	13.6 cB	13.8 dA
	Natural	14.6 aA	14.4 bB
12	5 °C / 70% RU	13.6 bB	14.2 aA
	12 °C / 70% RU	13.6 bB	13.8 cA
	18 °C / 65% RU	13.3 cB	13.4 dA
	Natural	14.1 aA	14.0 bB

[1]Means followed by the same lowercase letter in the column and uppercase letter in the line do not differ significantly at 5 % probability by Tukey test.

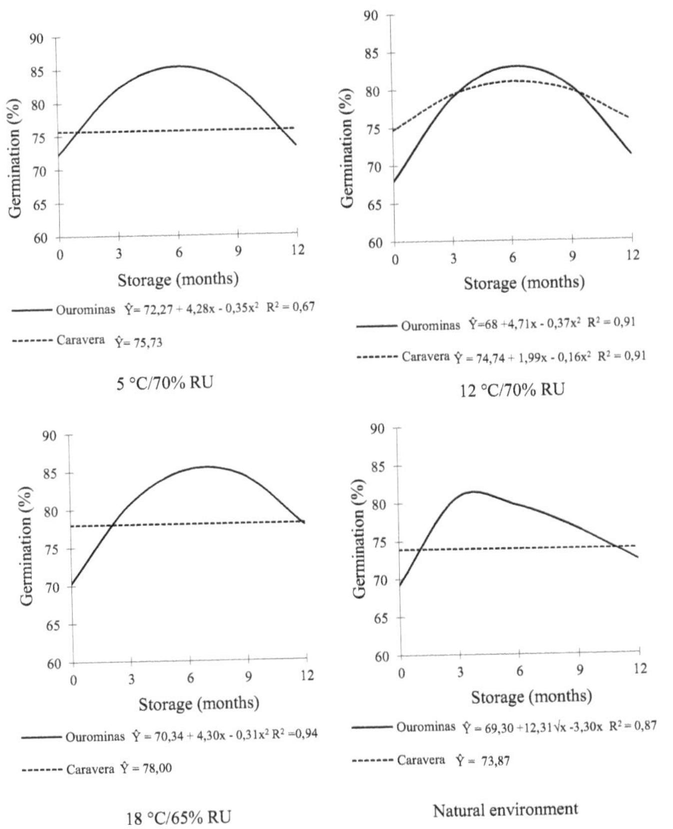

Figure 2. Germination of seeds of rice cultivars during storage in different environments.

Vieira et al. (2008) also found that rice seeds of irrigated cultivar Rio Grande stored in conventional environment in Lavras, MG, with a temperature between 15 and 28 °C and relative humidity ranging between 60 and 72%, have broken dormancy more rapidly during storage than stored seeds in a cold and dry room. These results show that seed dormancy decreases over the storage period, especially with increasing temperature.

The seeds of cultivar BRSMG Caravera have not had their germination reduced by storage at 5 °C, 18 °C and natural environment. At 12 °C, germination increased to 6.2 months of storage, suggesting that breaking dormancy (Figure 2) has occurred.

Only the seeds of cultivar BRS Ourominas, regardless of the environment, have maintained germination above the minimum required for commercialization until six months of storage.

In situations where there has been breaking dormancy, regardless of cultivar and the storage environment after reaching the maximum germination, a process of natural seed deterioration with a consequent reduction in germination vigor has begun.

Marques et al. (2014) have observed similar results studying three rice cultivars stored in four storage environments, namely, loss of germination vigor after reaching maximum germination and vigor. Macedo et al. (1999) when assessing rice seeds of cultivar IAC 165, stored for 12 months in Campinas, SP, with a relative humidity of 72% and an average temperature of 21 °C, have observed a gradual decrease in germination from the eighth month.

Theoretically, the deterioration starts at physiological maturity; however, it is most frequently detected during storage. Once the seed has reached a maximum physiological quality, namely, the point of maximum germination, vigor, and dry matter accumulation, a continuous and irreversible process of deterioration begins. The process of deterioration in seeds comprises a sequence of biochemical and physiological changes initiated immediately after physiological maturity, which lead to reduced vigor, culminating in the loss of germination capacity. This process cannot be avoided, but can be slowed down when the storage is done under favorable conditions, especially of temperature and relative humidity.

In this work, as the relative humidity of the storage environment was similar, reflecting in seeds with water content with little variation throughout the storage period (Table 1), the difference in seed germination was attributed to differences in temperatures. In general, seeds stored in room temperatures of 5, 12 and 18 °C have shown, at the end of storage, greater germination than the seeds stored in a natural environment. In this environment, for eight months of storage, temperatures above 25 °C could be seen, which certainly contributed to a further deterioration of seeds. The temperature significantly contributes to the conservation of seeds, directly affecting the speed of the biochemical processes, which also trigger other processes such as increasing the enzymatic activity (hydrolytic enzymes) and free fatty acids (Schwember and Bradford, 2010).

As to catalase enzyme activity (Figure 3), in the environment at 5 °C it was observed, in the seeds of cultivar BRS Ourominas, a linear increase in catalase activity during the storage period and, for seeds of cultivar BRSMG Caravera, an increased activity during the storage period corresponded to the square root model. At 12 °C it was found in the seeds of cultivar BRS Ourominas a gradual increase in the enzyme activity up to approximately nine months, followed by a larger increase until 12 months. In the seeds of cultivar BRSMG Caravera, the sharpest increase in activity began close to three months of storage and additions up to 12 months could be seen. At 18 °C, for both cultivars, the increase was gradual until about nine months, with highest increases at 12 months. In the natural environment, for the seeds of cultivar BRS Ourominas, enzyme activity remained at low levels until approximately six months, followed by a small increase up to 12 months of storage. For the seeds of cultivar BRSMG Caravera, there was a greater increase from approximately three months with increase up to 12 months of storage. Clearly, the longer the storage time, the greater the

catalase activity in seeds of rice cultivars for all treatments.

The increase in catalase activity has been related to the production of H_2O_2 in conditions of oxidative stress (Nasr et al., 2011). This result can be compared with the results of germination (Figure 2) during the storage period of rice seeds, namely, an advancement in the deteriorating process, with loss of germination, and the enzyme thereby acting as a protection mechanism. To prevent irreversible cell damage, enzymes of the antioxidant system get into action when levels of ROS (reactive oxygen species), preventing their formation and promoting the removal of reactive forms produced, exceed certain levels, according to each species (Moller et al., 2007 and Scandalios, 2005).

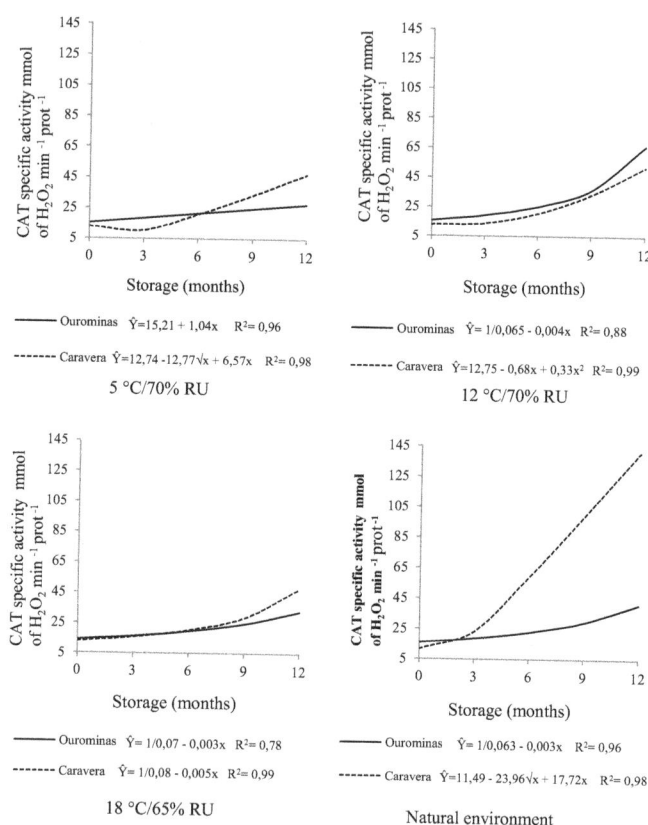

Figure 3. Specific activity of the catalase enzyme (CAT) (mmol of H_2O_2 min $^{-1}$μg of prot^{-1}) in rice cultivar seeds during storage in different environments.

Nonetheless, Yin et al. (2014) have observed in artificially aged rice seeds of *Oryza sativa* L. cv. (Wanhua number 11) a decreased catalase activity, resulting in a drop in the percentage of germination from 99 to 2%. Similar results were found by Chauhan et al. (2011), where the decrease of the activities of catalase and peroxidase in *Triticum aestivum* seeds during aging have resulted in lower viability and vigor.

The ascorbate peroxidase enzyme activity (Figure 4) at

the environment of 5 °C in seeds of two cultivars has linearly increased during storage. In the environment of 12 °C, in the seeds of cultivar BRSMG Caravera, it was observed that the activity of the enzyme remained at low levels during storage, and in the seeds of cultivar BRS Ourominas there was no difference during storage. At the environment of 18 °C in seeds of cultivar BRS Ourominas an increased activity during storage was observed, and in the seeds of cultivar Caravera there was no difference during storage. In the natural environment, for seeds of cultivar Ourominas, activity remained at low levels until approximately six months, followed by a slight increase up to 12 months. In cultivar BRSMG Caravera, the increase was very sharp from the three months of storage. These results demonstrate that an increased oxidative stress has occurred in this environment; therefore, there was a higher activity of catalase enzymes and ascorbate peroxidase.

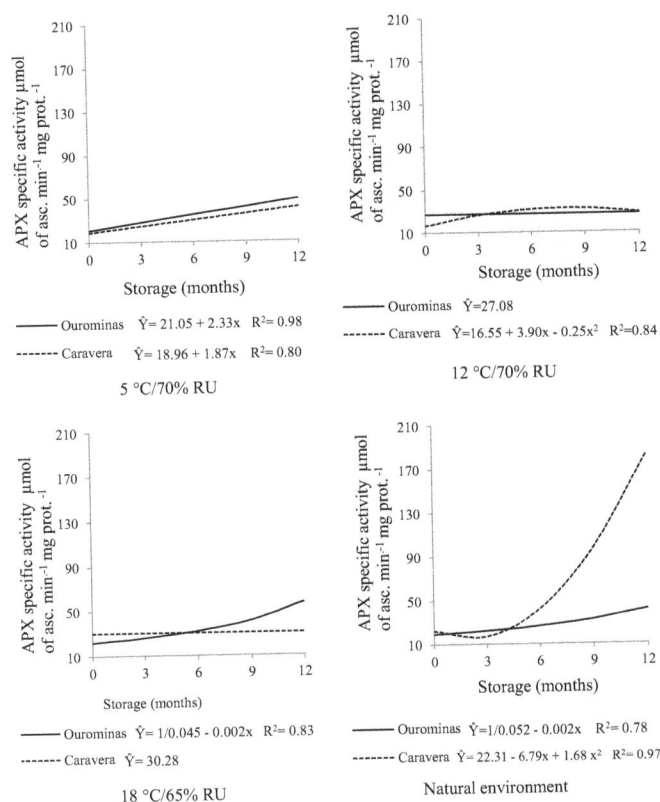

Figure 4. Specific activity of the enzyme ascorbate peroxidase (APX) (μmol de asc min^{-1} μg of prot^{-1}) in seeds of rice cultivars during storage in different environments.

Similar studies have been reported relating the increase of APX activity in different stress conditions on sunflower seeds (Carneiro et al., 2011) and bean seeds (Deuner et al., 2011).

However, Yin et al. (2014) have observed in artificially aged rice seeds of *Oryza sativa* L. cv. (Wanhua number 11), a reduction of ascorbate peroxidase activity, resulting in loss of seed viability.

The activity of the enzyme ascorbate peroxidase during storage in both cultivars, has shown a lower level than the route of catalase. The performance of the routes, involving the catalase and ascorbate peroxidase, is simultaneous, which can result in low activity for both, but performing an efficient detoxification. With the advancement in the process of deterioration during storage, an accumulation of toxic levels of H_2O_2 can occur in plant tissues. Different types of stresses may result in the combined action of CAT and APX to protect cells from the action of peroxides. However, these two enzymes have different affinities for H_2O_2. The APX has a high affinity (range μM), being responsible for the fine modulation in levels of EROs (Mittler, 2002).

However, it is necessary to emphasize that the in vitro enzymatic activities are limited because the activities in the tissue in vivo may be different because the plant is in other conditions. The in vitro antioxidant capacity assays are important to check whether there is or not a correlation between the antioxidants and the levels of oxidative stress (Huang et al., 2005).

By the α-amylase enzyme activity (Figure 5), it was observed, in the four environments, that the two cultivars have shown the same behavior, namely, a drastic reduction close to six months of storage, showing an exponential model. A small percentage of dormancy observed in the seeds of cultivars (Table 1), 17% in the seeds of BRS Ourominas and 5% in BRSMG Caravera, may be the reason for the high activity of α-amylase in the beginning of the storage period.

Livesley and Bray (1991); Vieira et al. (2008) have reported that in seeds with high dormancy, the activity of α-amylase has been little. Shaw and Ou-Lee, cited by Das and Sen-Mandi (1992), have reported that the α-amylase activity is essential for the germination of the rice seeds.

During the storage period, there was a reduced activity of α-amylase in the different environments. This result can be compared with the results of decrease of germination (Figure 2), namely, an advancement in the deteriorating process and a reduced activity of this enzyme. Decline in α-amylase activity has been reported in cereal seeds after aging. Ganguli and Sen-Mandi (1993) have observed that the scutellar amylase activity in wheat seeds was reduced in aged seeds. Livesley and Bray (1991) have observed that the α-amylase enzyme was synthesized at rates reduced by the aleurone layer of aged wheat seeds. According to the authors, the process of deterioration can interfere with enzymes in the aleurone layer during aging. Such changes may influence the production of α-amylase which, in turn, may interfere in germination.

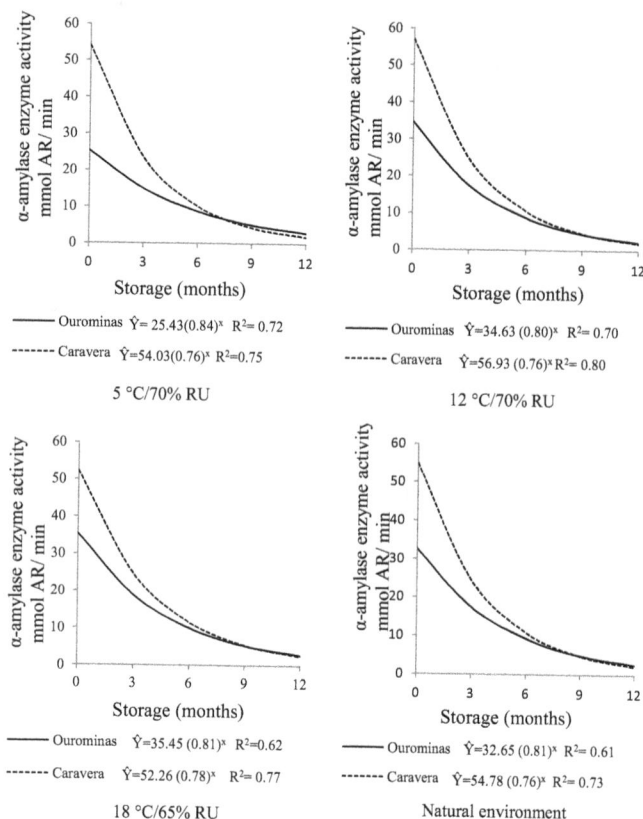

Figure 5. Activity of α-amylase enzyme (mmol of reducing sugar/min) in seeds of rice cultivars during storage in different environments.

Conclusions

The dormancy of seeds stored in the natural environment was surpassed by the shorter storage of seeds in cold room;

There was an increased activity of catalase and ascorbate peroxidase, most obviously in the seeds of cultivar BRSMG Caravera stored in a natural environment;

The activity of α-amylase enzyme has decreased during the storage period.

Acknowledgments

To CNPq for the doctoral scholarship to the first author and to FAPEMIG, for the financial support for the research project.

References

BEWLEY, J.D.; BLACK, M. *Seeds*: physiology of development and germination. New York: Plenum Press, 1994. 445p.

BRASIL. Ministério da Agricultura, Pecuária e Abastecimento. *Regras para análise de sementes*. Ministério da Agricultura, Pecuária e Abastecimento. Secretaria de Defesa Agropecuária. Brasília: MAPA/ACS, 2009. 395p.http://www.agricultura.gov.br/arq_editor/file/2946_regras_analise__sementes.pdf

CAKMAK, T.; ATICI, O.; AGAR, G.; SUNAR, S. Natural aging-related biochemical changes in alfafa (*Medicago sativa* L.) seeds stored for 42 years. *International Research Journal of Plant Science*, v.1, n.1, p.1-6, 2010.

CARNEIRO, M.M.L.C.; DEUNER, S.; OLIVEIRA, P.V.; TEIXEIRA, S.B.; SOUSA, C.P.; BACARIN, M.A.; MORAES, D.M. Atividade antioxidante e viabilidade de sementes de girassol após estresse hídrico e salino. *Revista Brasileira de Sementes*, v.33, n.4, p.754-763, 2011.www.scielo.br/pdf/rbs/v33n4/17.pdf

CARVALHO, L.F.; SEDIYAMA, C.S.; REIS, M.S.; DIAS, D.C.F.S.; MOREIRA, M.A. Influência da temperatura de embebição da semente de soja no teste de condutividade elétrica para avaliação da qualidade fisiológica. *Revista Brasileira de Sementes*, v.31, n.1, p.9-17, 2009.www.scielo.br/pdf/rbs/v31n1/a01v31n1.pdf

CHAUHAN, D.S.; DESWAL, D.P.; DAHIYA, D.P.; PUNIA, R.C. Change in storage enzymes activities in natural and accelerated aged seed of wheat (*Triticumaestivum*). *Indian Journal of Agricultural Sciences*, v.81, n.11, p.1037-1040, 2011.

DAS, G.; SEN-MANDI, S. Scutellar amylase activity in naturally aged accelerated aged wheat seeds. *Annals of Botany*, v.69, n.6, p.497-501, 1992.

DEMIRKAYA, M.; DIETZ, K.J.; SIVRITEPE, H.O. Changes in antioxidant enzymes during aging of onion seeds. *Notulae Botanicae Horti Agrobotanici*, v.38, n.1, p.49-52, 2010. <http://notulaebotanicae.ro/nbha/article/viewPDFInterstitial/4575/4417>

DEUNER, C.; MAIA, M.S.; DEUNER, S.; ALMEIDA, A.S.; MENEGHELLO, G.E. Viabilidade e atividade antioxidante de sementes de genótipos de feijão-miúdo submetidos ao estresse salino. *Revista Brasileira de Sementes*, v.33, n.4, p.711-720, 2011. www.scielo.br/pdf/rbs/v33n4/13.pdf

GANGULI, S.; SEN-MANDI, S. Effects of aging on amylase activity and scutellar cell structure during imbibition in wheat seed. *Annals of Botany*, v.71, n.5, p.411-416, 1993.

HUANG, D.; OU, B.; PRIOR, R.L. The chemistry behind antioxidant capacity assays. *Journal of Agricultural and Food Chemistry*, v.53, n.6, p.1841-1856, 2005.http://pubs.acs.org/doi/abs/10.1021/jf030723c

KETRING, A.L. Germination inhibitors. *Seed Science and Technology*, v.1, n.2, p.305-324, 1997.

LEHNER, A.; MAMADOU, N.; POELS, P.; CÔME, D.; BAILLY, C.; CORBINEAU, F. Changes in soluble carbohydrates, lipid peroxidation and antioxidant enzyme activities in the embryo during aging in wheat grains. *Journal of Cereal Science*, v.47, n.3, p.555-565, 2008. http://www.sciencedirect.com/science/article/pii/S0733521007001312

LIVESLEY, M.A.; BRAY, C.M. The effect of aging upon α-amylase production and protein synthesis by wheat aleurone layers. *Annals of Botany*, v.68, n.1, p.69-73, 1991.

MACEDO, E.C.; GROTH, D.; SOAVE, J. Influência da embalagem e do armazenamento na qualidade fisiológica de sementes de arroz. *Revista Brasileira de Sementes*, v.21, n.1, p.67-75, 1999. www.abrates.org.br/revista/artigos/1999/v21n1/artigo10.pdf

MARCOS-FILHO, J. *Fisiologia de sementes de plantas cultivadas*. Piracicaba: FEALQ, 2005. 495p.

MARQUES, E.R.; ARAÚJO, E.F.; ARAÚJO, R.F.; FILHO, S.M.; SOARES, P.C. Seed quality of rice cultivars stored in different environments. *Journal of Seed Science*, v.36, n.1, p.32-39, 2014.www.readcube.com/articles/10.1590/S2317-15372014000100004

MITTLER, R. Oxidative stress, antioxidants and stress tolerance. *Trends in Plant Science*, v.7, n.9, p.405-410, 2002.

MOLLER, I.M.; JENSEN, P.E.; HANSSON, A. Oxidative modifications to cellular components in plants. *Annual Review of Plant Biology*, v.58, p.459-481, 2007.http://www.annualreviews.org/doi/abs/10.1146/annurev.arplant.58.032806.103946

NASR, N.; CARAPETIAN, J.; HEIDARI, R.; ASRI REZAEI, S.; ABBASPOUR, N.; DARVISHZADEH, R.; GHEZELBASH, F. The effect of aluminum on enzyme activities in two wheat cultivars. *African Journal of Biotechnology*, v.10, n.17, p.3354-3364, 2011. http://www.ajol.info/index.php/ajb/article/viewFile/93401/82812

OLATOYE, S.T.; HALL, M.A. Interaction of ethylene and light on dormant weed seeds. In: HEYDECKER, W. (Ed.). *Seed Ecology*, Norwich: Pennsylvania State University, 1972. p.233-249.

ROBERTS, E.H. An investigation of inter varietal differences in dormancy and viability of rice seed. *Annals of Botany*, v.27, n.2, p.365-369, 1963.

SAEG – *Sistema para Análises Estatísticas*, Versão 9.1: Fundação Arthur Bernardes – UFV – Viçosa, 2007.www.ufv.br/saeg/download.htm

SANTOS, C.M.R.; MENEZES, N.L.; VILLELA, F.A. Alterações fisiológicas e bioquímicas em sementes de feijão envelhecidas artificialmente. *Revista Brasileira de Sementes*, v.26, n.1, p.110-119, 2004. www.scielo.br/scielo.php?script=sci_arttext&pid=S0101

SCANDALIOS, J.G. Oxidative stress: molecular perception and transduction of signals triggering antioxidant gene defenses. *Brazilian Journal of Medical and Biological Research*, v.38, n7, p. 995 -1014, 2005. http://www.scielo.br/scielo.php?pid=S0100-879X2005000700003&script=sci_arttext&tlng=pt

SCHWEMBER, A.; BRADFORD, K.J. Quantitative trait loci associated with longevity of lettuce seeds under conventional and controlled deterioration storage conditions. *Journal of Experimental Botany*, v.61, n.15, p.4423-4436, 2010. http://jxb.oxfordjournals.org/content/early/2010/08/06/jxb.erq248.short

TONIN, G.A.; PEREZ, S.C.J.G.A. Qualidade fisiológica de sementes de *Ocotea porosa* (Nees et Martius ex. Nees) após diferentes condições de armazenamento e semeadura. *Revista Brasileira de Sementes*, v.28, n.2, p.26-33, 2006. www.readcube.com/articles/10.1590/S0101-31222006000200004

VIEIRA, A.R.; OLIVEIRA, J.A.; GUIMARÃES, R.M.; VON PINHO, E.V.R.; PEREIRA, C.E.; CLEMENTE, A.C.S. Marcador isoenzimático de dormência em sementes de arroz. *Revista Brasileira de Sementes*, v.30, n.1, p.81-89, 2008. www.scielo.br/scielo.php?pid=S0101-31222008000100011&script

YIN, G.; XIN, X.; SONG, C.; CHEN, X.; ZHANG, J.; WU, S.; LI, R.; LIU, X.; LU, X. Activity levels and expression of antioxidant enzymes in the ascorbate-glutathione cycle in artificially aged rice seed. *Plant Physiology and Biochemistry*, v.80, p.1-9, 2014. www.sciencedirect.com/.../pii/S0981942814000849

Physiological and sanitary quality, and transmission of fungi associated with *Brachiaria brizantha* (Hochst. ex. A. Rich.) Stapf seeds submitted to thermal and chemical treatments

Cheila Cristina Sbalcheiro[1], Solange Carvalho Barrios Roveri José[1]*,
Jennifer Carine Rodrigues da Costa Molina Barbosa[2]

ABSTRACT - The Brazilian pastures establishment success depends on the use of seeds with high physiological and sanitary quality. The aim of this study was to evaluate the effect of thermal and chemical treatments on quality of *Brachiaria brizantha*, cv. BRS Piatã and pathogen transmission via seed. The treatments included the use of fungicides, detergent, sodium hypochlorite, sulfuric acid and alcohol, as well as thermal treatments. In addition to seeds without treatment, nine treatments were tested. The tests used to assess the physiological quality of the seeds were first and final count of germination test and root protrusion; germination and root protrusion speed index. A transmitting test and filter paper method had been performed for sanitary quality. Fungal lower incidence was observed in seeds treated with alcohol, however, this treatment reduced the physiological quality of the seeds and higher frequency of fungi transmission from seeds to seedlings was observed for *Bipolaris* sp., *Fusarium* sp. and *Phoma* sp. Treatment with sulfuric acid provided a better seed germination performance and reduced the incidence of fungi.

Index terms: germination, vigor, seed-seedling, incidence, forage.

Qualidade fisiológica, sanitária e transmissão de fungos associados às sementes de *Brachiaria brizantha* (Hochst. ex. A. Rich.) Stapf submetidas a tratamentos térmicos e químicos

RESUMO - O sucesso no estabelecimento das pastagens brasileiras depende da utilização de sementes com boa qualidade fisiológica e sanitária. Objetivou-se com o trabalho avaliar o efeito de tratamentos térmicos e químicos sobre a qualidade fisiológica e sanitária de sementes de *Brachiaria brizantha*, cultivar BRS Piatã e a transmissão de patógenos via semente. Os tratamentos incluíram o uso de fungicidas, detergente, hipoclorito de sódio, álcool e ácido sulfúrico, além de tratamentos térmicos. Os testes utilizados para a avaliação da qualidade fisiológica das sementes foram primeira contagem e contagem final do teste de germinação e de protrusão radicular; índice de velocidade de germinação e de protrusão radicular. Para a qualidade sanitária foi realizado o método de papel de filtro e o teste de transmissão. Menor incidência fúngica foi observada nas sementes tratadas com álcool, no entanto, esse tratamento reduziu a qualidade fisiológica das sementes e maior frequência de transmissão de fungos das sementes para as plântulas foi observada para *Bipolaris* sp., *Fusarium* sp. e *Phoma* sp. O tratamento com ácido sulfúrico proporcionou melhor desempenho germinativo das sementes e reduziu a incidência de fungos.

Termos para indexação: germinação, vigor, semente-plântula, incidência, forrageira.

Introduction

Brachiaria brizantha (Hochst. ex. A. Rich.) Stapf [syn. *Urochloa brizantha* (Hochst. ex. A. Rich.) R.D. Webster] is a species originated in Africa and among the cultivars used commercially the Marandu can be cited, a forage grass fairly planted in Brazil, especially in the North and Midwest regions and the BRS Piatã, cultivar noted for producing better quality forage and with a normally high production of seeds. By presenting different agronomic and adaptive characteristics, this cultivar is recommended to diversify grassland in various culture environments as an alternative to cv. Marandu (Almeida et al., 2009; Quadros et al., 2012).

Successful establishment of Brazilian pastures depends on the use of good quality seeds with high germination, vigor and good phytosanitary quality (Vechiato et al., 2010). However,

[1]Embrapa Recursos Genéticos e Biotecnologia, Caixa Postal 023725, 70770-917 - Brasília, DF, Brasil.

[2]Faculdade Anhanguera de Brasília, 72950-000 - Taguatinga, DF, Brasil.
*Corresponding author <solange.jose@embrapa.br>

Physiological and sanitary quality, and transmission of fungi associated with Brachiaria brizantha...

123

most species of tropical forage grasses presents factors that impair the production of good quality seeds, such as the low number of fertile seeds, disuniformity in the emission of inflorescence, irregular flowering within of panicles, high natural shattering, seed dormancy and emergence of weeds in pastures (Martins and Silva, 2003; Bonome et al., 2006). Thus, the presence of seed dormancy is a limiting factor in the germination in the field and under controlled conditions (Garcia and Cícero, 1992; Lacerda et al., 2010).

It is understood by dormancy, the phenomenon in which seemingly viable seeds do not germinate even under favorable environmental conditions (Brasil, 2009; Lacerda et al., 2010). It is remarkable the presence of dormancy in freshly harvested seeds in several species of tropical grasses (Usberti and Martins, 2007; Costa et al., 2011), associated to physiological causes and which are progressively removed during storage (Almeida and Silva, 2004). The recommendations described in the Rules for Testing Seeds - RAS (Brasil, 2009), as well as in most studies conducted to break dormancy in seeds of *Brachiaria*, the use of chemical scarification with sulfuric acid and thermal treatment are mentioned (Almeida and Silva, 2004; Lacerda et al., 2010; Costa et al., 2011; Custódio et al., 2012).

Alternatives to overcome dormancy can benefit seed quality assessment in laboratory and the development of methods that, being frequently used, allow the marketing of seeds with dormancy partially or totally eliminated (Martins and Silva, 2006).

Besides the physiological quality, seed sanitary quality should be considered. The forage seed production system is threatened by the presence of pathogens, mainly due to the absence of sanity standards to the domestic market, tolerance to seeds infection by pathogens and seed treatments effective techniques (Vechiato et al., 2010).

Thus, the seeds become an efficient mechanism for the introduction and dispersion of various pathogens, which may cause damage from the stand establishment until harvest. The presence of contaminating microorganisms in storage or after the point of physiological maturity also presents a major threat to seed health (Marchi et al., 2007; Marchi et al., 2008; Marchi et al., 2010), therefore it is important to study pathogen transmission via seed to seedling.

Potentially pathogenic fungi such as *Bipolaris* spp., *Curvularia* sp., *Fusarium* sp. and *Phoma* sp., and storage fungi such as *Aspergillus* sp., *Penicillium* sp., *Rhizopus* sp. associated with *Brachiaria* seeds have been reported as damage cause to the quality and establishment of forage plants (Marchi et al., 2010; Santos et al., 2014).

Thermotherapy with or without chemical treatment has been used for the eradication of pathogens associated with

seeds, and it may reduce pesticide use in vegetable material, in addition to promoting seed germination. This strategy reduces the introduction of new pathogens species in free areas and it can bring benefits to farmers, who can use this practice in the prevention and control of pathogens associated with seed (Tenente et al., 2005).

Thus, in order to the enterprise succeed in deploying pastures, using high physiological and sanitary quality seed is needed. However, information about the physiological and sanitary quality of forage seeds are scarce and lack studies that investigate procedures that benefit getting lots of seeds with higher germination and low infection rate.

Thus, the objective of this study was to evaluate the effect of thermal and chemical treatments on physiological and sanitary quality of *Brachiaria brizantha* cv. BRS Piatã from two harvests, as well as to analyze the pathogens transmission via seeds to seedlings.

Material and Methods

The experiments were performed in the Seed Laboratory of Embrapa Genetic Resources and Biotechnology (Brasília, DF). Seeds of *Brachiaria brizantha* cv. BRS Piatã, 2010 and 2011 harvests, produced in the area of seed production of Sucupira farm (Brasília, DF) were used, and crop seeds system was on bunches. The seeds were received in the laboratory in August 2011 and stayed for 15 days in cold storage to 10 °C and relative humidity of 30% when the experiment was installed. The seeds humidity at the beginning of the experiment was 7.9% and 9.8% and germination of 29% and 42% for seeds harvested in 2010 and 2011, respectively. Seeds were subjected to the following treatments: 1) Profile: untreated seeds; 2) Control: only seeds moistened with ultrapure water at a ratio of 6:1 (seed: water) to simulate the same conditions as other treatments; 3) Detergent: seeds immersion in a solution of neutral detergent (10 drops/400 mL water) for 5 minutes under constant agitation and later rinsing; 4) Carbendazim + Thiram (150+350 SC): seeds moistened with ultrapure water (6:1) were treated with the product dosage of 300 mL/100 kg seed; 5) Captan PM 500 (50/kg seed): seeds moistened with ultrapure water (6:1) were treated with the product dosage of 300 g/100 kg of seeds; 6) Alcohol 70%: immersing the seeds in 70% alcohol for 30 minutes and further rinsing with ultrapure water; 7) Hot hypochlorite: the seeds were placed in Erlenmeyer flask containing warm sodium hypochlorite (1%) solution and kept in a water bath at 65°C for 5 min and then rinsed with water; 8) Hypochlorite paper: blotter paper was moistened with a solution of sodium hypochlorite (0.5%), except for the transmission test in which the seeds were placed in a flask containing sodium hypochlorite at 0.5% in the ratio 2:1 (hypochlorite:seed), kept under constant stirring for 1 minute and rinsed with ultrapure water three times for 2 minutes each and being called hypochlorite solution treatment; 9) Hot water: the seeds were placed in Erlenmeyer flask containing hot ultrapure water and kept in water bath at 65 °C for 5 minutes; 10) Sulfuric acid: seeds immersion in sulfuric acid for 5 minutes, later rinsed with plentiful ultrapure water and final immersion in ultrapure water for 30 minutes. After the treatments, the seeds were subjected to surface drying in a laminar flow cabinet (sterile environment), except seed treatment profile.

Germination test: for each treatment four replicates of 100 seeds were distributed in four gerbox type boxes, containing two sheets of blotting paper moistened with ultrapure water (except for hypochlorite treatment paper), were used. After sowing, the seeds remained in germination BOD chamber at 20-35 °C with 16 hours light. The first germination test count was carried out at the third day and the final count after 21 days of sowing by computing the number of normal seedlings, dormant and dead seeds, as recommended by the Rules for Seed Testing (Brasil, 2009). Together with the germination test, the seeds with protruding primary root were count, the germination speed index (GSI) and the root protrusion speed index (RPSI) were performed according to the formula of Maguire (1962).

Health test: for pathology test, four replicates of 400 seeds from each treatment were distributed in gerbox type boxes, containing three sheets of blotting paper moistened with sterile distilled water (except for hypochlorite treatment paper) in laminar flow cabinet. The seeds remained in a germination chamber BOD at 25 °C with 12 hours light. The sanitary examination was performed seven days after sowing, each seed was individually evaluated with stereomicroscope, observing the occurrence of typical fruiting of fungal growth, and when necessary, the identification was made under an optical microscope. The results were expressed as a percentage of fungi occurrence relating to the number of seeds (incidence) and the rate of fungal incidence control in relation to the profile for each treatment. The sum of all fungi incidences detected in seeds from each treatment and control of the total fungal incidence in relation to treatment profile was adopted as the total fungal incidence.

Pathogen transmission via seed to seedling test: the same treatments of seed pathological analysis were used, replacing "hypochlorite paper" with "hypochlorite solution". The seeds from each treatment were sown in trays containing sterile sieved sand, moistened as its field capacity. 100 seeds per tray were distributed to each treatment, being four replicates of 25 seeds each, and, after sowing, they remained in the germination and BOD chamber at 25 °C with 12 hours light. Assessments occurred at seven and 14 days after sowing, noting the percentage of seedling emergence and the presence of fungus signs or symptoms in the aerial part of the seedling. All seedlings emerged after 14 days were placed in moist chamber to fungal structures observation.

Statistical Analysis: the experimental design was randomized to the physiological and sanitary analyzes (filter paper) using the factorial (10x2), 10 treatments x 2 harvests. Data were evaluated by analysis of variance and means, compared by Tukey's test at 5%. Analyses were performed in Sisvar (System Analysis of Variance) statistical program (Ferreira, 2011).

Results and Discussion

There was significant interaction between the harvesting factors and treatments for all tests done, except for the percentage of dormant seeds. The average results are presented in Tables 1 to 6. In general, greater vigor was observed in the treated seeds of the 2010 harvest, compared to the 2011 harvest, in the first count of germination and root protrusion test (Tables 1 and 2), ie, at three days of sowing a higher percentage of normal seedlings and root protrusion on the seeds from the 2010 harvests was found, to most treatments.

The seeds germination of both harvests was similar only when they were treated with sulfuric acid, hypochlorite paper and 70% alcohol, and the treatment with sulfuric acid was the one that most improved the seeds germination (Tables 1 and 2). The dormancy breaking of the 2011 harvest seeds was only observed in the treatment with sulfuric acid, unlike the seeds of the 2010 harvest, which in addition to sulfuric acid, hot hypochlorite treatment exceeded the germination of untreated seeds (profile).

A similar tendency was observed in the behavior of seeds in the final count of germination test, both in the percentage of normal seedlings (Table 1) as in the root protrusion (Table 2), in the 2011 harvest. Treatment with sulfuric acid outperformed the others, promoting greater germination of the seeds of the newest harvest, 2011. Almeida and Silva (2004) researching the behavior of dormancy in seeds of *B. dictyoneura* found that after thermal treatment of the seeds at a temperature of 85 °C and sulfuric acid treatment there was a reduction in seed dormancy and improvement in their performance in a similar way for both treatments. However, when the seeds were treated with sulfuric acid and stored for six months, the authors found no increase in normal seedlings or a reduction of dormant seeds, unlike the results observed in this study, in which the seeds were not subjected to any treatment before of the experiment. It was also observed in seeds of the 2010 harvest, that the

Physiological and sanitary quality, and transmission of fungi associated with Brachiaria brizantha...

125

hot hypochlorite treatment, hot water, hypochlorite paper and detergent provided the same performance as the treatment with sulfuric acid and superior performance to profile treatment, with better results on germination (Table 1). Lacerda et al. (2010) found that boiling water for 1-3 minutes was more effective in overcoming dormancy of *B. brizantha*, compared to treatments with gibberellic and sulfuric acid, which does not differ from the control.

Table 1. Mean values of normal seedlings (%) obtained at first count germination (FCG) and at final count germination (FG) of *B. brizantha* cv. BRS Piatã seeds from two harvests.

Treatment	FCG		FG	
	2010	2011	2010	2011
Sulfuric acid	42 aA	47 aA	53 bA	63 aA
Detergent	20 aBC	1 bB	48 aAB	32 bB
Hypochlorite paper	3 aE	0 aB	43 aABC	44 aB
Hot water	22 aBC	0 bB	43 aABC	33 bB
Hot hypochlorite	29 aB	0 bB	4 aABC	36 aB
Captan	10 aDE	0 bB	40 aABCD	38 aB
Carbendazim+Thiram	14 aCD	0 bB	35 bBCD	45 aB
Control	20 aBC	0 bB	30 bCD	38 aB
Profile	19 aC	0 bB	29 bD	42 aB
Alcohol 70%	1 aE	0 aB	1 aE	1 aC

Means followed by the same lowercase letter on the line, and capitalized on the column; do not differ statistically by the Tukey's test at 5% probability.

Table 2. Mean values of root protrusion (%) obtained at first count (RPFC) and at final count (RP) of *B. brizantha* cv. BRS Piatã seeds germination test from two harvests.

Treatment	RPFC		RP	
	2010	2011	2010	2011
Sulfuric acid	42 aA	47 aA	57 aA	64 aA
Detergent	20 aBC	5 bB	49 aABC	32 bC
Hypochlorite paper	3 aE	0 aB	45 aABC	45 aBC
Hot water	22 aBC	5 bB	47 aABC	33 bBC
Hot hypochlorite	29 aB	3 bB	50 aAB	37 bBC
Captan	10 aDE	0 bB	47 aABC	39 bBC
Carbendazim+Thiram	14 aCD	1 bB	41 aBC	46 aB
Control	20 aBC	2 bB	40 aBC	39 aBC
Profile	19 aC	1 bB	37 aC	42 aBC
Alcohol 70%	1 aE	0 aB	2 aD	1 aD

Means followed by the same lowercase letter on the line, and capitalized on the column; do not differ statistically by the Tukey's test at 5% probability.

Costa et al. (2011) found that storage for 21 months was more effective for overcoming *B. humidicola* seed dormancy than acid scarification or germination promoting substances

application. However, seeds stored for periods longer than 12 months may have lower quality and treatments with sulfuric acid does not promote germination increase, as evidenced by Custódio et al. (2012), in *B. brizantha* seeds lots. Paniago et al. (2014) found that the storage time did not provide *B. humidicola* seed quality increase nor reduced the percentage of dormant seeds. However, the study period was only 90 days of storage. In this study, despite the seeds germination of the 2010 harvest was low, treatment with sulfuric acid provided an 82% percentage increase of normal seedlings compared to untreated seeds (profile), with 29% germination. For seeds of the 2011 harvest, this increase was 50% (Table 1). These results corroborate those obtained by Câmara and Stacciarini-Seraphin (2002), who observed that seeds of *B. brizantha* stored for sixteen months at laboratory conditions showed poor germination, around 23%, but maintained viability. It was expected that the percentage of dormant seeds in 2011 harvest was higher, since the seeds are the newest crop. However, the percentage of dormant seeds of the 2011 harvest was 25% and for the 2010 harvest, that number was 46%. A higher percentage of dead seeds of the 2011 harvest may have contributed to reduce the proportion of dormant seeds of this harvest (Table 3). It is clear that the storage time may promote dormancy breaking, but the history of seeds must be considered.

Table 3. Mean values of dormant seeds (DorS) and dead seeds (DS) obtained at *B. brizantha* cv. BRS Piatã seeds final germination.

Treatment	DorS (%)*	DS (%)	
		2010	2011
Sulfuric acid	19 C	15 bAB	25 aDE
Detergent	32 B	9 bABC	43 aABC
Hypochlorite paper	27 BC	9 bABC	39 aBC
Hot water	29 B	9 bABC	50 aA
Hot hypochlorite	31 B	4 bC	44 aAB
Captan	33 B	7 bBC	36 aBC
Carbendazim+Thiram	30 B	14 bABC	33 aCD
Control	31 B	19 bA	39 aABC
Profile	35 B	11 bABC	37 aBC
Alcohol 70%	86 A	5 bBC	21 aE
Mean DorS (%) harvest 2010	46 A		
Mean DorS (%) harvest 2011	25 B		

*There was no interaction between harvests and treatments for this variable. Means followed by the same lowercase letter on the line, and capitalized on the column; do not differ statistically by the Tukey's test at 5% probability.

In Table 4 the data obtained for the germination speed rate can be observed, measured by the germination speed index (GSI) and root protrusion (RPSI). Higher values for the index means more vigorous seeds. Treatment with sulfuric acid provided the best results, outperforming the other treatments, and these rates were higher in the seeds of the 2011 harvest, confirming the results obtained by Dias and Toledo (1993), Lago and Martins (1998) and Martins and Silva (2006) who also found that the best performance in germination and dormancy breaking of B. brizantha seeds was using sulfuric acid treatment, even during the storage for six months (Sallum et al., 2010).

Treatment with 70% alcohol reduced physiological seed quality, and on germination test the values observed in the two harvests were statistically equal, within 1% normal seedlings (Table 1). Seed vigor was also negatively affected by this treatment, regardless of the harvest, providing the worst results (Table 4). Although treatment with sulfuric acid is effective in breaking dormancy, it can also cause injuries to the seeds (Macedo et al., 1994; Meschede et al., 2004). As mentioned by Garcia and Cícero (1992) and Dias and Alves (2008), responses to different treatments to overcome dormancy vary with the history and age of seeds, storage conditions, among other factors, and may not cause the expected effect. Undesirable effect can also be observed when the seed lot to be treated presents dormant and not dormant seeds. In this research, although the seeds are from different harvests, ie, different ages, treatment with sulfuric acid caused no injuries to them, favoring both germination and seed vigor.

Regarding the sanitary quality of seeds, on the 2010 harvest the incidence of the following fungi were observed: Alternaria spp., Aspergillus flavus, Aspergillus niger, Aspergillus ochraceus, Bipolaris spp., Chaetomium sp., Cladosporium sp., Curvularia sp., Fusarium sp., Penicillium sp., Phoma sp., Pithomyces sp., Rhizopus sp. and Trichoderma sp. (Table 5). On the 2011 harvest, the following fungi were observed: Alternaria spp., A. flavus, A. niger, Bipolaris spp., Chaetomium sp., Curvularia sp., Fusarium sp., Penicillium sp., Phoma sp. and Rhizopus sp. When comparing the two harvests, higher total fungal incidence in seeds of the 2011 season was observed, including the untreated seeds (profile), which may have provided an increase in the percentage of dead seeds in this harvest, as seen in Table 3. Among the fungi identified in this study, six were identified by Mallmann et al. (2013) in B. brizantha cv. BRS Piatã: Alternaria sp., Bipolaris sp., Cladosporium sp., Curvularia sp., Fusarium sp. and Phoma sp. and 10 were observed by Santos et al. (2014): Aspergillus niger, Bipolaris sp., Chaetomium sp., Curvularia sp., Fusarum sp., Penicillium sp., Phoma sp., Pithomyces sp., Rhizopus sp., Trichoderma sp.

Table 4. Mean values of germination speed index (GSI) and root protrusion speed index (RPSI) obtained at B. brizantha cv. BRS Piatã seeds germination test from two harvests.

Treatment	GSI		RPSI	
	2010	2011	2010	2011
Sulfuric acid	16 bA	26 aA	16 bA	26 aA
Detergent	6 aB	5 bB	6 aB	7 aB
Hypochlorite paper	5 aB	6 aB	5 bB	7 aB
Hot water	6 aB	5 bB	6 aB	6 aB
Hot hypochlorite	6 aB	5 aB	7 aB	6 aB
Captan	5 aB	5 aB	6 aB	6 aB
Carbendazim+Thiram	5 aB	6 aB	5 bB	7 aB
Control	4 aB	5 aB	5 bB	7 aB
Profile	4 aB	6 aB	5 bB	7 aB
Alcohol 70%	0 aC	0 aC	0 aC	0 aC

Means followed by the same lowercase letter on the line, and capitalized on the column; do not differ statistically by the Tukey's test at 5% probability.

Table 5. Fungal incidence and incidence control in B. brizantha cv. BRS Piatã seeds from two harvests.

Treatment	Total Fungal Incidence (%)			Incidence control*(%)		
	2010	2011	Means	2010	2011	Means
Profile	38 bB	118 aB	78 B			
Control	39 bB	123 aB	81 B	0 aD	0 aE	0 E
Detergent	60 bA	124 aB	92 A	0 aD	0 aE	0 E
Carbendazim+Thiram	2 bD	42 aD	22 DE	94 aA	64 bC	79 B
Captan	3 bD	20 aE	11 FG	92 aA	83 bB	88 AB
Alcohol 70%	7 aD	3 aF	5 G	82 bA	97 aA	90 A
Hot hypochlorite	21 aC	11 bEF	16 EF	44 bC	90 aAB	67 C
Hypochlorite paper	48 bAB	137 AA	93 A	0 aD	0 aE	0 E
Hot water	12 bCD	78 AC	45 C	67 aB	34 bD	50 D
Sulfuric acid	13 bCD	41 aD	27 D	65 aB	65 aC	65 C
Means	24 b	70 a		44 a	43 a	

*Total fungal incidence control in relation to profile treatment. Means followed by the same lowercase letter on the line, and capitalized on the column; do not differ statistically by the Tukey's test at 5% probability.

Physiological and sanitary quality, and transmission of fungi associated with Brachiaria brizantha...

127

The lowest total fungal incidences in the seeds of the 2010 harvest were observed in the treatments with Carbendazim + Thiram (2%), Captan (3%), alcohol 70% (7%), hot water (12%) e sulfuric acid (13%). For the seeds of the 2011 harvest, the lowest values were 3%, 11% and 20% for treatments with alcohol, hot hypochlorite and Captan, respectively. As for controlling the fungi incidence on seeds in relation to treatment profile, it was observed that the alternative alcohol treatment had the highest control percentage on seeds of the 2011 harvest and did not differ statistically from fungicide treatments in 2010 harvest (Table 5). Comparing the results in Table 5 with those shown in Tables 1 and 2, treatments with hot water and sulfuric acid were effective for improving the physiological quality, assessed at the final count of testing and seed of the 2010 harvest sanitary quality.

According to Martins et al. (2001) treatment with sulfuric acid reduces the incidence of plant pathogens. Similar results were observed in this study, for the two crops analyzed, however, seed treatment with sulfuric acid was not superior to treatment with fungicides and 70% alcohol in relation to the incidence of pathogens.

Although alcohol treatment has been effective in controlling plant pathogens, in two harvests, this treatment has damaged seed performance in tests that assessed the physiological quality of seeds. Observing the values obtained in the germination test (Table 1), which is the standard for seed quality evaluation for marketing purposes, it can be seen that, except for alcohol, other treatments did not reduce seed germination, compared to seed treatment profile. However, seed treatment with sulfuric acid provided higher germination and reduced overall incidence of fungi by 65% for the two harvests, being a viable alternative for improving the physiological and sanitary quality of the seeds. This treatment reduced the incidence of the following fungi: *Alternaria* spp., *A. ochraceus*, *Bipolaris* sp., *Chaetomium* sp., *Cladosporium* sp., *Curvularia* sp., *Fusarium* spp., *Phoma* sp., *Rhizopus* sp. and *Trichoderma* sp., in relation to profile treatment seed and in a similar way to treatments with fungicides, and it was more effective in controlling *Alternaria* spp. and *Bipolaris* sp., than fungicides. Probably, sulfuric acid acted in reducing the inoculum potential present in the *B. brizantha* cv. BRS Piatã seed, becoming an alternative method for controlling pathogens.

In the transmission of pathogens via seed sowed on sand testing (Table 6), the occurrence of *Aspergillus* sp., *Bipolaris* sp., *Curvularia* sp., *Fusarium* spp., *Penicillium* sp., *Phoma* sp. e *Rhizopus* sp. pathogens was observed, confirming the transmission of pathogens present in the seeds to seedlings since those fungi were detected in seed pathology test. These fungi, for having the ability to infest/infect and be transmitted by seed, besides reducing germination, they become a risk to the forage crops, especially under favorable environmental conditions and in the absence of an efficient controlling method. Santos et al. (2014) observed the transmission of *Bipolaris* sp. and *Curvularia* sp. from *B. brizantha* seeds for seedlings and found that the *Bipolaris* sp. fungus, is damaging the seedlings of the *Brachiaria*, *Crotalaria* and *Panicum* genera.

In 2010 harvest (Table 6), seed treatment with hot water and sulfuric acid reduced the percentage of seedlings with pathogens, as well as treatments with Carbendazim+Thiram and Captan fungicides. Similar results were observed for fungal incidence (Table 5) to the same treatments. Seed treatment with 70% alcohol interfered with seedling emergence in sand, corroborating the results obtained in this study on the germination test. However, the two seedlings that emerged from this treatment had *Bipolaris* sp. fungus, ie, even being the best treatment to control pathogens associated with seed in filter paper test, the seedlings that emerged were seen to be infected by fungi.

In the 2011 harvest treatment of seeds with hot sodium hypochlorite reduced the percentage of pathogens infected seedling in relation to the profile (Table 6). It was observed that the treatment of seeds with fungicides was not efficient in controlling *Bipolaris* sp. It is important to mention that the active principles Captan and Carbendazim do not present the toxic action to *Bipolaris*, *Curvularia*, *Alternaria* and *Drechslera*, according to Reis et al. (2001).

Differences in results for the same treatments between harvests were observed, indicating that thermal treatment associated or not with chemical treatment make the fungi incidences associated with *B. brizantha* seeds be variable in different harvests, possibly due to the position, way of infection and colonization of the microorganism in the seed, initial inoculum amount and time exposed to the treatment.

Although sulfuric acid is effective in promoting germination, as well as at the fungi incidence control, this treatment poses risks to the health, and handling precautions should be taken. Therefore, the study of alternative and effective methods should be investigated. In the present study, it was found that the thermal treatment with sodium hypochlorite and water may be an alternative to improve the seeds physiological and sanitary quality.

Table 6. Pathogen transmission via seed for de *B. brizantha* cv. BRS Piatã seedlings from two harvests.

Treatments	Harvest	Emerging seedlings (%)	% of seedlings with pathogen						
			Asp.*	Bip.	Cur.	Fus.	Pen.	Pho.	Rhi.
Profile	2010	45	16	4	0	29	0	7	0
	2011	29	0	55	7	38	0	31	0
Control	2010	49	0	0	0	80	0	31	2
	2011	34	0	32	0	44	0	18	0
Detergent	2010	20	0	0	0	10	10	0	15
	2011	29	3	48	0	31	0	24	0
Carbendazim+Thiram	2010	44	0	7	0	2	0	0	0
	2011	41	0	98	0	15	0	5	0
Captan	2010	49	2	0	0	0	0	0	0
	2011	42	0	38	0	26	0	26	0
Alcohol 70%	2010	2	0	100	0	0	0	0	0
	2011	4	0	50	0	0	0	25	0
Hot hypochlorite	2010	42	12	5	0	0	0	0	0
	2011	43	2	23	2	7	0	14	0
Hypochlorite solution	2010	31	0	71	0	0	0	0	0
	2011	32	0	53	0	44	0	25	0
Hot water	2010	37	0	0	0	3	0	0	0
	2011	31	0	29	0	48	0	7	0
Sulfuric acid	2010	57	0	0	0	0	16	0	0
	2011	53	13	34	0	30	0	42	0

*Asp.= *Aspergillus* sp.; Bip.= *Bipolaris* sp.; Cur.= *Curvularia* sp.; Fus.= *Fusarium* sp.; Pen.= *Penicillium* sp.; Pho.= *Phoma* sp.; Rhi.= *Rhizopus* sp.

Conclusions

The treatment with sulfuric acid provides better seed germination performance and contributes to reducing the fungi incidence.

Minor fungal incidences are observed in seeds treated with alcohol, but this treatment is highly detrimental to physiological seed quality.

Bipolaris sp., *Fusarium* sp. and *Phoma* sp. fungi have higher seed to seedling transmission frequency in *Brachiaria brizantha* cv. Piatã.

References

ALMEIDA, R.G.; COSTA, J.A.A.; KICHEL, A.N.; ZIMMER, A.H. Taxas e Métodos de Semeadura para *Brachiaria brizantha* cv. BRS Piatã em Safrinha. *Comunicado Técnico*, 113, Campo Grande, 2009. http://ainfo.cnptia.embrapa.br/digital/bitstream/CNPGC-2010/13218/1/COT113.pdf

ALMEIDA, C.R.; SILVA, W.R. Comportamento da dormência em sementes de *Brachiaria dictyoneura* cv. Llanero submetidas às ações do calor e do ácido sulfúrico. *Revista Brasileira de Sementes*, v.26, n.1, p.44-49, 2004. http://www.scielo.br/pdf/rbs/v26n1/a07v26n1.pdf

BONOME, L.T.S.; GUIMARÃES, R.M.; OLIVEIRA, J.A.; ANDRADE, V.C.; CABRAL, P.S. Efeito do condicionamento osmótico em sementes de *Brachiaria brizantha* cv. Marandu. *Ciência e Agrotecnologia*, v.30, n.3, p.422-428, 2006. http://www.scielo.br/pdf/cagro/v30n3/v30n3a06.pdf

BRASIL. Ministério da Agricultura, Pecuária e Abastecimento. *Regras para análise de sementes*. Ministério da Agricultura, Pecuária e Abastecimento. Secretaria de Defesa Agropecuária. Brasília: MAPA/ACS, 2009. 395p. http://www.agricultura.gov.br/arq_editor/file/2946_regras_analise__sementes.pdf

CÂMARA, H.H.L.L.; STACCIARINI-SERAPHIN, E. Germinação de sementes de *Brachiaria brizantha* cv. Marandu sob diferentes períodos de armazenamento e tratamento hormonal. *Pesquisa Agropecuária Tropical*, v.32, n.1, p.21-28, 2002. http://www.revistas.ufg.br/index.php/pat/article/view/2436

COSTA, C.J.; ARAÚJO, R.B.; VILLAS BÔAS, H.D. C. Tratamentos para a superação de dormência em sementes de *Brachiaria humidicola* (Rendle) Schweick. *Pesquisa Agropecuária Tropical*, v.41, n.4, p.519-524, 2011. http://www.scielo.br/pdf/Pat/v41n4/a11v41n4.pdf

CUSTÓDIO, C.C.; DAMASCENO, L.R.; NETO, N.B.M. Imagens digitalizadas na interpretação do teste de tetrazólio em sementes de *Brachiaria brizantha*. *Revista Brasileira de Sementes*, v.34, n.2, p.334-341, 2012. http://www.scielo.br/pdf/rbs/v34n2/20.pdf

DIAS, M.C.L.L.; ALVES, S.J. Avaliação da viabilidade de sementes de *Panicum maximum* Jacq pelo teste de tetrazólio. *Revista Brasileira de Sementes* v.30, n.3, p.152-158, 2008. http://www.scielo.br/pdf/rbs/v30n3/20.pdf

DIAS, D.C.F.S.; TOLEDO, F.F. Germinação e incidência de fungos em testes com sementes de *Brachiaria brizantha* (Hochst ex A. Rich.) Stapf. *Scientia Agricola*, v.50, n.1, p.68-76, 1993. http://www.scielo.br/pdf/sa/v50n1/11.pdf

FERREIRA, D.F. SISVAR: a computer statistical analysis system. *Ciência e Agrotecnologia*, v.35, n.6, p.1039-1042, 2011. http://www.scielo.br/pdf/cagro/v35n6/a01v35n6

GARCIA, J.; CÍCERO, S.M. Superação de dormência em sementes de *Brachiaria brizantha* Marandu. *Scientia Agricola*, v.49, n.1, p.9-13, 1992. http://www.scielo.br/pdf/sa/v49nspe/02.pdf

LACERDA, M.J.R.; CABRAL, J.S.R.; SALES, J.F.; FREITAS, K.R.; FONTES, A.J. Superação da dormência de sementes de *Brachiaria brizantha* cv. "Marandu". *Semina: Ciências Agrárias*, v.31, n.4, p.823-828, 2010. http://www.uel.br/portal/frm/frmOpcao.php?opcao=http://www.uel.br/revistas/uel/index.php/semagrarias

LAGO, A.A.; MARTINS, L. Qualidade fisiológica de sementes de *Brachiaria brizantha*. *Pesquisa Agropecuária Brasileira*, v.33, n.2, p.199-204, 1998. http://seer.sct.embrapa.br/index.php/pab/article/view/4832/6950

MACEDO, E.C.; GROTH, D.; LAGO, A.A. Efeito de escarificação com ácido sulfúrico na germinação de sementes de *Brachiaria humidicola* (RENDLE) SCHWEICK. *Pesquisa Agropecuária Brasileira*, v.29, n.3, p.455-460, 1994.

MAGUIRE, J.D. Speeds of germination and seedling emergence and vigor. *Crop Science*, v.2, n. 2, p.176-177, 1962.

MALLMANN, G.; VERZIGNASSI, J.R.; FERNANDES, C.D.; SANTOS, J.M.; VECHIATO, M.H.; INÁCIO, C.A.; BATISTA, M.V.; QUEIROZ, C.A. Fungos e nematoides associados a sementes de forrageiras tropicais. *Summa Phytopathologica*, v.39, n.3, p.201-203, 2013. http://www.scielo.br/pdf/sp/v39n3/a10v39n3.pdf

MARCHI, C.E.; FERNANDES, C.D.; BORGES, C.T.; SANTOS, J.M.; JERBA, V.F.; TRENTIN, R.A.; GUIMARÃES, L.R.A. Nematofauna fitopatogênica de sementes comerciais de forrageiras tropicais. *Pesquisa Agropecuária Brasileira*, v.42, n.5, p.655-660, 2007. http://www.scielo.br/pdf/pab/v42n5/07.pdf

MARCHI, C.E.; FERNANDES, C.D.; ANACHE, F.C.; JERBA, V.F.; FABRIS, L.R. Quimio e termoterapia em sementes e aplicação de fungicidas em *Brachiaria brizantha* como estratégias no manejo do carvão. *Summa Phytopathologica*, v.34, n.4, p.321-325, 2008. http://www.scielo.br/pdf/sp/v34n4/v34n4a04.pdf

MARCHI, C.E.; FERNANDES, C.D.; BUENO, M.L.; BATISTA, M.V.; FABRIS, L.R. Fungos veiculados por sementes comerciais de braquiária. *Arquivo Instituto Biológico*, v.77, n.1, p.65-73, 2010. http://www.biologico.sp.gov.br/docs/arq/v77_1/marchi.pdf

MARTINS, L.; SILVA W.R.; ALMEIDA, R.R. Sanidade em sementes de *Brachiaria brizantha* (Hochst. ex. A. Rich) Stapf submetidas a tratamentos térmicos e químico. *Revista Brasileira de Sementes*, v.23, n.2, p.117-120, 2001. http://www.abrates.org.br/revista/artigos/2001/v23n2/artigo17.pdf

MARTINS, L.; SILVA, W.R. Efeitos imediatos e latentes de tratamentos térmico e químico em sementes de *Brachiaria brizantha* cultivar Marandu. *Bragantia*, v.62, n.1, p.81-88, 2003. http://www.scielo.br/pdf/brag/v62n1/18504.pdf

MARTINS, L.; SILVA, W.R. Ações fisiológicas do calor e do ácido sulfúrico em sementes de *Brachiaria brizantha* cultivar Marandu. *Bragantia*, v.65, n.3, p.495-500, 2006. http://www.scielo.br/pdf/brag/v65n3/a16v65n3.pdf

MESCHEDE, D.K. SALES, J.G.C.; BRACCINI, A.L.; SCAPIM, C.A. SCHUAB, S.R.P.. Tratamentos para superação da dormência das sementes de capim-braquiária cultivar Marandu. *Revista Brasileira de Sementes*, v.26, n.2, p.15-19, 2004. http://www.scielo.br/pdf/rbs/v26n2/24492.pdf

PANIAGO, B. C.; PEREIRA, S.R.; RODRIGUES, A.P.D´A.C.; LAURA, V.A, Dormência pós-colheita de sementes de *Urochloa humidicola* (Rendle) Morrone e Zuloaga. *Informativo ABRATES*, v.24, n.1, p.22-26, 2014. http://www.abrates.org.br/portal/images/Informativo/v24_n1/004_2014_Silvia_Rahe.pdf

QUADROS, D.G.; ANDRADE, A.P.; OLIVEIRA, G.C.; OLIVEIRA, E.P.; MOSCON, E.S. Componentes da produção e qualidade de sementes dos cultivares marandu e xaraés de *Brachiaria brizantha* (Hochst. ex A. Rich.) Stapf colhidas por varredura manual ou mecanizada. *Semina: Ciências Agrárias*, v.33, n.5, p.2019-2028, 2012. http://www.uel.br/revistas/uel/index.php/semagrarias/article/view/9712/11581

REIS, E.M.; FORCELINI, C.A.; REIS, A.C. *Manual de fungicidas:* Guia para o controle químico de doenças de plantas. Florianópolis: Insular, 2001. 176p.

SALLUM, M.S.S.; ALVES, D.S.; AGOSTINI, E.A.T.; MACHADO NETO, N.B. Neutralização da escarificação química sobre a germinação de sementes de *Brachiaria brizantha* cv. Marandu. *Revista Brasileira de Ciências Agrárias*, v.5, n.3, p.315-321, 2010. http://agraria.pro.br/sistema/index.php?journal=agraria&page=article&op=view&path%5B%5D=agraria_v5i3a603&path%5B%5D=748

SANTOS, G.R.; TSCHOEKE, P.H.; SILVA, L.G.; SILVEIRA, M.C.A.C.; REIS, H.B.; BRITO, D.R.; CARLOS, D.S. Sanity analysis, transmission and pathogenicity off fungi associated with forage plant seeds in tropical regions of Brazil. *Journal of Seed Science*, v.36, n.1, p.54-62, 2014. http://www.scielo.br/pdf/jss/v36n1/a07v36n1.pdf

TENENTE, R. C. V.; GONZAGA, V.; SOUSA, A.I.M.; SANTOS, D.S. *Aplicação de tratamentos térmicos e químicos em sementes de beterraba importada, na erradicação de Ditylenchus dipsaci (Kuhn, 1857) Filipjev, 1936*. Embrapa Recursos Genéticos e Biotecnologia, 2005. 8p. (Circular Técnica 36). http://www.cenargen.embrapa.br/clp/publicacoes/cit/2005/cit036.pdf

USBERTI, R.; MARTINS, L. Sulphuric acid scarification effects on *Brachiaria brizantha, B. humidicola* and *Panicum maximum* seed dormancy release. *Revista Brasileira de Sementes*, v.29, p.143-147, 2007. http://www.scielo.br/pdf/rbs/v29n2a20.pdf

VECHIATO, M.H.; APARECIDO, C.C.; FERNANDES, C.D. Frequência de fungos em lotes de sementes comercializadas de *Brachiaria* e *Panicum*. *Arquivos Instituto Biológico*. Documento Técnico n.4, p.1-11, 2010. http://www.biologico.sp.gov.br/docs/dt/DT_07_2010.pdf

18

Heat treatment and germination of seeds of interspecific hybrid between American oil palm (*Elaeis oleifera* (H.B.K) Cortes) and African oil palm (*Elaeis guineensis* Jacq.)

Wanderlei Antônio Alves Lima[1*], Ricardo Lopes[2], Márcia Green[1], Raimundo Nonato Vieira Cunha[1], Samuel Campos Abreu[1], Alex Queiroz Cysne[1]

ABSTRACT – The oil palm (*E. guineensis*) is the African origin and the world's leading source of vegetable oil. The interspecific hybridization of the African oil palm (*E. guineensis*) with American oil palm (*E. oleifera*) aims to improve resistance to diseases, to improve oil quality and lower plant height. EMBRAPA (Empresa Brasileira de Pesquisa Agropecuária, Brazilian Corporation of Agricultural Research) has developed the first Brazilian interspecific hybrid cultivar (HIE) between American oil palm and African oil palm. The procedures adopted for commercial seed germination assessment have shown an average germination rate of 32%. The objective of this work was to assess the effect of the period of heat treatment and seed water content that are ideal for breaking dormancy and obtaining maximum germination. A completely randomized design was adopted, in a 4 x 3 factorial design, with four ranges of moisture contents: 18-19; 19-20; 20-21 and 21-22%, and three periods of heat treatment: 55, 75 and 100 days, with three replicates of 500 seeds. The percentage of germination, the first count and the germination speed index were assessed. To break dormancy and germination, the hybrids seeds of HIE, *oleifera* versus *guineensis*, should have their water content adjusted to values between 19 and 22%, and be subjected to heat treatment at a temperature of 39 ± + 1 °C and relative humidity of approximately 75% for 75 days.

Index terms: *Elaeis guineensis*, *Elaeis oleifera*, oil palm, breaking dormancy.

Tratamento térmico e germinação de sementes do híbrido interespecífico entre o caiaué e o dendezeiro

RESUMO - O dendezeiro, palmeira de origem africana, representa a principal fonte mundial de óleo vegetal. A hibridação interespecífica do dendezeiro com o caiaué, de origem americana, tem como objetivos a resistência a doenças, melhoria na qualidade do óleo e menor altura de plantas. A Embrapa desenvolveu a primeira cultivar brasileira híbrida interespecífica (HIE) entre o caiaué e o dendezeiro. Os procedimentos adotados para avaliar a germinação das sementes comerciais dessa cultivar têm apresentado valores médios de 32%. O objetivo desse trabalho foi avaliar o efeito do período de tratamento térmico e do teor de água nas sementes, ideais para superação da dormência e obtenção da germinação máxima. Adotou-se o delineamento inteiramente casualizado, em esquema fatorial 4 x 3, sendo quatro intervalos de graus de umidade: 18-19; 19-20; 20-21 e 21-22% e três períodos de tratamento térmico: 55, 75 e 100 dias, com três repetições de 500 sementes. Foram avaliadas a porcentagem de germinação, a primeira contagem e o índice de velocidade de germinação. Para superação da dormência e germinação, as sementes híbridas do HIE: *oleifera* x *guineensis* devem ter seu teor de água ajustado para valores entre 19 e 22% e ser submetidas ao tratamento térmico sob temperatura de 39 ± 1 °C e umidade relativa do ar de aproximadamente 75% por 75 dias.

Termos para indexação: *Elaeis guineensis*, *Elaeis oleifera*, palma de óleo, superação de dormência.

Introduction

The oil palm (*Elaeis guineensis* Jack.) is of African origin and the world's leading source of vegetable oil (USDA, 2014). American oil palm (*Elaeis oleifera* H.B.K. Cortés), a semi-domesticated species of South American origin, is from the same genus as the African oil palm, but has no importance in itself in the commercial production of oil. These two species can be crossed with ease and interspecific hybridization has been used for genetic improvement, seeking to associate the

[2]Embrapa Amazônia Ocidental, Caixa Postal 319, 69010-970 – Manaus, AM, Brasil.
*Corresponding author < wanderlei.lima@embrapa.br

Heat treatment and germination of seeds of interspecific hybrid between American oil palm...

131

high productivity of the African oil palm with characteristics of the American oil palm, such as smaller height, improved oil quality (greater unsaturation) and mainly resistance to diseases, particularly to fatal yellowing (FY) (Cunha et al., 2012), a disease of unknown etiology that has already claimed thousands of hectares of plantations of the African species (Boari, 2008).

From studies started in the 1990s, the first national interspecific hybrid cultivar was developed by Embrapa between American oil palm and African oil palm (HIE OxG), called BRS Manicoré, which, in addition to a high productivity, is resistant to FY, presents a reduced height growth and a more unsaturated oil than the tenera cultivars (intraspecific hybrid) of the African species (Cunha and Lopes, 2010).

In commercial seed production of interspecific hybrid, BRS Manicoré has been showing low rates of germination, about 30 to 35%, according to a personal communication from the business office Escritório de Negócios da Amazônia - Embrapa Produtos e Mercado. Increasing the germination rate of HIE OxG reduces the cost of production and, consequently, the cost of selling to the producer, as well as increasing the supply capacity of germinated seeds of this material, without expanding fields of seed production. According to Guerrero et al. (2011), to the arrival of new OxG cultivars are added new problems and among them stands out the low germination of these seeds. The authors have reported that four Colombian producers: Hacienda La Cabaña, Meta; Unipalma, Meta; Indupalma, Cesar y Corpoica El Mira, Nariño and Embrapa report germination rates of HIE OxG below 30%.

According to Hussey (1958), the African oil palm seeds have low germination rates due to the dormancy that they present after physiological maturity, caused by the mechanical resistance of the endocarp, which has a hard and dense consistency, and the absorption of oxygen, preventing the growth of the embryo. Under natural conditions, as is common in the *Elaeis* genus, germination can take years and is generally non-uniform and low. Studies related to commercial seed germination of African oil palm were developed by Hussey (1958) and Rees (1962), who observed the need to subject seeds to heat treatment to break dormancy. Since then, several procedures have been developed for germination of African oil palm seeds based on the thermal treatment (heating) to break dormancy.

For germination of seeds of HIE OxG, Embrapa has used the method described by Corrado and Wuidart (1990), which consists, in the storage of seeds with moisture content between 7-10% (fresh weight of almond); hydration of seeds by soaking in water for seven days; heat treatment of the seeds in Heat Room Seed Germinator at a temperature from 39 to 40 °C, for 100 days; and at the end, a new hydration

for seven days and then seed germination at a temperature of about 27 °C. This process requires approximately 150 days and the results are not satisfactory, because the germination percentage obtained is low.

New materials have been produced by research to meet the different demands of national African oil palm culture, including HIEs, families of American oil palm and of African oil palm selected in germination germplasm bank. Simultaneously, the production, storage and germination procedures should be improved so that the demand of the farmers for the seeds of the new materials are met in terms of quality and quantity.

The aim of this study was to determine the water content of the seeds and the period of optimal heat treatment to break dormancy and germination of seeds of HIE OxG BRS Manicoré.

Material and Methods

The experiment was conducted at Embrapa Amazônia Ocidental, Campo Experimental do Rio Urubu (CERU), Rio Preto da Eva, AM (2°27'08.44" S, 59°34'13.69" W) and in the laboratory Laboratório de Dendê e Agroenergia, km 29 da Rodovia AM, in Manaus, AM.

Seeds of cultivar BRS Manicoré (Cunha and Lopes, 2010) were used, interspecific hybrid between species *E. oleifera* (origin: Manicoré) and *E. guineensis* (origin: La Mé), produced by Embrapa Amazônia Ocidental. The seeds were produced by controlled pollination, and the bunches were harvested when they had three to five mature fruits naturally detached from the bunch (physiological maturity). After harvesting, the fruits were manually removed from the bunch and mesocarp was extracted in an electric centrifugal depulper. Then the seeds (endocarp, endosperm and embryo) were dried eliminating the ones deformed or damaged by processing, performing a fungicide treatment and then homogenization of seeds and formation of lots to be used in the study.

A completely randomized design was adopted, with three replications of 500 seeds for each treatment, in a 4 x 3 factorial design, with four ranges of moisture content of the seeds and three periods of stay in the Heat Room Seed Germinator (heat treatment). The periods of stay of seeds in the heat treatment (39 ± 1 °C and relative humidity of approximately 75%) were of 55, 75 and 100 days and the range of seed moisture contents of 18 to 19%; 19 to 20%; 20 to 21% and 21 to 22%. The determination of moisture content of the seeds was performed by the oven method at 105 °C ± 2 °C, for 24 hours, according to Brasil (2009), using as sampling four replicates of 10 seeds for each lot of 500 seeds. The ranges of moisture content were obtained and set using the formulas:

$$U(\%) = \left[\frac{(A_1 - A_2)}{A_2}\right] \times 100;$$

$$U_m(\%) = \frac{U_1(\%) + U_2(\%) + U_3(\%) + U_4(\%)}{4};$$

$$PS = \left[\frac{PFT}{(100 + U_m\%)}\right] \times 100;$$

$$U_T(\%) = \left[\frac{P}{PS - 1}\right] \times 100;$$

where:

$U(\%)$ = moisture content of the treatment;

A_1 = initial weight of the moist sample, in grams;

A_2 = final weight of the dry sample (after 24 h in the oven), in grams;

$U_m(\%)$ = average moisture content of the treatment;

$U_1(\%), U_2(\%), U_3(\%), U_4(\%)$ = moisture contents of samples 1, 2, 3 and 4, respectively;

PS = original dry weight of the treatment;

PFT = fresh weight of the treatment;

$U_T(\%)$ = present moisture of the seeds of the treatment.

After adjustment of the moisture content, the seeds of each moisture range were packed in polyethylene bags of 65 x 50 cm and a thickness 0.2 mm, sealed, containing air volume at least equal to the volume of seeds and placed in a Heatbox Seed Germinator (heating chamber controlled by electric resistance, system of digital temperature setting, forced air circulation and monitoring of relative humidity) to be submitted to different periods of heat treatment.

At the end of each period of thermal treatment, the seeds were rehydrated by immersion in water tanks, under oxygen for eight days. Then, the seeds were placed in polyethylene bags (65 cm x 50 cm, thickness 0.2 mm), tightly closed and kept for germination in a room with controlled temperature between 27 e 30 ºC.

The count of germinated seeds was performed four times,

the first being held 15 days after preparation of the seeds in the germination room and the others weekly. The seed with visible protrusion of the hypocotyl-radicle axis was considered germinated. To calculate the percentage of germination, the seeds discarded by contamination and non-germinated were considered as non-germinated. At 15 days the first count was assessed (FC) and after 35 days the percentage of germination was assessed (GERM). At the end of the germination period, the Emergence Speed Index (ESI) was calculated using the formula described by Nakagawa (1999), with adaptations:

$$ESI = G_1/N_1 + G_2/N_2 + ... + G_n/N_n$$

where:

G_n = number of germinated seeds in the nth week.

N_n = n^{th} assessment week of germination.

The data for these variables was tested using the Lilliefors normality test and analysis of variance and means were compared by Tukey test (5 % probability). Statistical analyses were performed using the software ASSISTAT (Silva and Azevedo, 2002).

Results and Discussion

All variables were normally distributed, with the original data submitted to analysis of variance (Table 1). Significant effects of the factors of the period of stay in a heat treatment and moisture content of the seeds were observed, as well as the interaction among them in the variables related to FC and ESI force. In the GERM variable, the effects of the factors were significant and the interaction among them was not significant. The results show that there is a complex interaction in the response of FC and ESI, and the simple effects of the factors should be analyzed, i.e., for each residence time in the heat treatment the answer should be analyzed in relation to changes in moisture content, and vice versa. In the case of GERM, as the interaction effect was not significant, the main effects of the factors were analyzed.

Table 1.   Summary of analysis of variance of germination (GERM), first count (FC) and emergence speed index (ESI) assessed for cultivar BRS Manicoré seeds.

SV	DF	GERM		FC		ESI	
		AS	F	AS	F	AS	F
Heat treatment (F1)	2	1532.97	24.22**	814.68	13.14**	277.80	7.89**
Moiture content (F2)	3	3403.46	53.77**	6048.84	97.55**	1955.36	55.57**
Interaction F1 x F2	6	77.52	1.22ns	317.95	5.13**	120.67	3.43**
Treatment	11	13741.47	19.74**	1971.23	31.79**	649.61	18.46**
Residue	24	63.29		62.01		35.19	
Mean		60.94		39.94		25.42	
CV (%)		13.05		19.71		23.33	

ns – non significant (p ≥ 0.05); ** significant at 1 % probability by the F test; SV – Source of variation; F1 – Factor 1; F2 – Factor 2; DF – Degree of freedom; CV – Coefficient of variation; AS – Average square.

Heat treatment and germination of seeds of interspecific hybrid between American oil palm...

133

The germination of seeds subjected to 55 days of stay in the heat treatment was lower than that obtained in periods of 75 and 100 days, which did not differ statistically with each other (Table 2). The results indicate that the period of 75 days of treatment was sufficient to break dormancy and promote a high germination rate of seeds (68%). Seeds with 21-22% moisture had a germination of 76%, a value with no statistical difference of the 72% of germination of seeds with 20-21% moisture. In the range of 19-20% moisture, seed germination was 63% and did not differ statistically from seed germination with 20-21% moisture (72%). Seed germination with 18 to 19% (33%) was lower than the other treatments. The results indicate that to obtain the highest rates of germination seed, moisture content should be between 21 and 22%.

In Figure 1, it can be seen that in the heat treatment period of 75 days after germination of seeds from treatments with higher contents of moisture (19-20%, 20-21% and 21-22%), it ranged from 71.1 to 84.9%, values that did not differ statistically with each other and were superior to the interval with lower moisture content (18-19%). Similar behavior was obtained for 100 days of heat treatment, i.e., the range of moisture 18-19% was statistically inferior to the others, and

the values of germination of seeds with higher moisture were not statistically different from each other. It was also found that the germination rates obtained in the three upper ranges of moisture content were not statistically different between the periods of heat treatment of 75 and 100 days.

Table 2. Germination (%) of seeds of the interspecific hybrid between *Elaeis oleifera* and *E. guineensis* (cultivar BRS Manicoré) according to the period of heat treatment (39 °C ± 1 °C) and moisture content of the seeds.

Heat treatment period (days)	Germination (%)
55	48 b
75	68 a
100	67 a
Moisture content of seeds (%)	Germination (%)
18-19	33 c
19-20	63 b
20-21	72 ab
21-22	76 a

Means followed by the same letter in the column do not differ among themselves by the Tukey test at 5% probability.

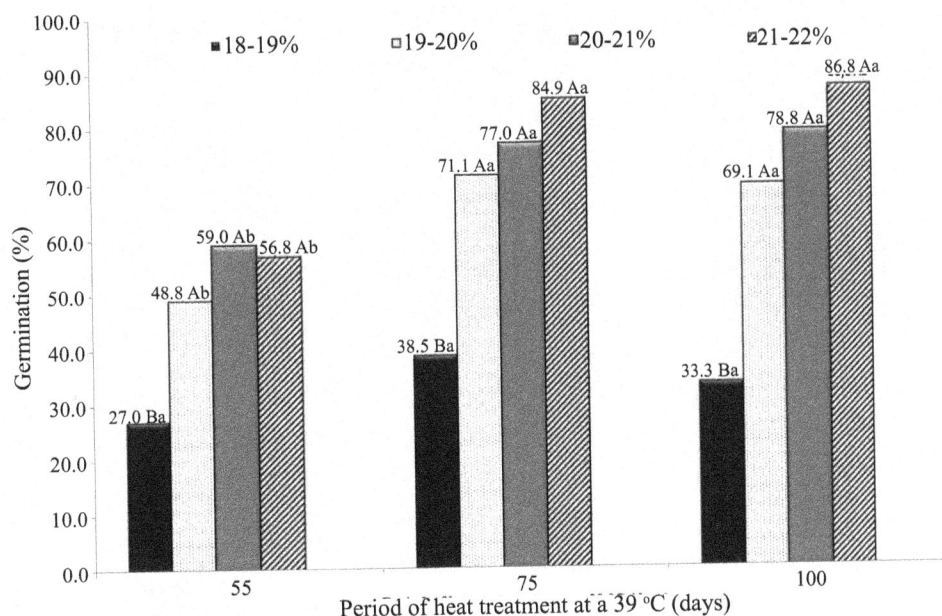

Figure 1. Germination (%) of seeds of BRS Manicoré in different ranges of moisture content (18-19%, 19-20%, 20-21% and 21-22%), subjected to periods of stay in the heat treatment of 55, 75 and 100 days. Means followed by equal uppercase letters within the periods of stay, and by equal lowercase letters within the same range of moisture, do not differ with each other by the Tukey test (p < 0.05).

To maximize the germination process of commercial seeds, one must consider the percentage of germination obtained with the combination of different treatments and

costs, mainly energy spent for heating and maintaining the heat room and specific worker for weekly reviews of lots (exchange air from the bags, check incidence of fungi, assessment of germination). Considering these aspects, as no significant differences between the germination of the seeds were observed in three ranges of higher contents of moisture and between the periods of heat treatment of 75 to 100 days, the minimum time should be used, i.e., 75 days, by adjusting seed moisture content to values between 19-22 %, conditions in which germination values higher than 71% were obtained. The water content of the seeds to be subjected to heat treatment is directly related to the probability of incidence of fungi, since the seeds are exposed for more than two months at ambient temperature and relative humidity conducive to growth of microorganisms, namely, 39 °C and 75%, respectively.

In the process of these seeds after heat treatment, weekly assessments are conducted in order to separate and count the germinated seeds. Information obtained in these counts can be used to calculate the speed of germination, which are related to seed vigor.

Significant differences between FC and ESI variable averages were observed, by the effect of moisture and by the heat treatment time (Figures 2 and 3). Significantly higher values for FC and ESI were observed in the highest ranges of moisture, especially during the heat treatment of 75 and 100 days, in which the averages for the maximum range of moisture (21-22%) were higher than the other treatments and did not differ statistically with each other.

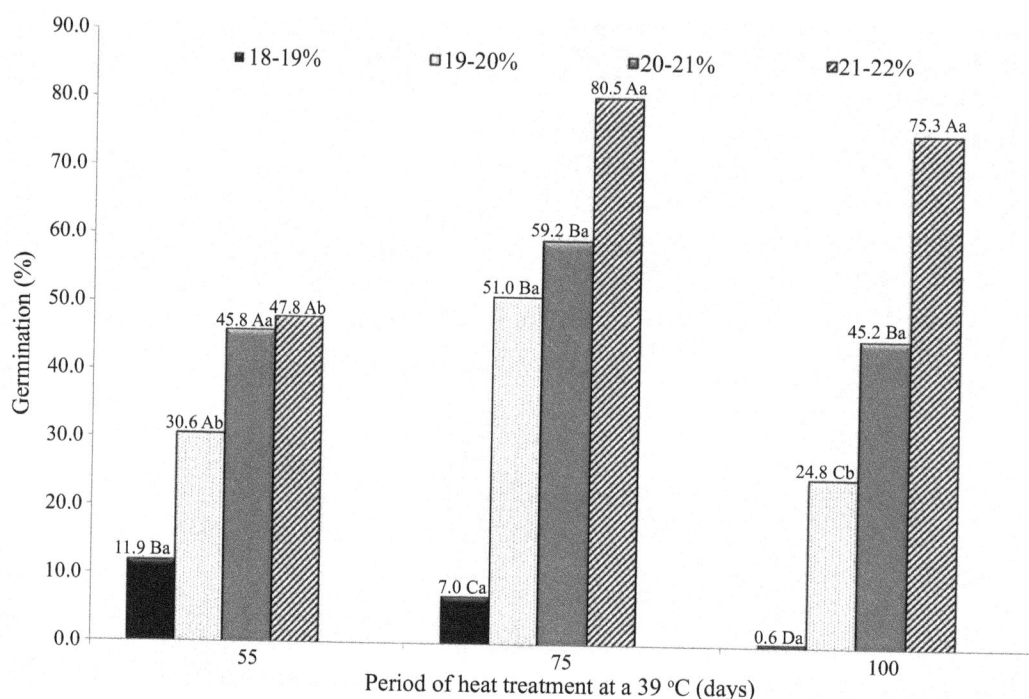

Figure 2. Germination in the first count (%) of seeds of BRS Manicoré in different ranges of moisture content (18-19%, 19-20%, 20-21% and 21-22%), subjected to periods of stay in the heat treatment of 55, 75 and 100 days. Means followed by equal uppercase letters within the periods of stay, and by equal lowercase letters within the same range of moisture, do not differ with each other by the Tukey test (p < 0.05).

Methods for rapid and uniform germination of African oil palm seeds, in commercial scale, go back to the middle of the last century, but there is still the need to investigate new practices and/or techniques involved in the germination process, taking into account the retelling of the period required for heat treatment to break dormancy with the market launch of genetically distinct materials. Recent example is the work done by Green et al. (2013), who studied the effect of heating periods (40, 50, 60 e 80 days) on the germination of six cultivars of tenera of African oil palm Deli x La Mé produced by Embrapa,

in which the results showed different responses, depending on the cultivar, and the maximum value can be obtained germination between 45 and 80 days of heat treatment, with varying percentages of germination of 70-92%. Fondom et al. (2010), also analyzing the germination of seeds of ten progenies of African oil palm, found that seeds with moisture content of 18% germinated better when subjected to 39 ± 1 °C for 60 days, compared to periods of 80, 100 and 120 days.

Even with the growing demand for HIEs, especially in Latin America, most of the required methodology for germination of these hybrids are still based on procedures for the African oil palm. The specific studies on the germination of HIE OXG are recent. Guerrero et al. (2011) assessed the effect of varying moisture content of the seeds at the beginning of the heat treatment and different periods of heat treatment. These authors have obtained better results with seeds with 20% and 22% moisture and heat treatment for 70 days at 39 °C, reaching an average germination of 40.4

and 36.3%, respectively, which are lower values compared to the germination of hybrid seeds of HIEs BRS Manicoré in this study, where germination above 70% was obtained. The authors have attributed the low percentage of germination to endogenous factors (sterility and immature embryo), mainly because it is an artificial cross between two different species and exogenous factors (latency and chemical inhibitors) that induce prolonged latency of seeds.

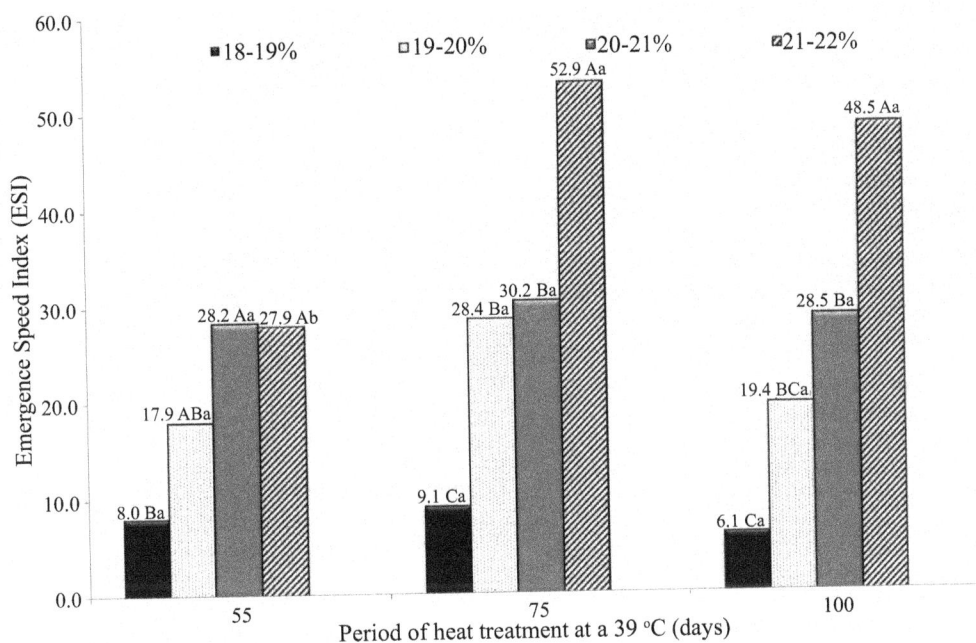

Figure 3. ESI of seedlings of BRS Manicoré in different ranges of moisture content (18-19%, 19-20%, 20-21% and 21-22%), subjected to periods of stay in the heat treatment of 55, 75 and 100 days. Means followed by equal uppercase letters within the periods of stay, and by equal lowercase letters within the same range of moisture, do not differ with each other by the Tukey test (p < 0.05).

The study by Guerreiro et al. (2011) was developed with the same limits of moisture content when entering in the heat treatment as in the present study, of 18 % and 22%; however, dividing the treatments into three moisture levels (18, 20 and 22%), whereas in the present four ranges of moisture were used (18-19%, 19-20%, 20-21% and 21-22%). Guerrero et al. (2011) recommended in their study that moisture levels should be explored next to the ones where they have obtained the best results, 20-22%, and in this range of seed moisture the highest values of germination were obtained, in this study. With respect to the period of stay of the seeds in the heat treatment, Guerrero et al. (2011) used five treatments (60, 70, 80, 90 and 100 days) and obtained a better germination in 70 days, an amount next to the one identified as the best treatment in this study, of 75 days. It is noteworthy that the female progenitors matrices (American oil palm) of the

two studies are of different origins, Colombian in the case of Guerreiro et al. (2011) and Brazilian in this study. As a certainty, the results obtained by Guerreiro et al. (2011) and the one obtained in this study show that it is possible to increase the rate of germination of HIEs, reduce spending on the germination process and optimize the use of facilities by testing combinations of exposure time and moisture content of the seeds in the treatment of dormancy break.

Conclusions

To break dormancy and germination of hybrid seeds of HIE OxG BRS Manicoré, seeds with water content adjusted to values between 19 and 22% should be used, and heat treatment must be conducted at temperatures of 39 ± 1 °C and relative humidity of approximately 75% for 75 days.

Acknowledgments

To the workers at the 'Laboratório de Dendê e Agroenergia' da Embrapa Amazônia Ocidental, scholarship fellows and co-workers of the African oil palm staff for their support in carrying out the work.

References

BOARI, A.J. Estudos realizados sobre o amarelecimento fatal do dendezeiro (*Elaeis guineensis* Jacq) no Brasil. Belém, 2008. 66p. (Embrapa-CPATU. Documentos, 348). http://ainfo.cnptia.embrapa.br/digital/bitstream/item/27984/1/Doc348.pdf

BRASIL. Ministério da Agricultura, Pecuária e Abastecimento. *Regras para análise de sementes*. Ministério da Agricultura, Pecuária e Abastecimento. Secretaria de Defesa Agropecuária. Brasília: MAPA/ACS, 2009. 395p. http://www.agricultura.gov.br/arq_editor/file/2946_regras_analise__sementes.pdf

CORRADO, F.; WUIDART, W. Germination des graines de palmier à huile (*E. guineensis*) em sacs de polyétylène. Méthode par 'charleur sèche'. *Oléagineux*, v.45, n.11, p.511-514, 1990. http://www.cabdirect.org/abstracts/19920312796.html;jsessionid=42a420545b790911114d263e398118f5?freeview=true

CUNHA, R.N.V.; LOPES, R.; ROCHA, R.N.C.; LIMA, W.A.A.; TEIXEIRA, P.C.; BARCELOS, E.; RODRIGUES, M.R.L; RIOS, S.A. Domestication and Breeding of the American Oil Palm. In: BORÉM, A.; LOPES, M.T.G.; CLEMENT, C.R. (Ed.). *Domestication and Breeding*: Amazonian species. Viçosa: UFV, p.275-296, 2012.

CUNHA, R N.V.; LOPES, R. *BRS Manicoré*: híbrido interespecífico entre o caiaué e o dendezeiro africano recomendado para áreas de incidência de amarelecimento-fatal. Manaus, 2010. 4p. (Embrapa Amazônia Ocidental. Comunicado Técnico, 85). http://www.snt.embrapa.br/publico/usuarios/produtos/85-Anexo1.pdf

FONDOM, N.Y.; ETTA, C.E.; MIH, A.M. Breaking Seed Dormancy: Revisiting heat-treatment duration on germination and subsequent seedling growth of oil palm (*Elaeis guineensis* Jacq.) progenies. *Journal of Agricultural Science*, v.2, n.2, p.101-110, 2010. http://www.cabdirect.org/abstracts/20113238443.html

GUERRERO, J.; BASTIDAS, S.; GARCÍA, J. Estandarización de uma metodologia para germinar semillas del híbrido interespecífico *Elaeis oleifera* H.B.K. x *Elaeis guineensis* J. *Revista Ciências Agrárias*, v.1, n.1, p.132-146, 2011. http://www.researchgate.net/publication/233997924_Estandarizacin_de_una_metodologa_para_germinar_semillas_del_hbrido_interespecfico_Elaeis_oleifera_H.B.K._x_Elaeis_guineensis_Jaq

GREEN, M.; LIMA, W.A.A.; FIGUEIREDO, A.F.; ATROCH, A.L.; LOPES, R.; CUNHA, R.N.V.; TEIXEIRA, P.C. Heat treatment and seed germination of oil palm (*Elaeis guineensis* Jacq.). *Journal of Seed Science*, v. 35, n.3, p.296-301, 2013. http://www.scielo.br/scielo.php?pid=S2317-15372013000300004&script=sci_arttext

HUSSEY, G. An analysis of the factors controlling the germination of seed of the oil palm, *Elaeis guineensis* (Jacq.). *Annals of Botany*, v.22, n.86, p.259-284, 1958. http://aob.oxfordjournals.org

NAKAGAWA, J. Testes de vigor baseados no desempenho das plântulas In: KRZYZANOWSKI, F. C.; VIEIRA, R. D.; FRANÇA-NETO, J. B. (Eds.). *Vigor de sementes:* conceitos e testes. Londrina: ABRATES, p. 2.1-2.24, 1999.

REES, A.R. High-temperature pre-treatment and germination of seed of oil palm, *Elaeis guineensis* (Jacq.). *Annals of Botany*, v.26, n.4, p.569-581, 1962. http://aob.oxfordjournals.org/search?fulltext=Rees%2C+A.R.+v.26%2C+n.104%2C+p.569-581%2C+1962&submit=yes&x=0&y=0

SILVA, F.A.S.; AZEVEDO, C.A.V. Versão do programa computacional Assistat para o sistema operacional Windows. *Revista Brasileira de Produtos Agroindustriais*, v.4, n.1, p.71-78, 2002. http://www.deag.ufcg.edu.br/rbpa/rev41/Art410.pdf

USDA – United States Department of Agriculture. *Oil Seeds*: World Markets and Trade. Foreign Agricultural Service. http://apps.fas.usda.gov/psdonline/circulars/oilseeds.pdf. Accessed on: Jul,03rd, 2014.

19

SGlu2 gene expression in coats of soybean seeds

Carlos André Bahry[1]*, Paulo Dejalma Zimmer[2]

ABSTRACT – Glucanases can act in plant defense against biotic factors. Despite its importance, research to study the expression of genes encoding glucanases in soybean seed coats is limited. The aim of this study was to assess the relative expression of the SGlu2 gene (β-1.3-Glucanase 2), possibly involved in defense against biotic factors, in coats of seeds of four soybean genotypes. Two genotypes of black seed coats, IAC and TP, and two of yellow seed coats, BMX Potência RR and CD 202 were used. Seeds were multiplied in a greenhouse at Embrapa Clima Temperado – ETB, and the gene expression assay was performed at the Laboratório de Sementes e Biotecnologia, UFPel. Seed coat gene expression was assessed by qPCR technique in four development stages: 40, 45, 50 and 55 days after anthesis. The SGlu2 gene shows more expression in the BMX Potência RR genotype compared to other genotypes. The gene expression in the seed coat is constant in different development stages of CD 202 cultivar and IAC and TP strains, except at 45 DAA (days after application) for this latter genotype.

Index terms: *Glycine max*, seed development stages, qPCR technique.

Expressão do gene SGlu2 em tegumentos contrastantes de sementes de soja

RESUMO – As glucanases podem atuar na defesa das plantas contra fatores bióticos. Apesar da sua importância, pesquisas visando estudar a expressão de genes que codificam glucanases em tegumento de sementes de soja são limitadas. O objetivo do trabalho foi avaliar a expressão relativa do gene SGlu2 (β-1,3-Glucanase 2), possivelmente envolvido na defesa contra fatores bióticos, em tegumento de sementes de quatro genótipos de soja. Foram utilizados dois genótipos de tegumentos pretos, IAC e TP, e dois de tegumentos amarelos, BMX Potência RR e CD 202. As sementes foram multiplicadas em casa de vegetação na Embrapa Clima Temperado – ETB, sendo o ensaio de expressão gênica realizado no Laboratório de Sementes e Biotecnologia da UFPel. A expressão do gene no tegumento das sementes foi avaliada pela técnica qPCR, em quatro fases de desenvolvimento: 40, 45, 50 e 55 dias após a antese. O gene SGlu2 apresenta maior expressão no genótipo BMX Potência RR em relação aos demais genótipos. A expressão do gene no tegumento das sementes é constante nas diferentes fases de desenvolvimento da cultivar CD 202 e das linhagens IAC e TP, com exceção aos 45 DAA para este último genótipo.

Termos para indexação: *Glycine max*, fases de desenvolvimento de semente, qPCR.

Introduction

Coats perform many important functions during the development of soybean seeds, such as temporary storage and supply of nutrients to an embryo being formed, modulation of interactions among the internal structures of the seed and the external environment, process control of germination and dormancy, among others (Marcos-Filho, 2005; Moise et al., 2005; Weber, 2005).

As a structure protection, seed coats play a unique role against biotic and abiotic factors that may compromise the physiological, physical and sanitary qualities of the seeds, especially in the final stages of development and after physiological maturity, which involves the expression of many genes (Moise et al., 2005; Senda et al., 2004). This expression is variable among genotypes, among tissues and among stages of development of each tissue and also suffers direct influence of external factors.

Among the expressed gene families in plant tissues that can act to protect these ones against biotic and abiotic factors are the β-1.3-glucanases, which have, among others, the function of defense against invading pathogens, particularly in synergism with chitinases, degrading the homopolymer β -1.4-N-acetylglucosamine, an abundant cell wall component of many pathogens (Bishop et al., 2005; Jin et al., 1999).

[1]Departamento de Ciências Agrárias, UTFPR, 85.503-390 – Pato Branco, PR, Brasil.

[2]Departamento de Fitotecnia, UFPel, Caixa Postal 354, 96010-900 – Capão do Leão, RS, Brasil.
*Corresponding author: <carlosbahry@hotmail.com>

Despite the recognized importance of glucanases for vegetables, there are few studies related to its expression in seed coats of soybean seeds. Based on this, the application of specific molecular biology techniques can contribute to the development of superior constitutions, reflecting the quality of the seeds produced, especially when the research emphasizes the study of soybean genotypes with contrasting seed coats, which have a higher genetic variability (Liu et al., 2007; Mertz et al., 2010; Ranathunge et al., 2010; Tuteja et al., 2004).

Therefore, this study aimed to assess the relative expression of the SGlu2 gene, possibly involved in the defense against biotic and abiotic factors, by means of the qPCR technique, in seeds of four soybean genotypes with contrasting seed as to the color of the coats and at different stages development.

Materials and Methods

Four genotypes of soybeans, TP and IAC strains, both with seeds featuring black coats and CD 202 (conventional) and BMX Potência RR (GM) cultivars were used, both with seeds of yellow coat.

The multiplication of the plant material was performed in a greenhouse at Embrapa Clima Temperado – Estação Terras Baixas (CPACT/ETB), located in the Brazilian city of Capão do Leão/RS, in harvest 2012/2013.

From the anthesis, the marking of flowers was performed, so that all the seeds sampled were at the same stage of development. Four samples of vegetables were collected every five days (40, 45, 50 and 55 days after anthesis), for each genotype.

Immediately after each collection at Laboratório de Sementes e Biotecnologia, Universidade Federal de Pelotas, the coat of about fifteen seeds was removed with the aid of sterile blades, and care was taken to keep the plant tissue free of impurities. Once separated from the seeds, the coats of each genotype were stored in Ultrafreezer at -80 °C until the procedure for obtaining RNA.

RNA was extracted using *Concert Plant RNA Reagent* (Invitrogen™). The extraction was performed at the same time for all treatments (sampling dates of coats and contrasting genotypes). After extraction, RNA samples were treated with DNase and their purity and integrity were assessed by analysis of absorbance (260/280 nm) and electrophoresis in 1% agarose gel. RNA extraction and cDNA synthesis were performed using three biological replicates and each replicate consisted of a mixture of coats of seeds in each stage assessed.

Single-stranded cDNAs were synthesized by reverse transcription using the *SuperScript III®* (Invitrogen™) enzyme, Single-stranded cDNAs, according to the manufacturer's instructions. To assess the quality of the cDNA, semi-quantitative PCR reaction was performed using *Master mix Go Taq*, cDNA of each sample, water and ß-actin. Purity and integrity of the cDNA were also measured to ensure the quality of the material used.

For the design of the primer pair for amplification of the gene SGlu2 (β-1.3-Glucanase 2), access AF034107.1, (sense: 5' CGGCGTGTGTTATGGAAGACTTGG 3' and antisense: 5' CTGAAAC GTATCTGAATCTGACATTGTTGGCATAG 3') was carried out by a search of the EST sequences of proteins corresponding to genes at the *National Center for Biotechnology Information* (NCBI) database.

Primers were designed with the aid of the program Vector NTI Advance 11.0 (Invitrogen™, 2008), observing the parameters of the annealing temperature, primer size, percentage of GC (40-60%), size of the amplicon, absence of dimerization and absence of secondary annealing sites.

Five normalizing genes were tested preliminarily, ACT11, SKIP16, UKN1, UKN2 and ß-ACTIN, opting by normalizers ACT11 and SKIP16, which showed lower variation of expression.

After the relative quantitation, the quality of the amplified product was found by means of the dissociation curve at the end of qPCR (Bustin et al., 2009) by the gradual increase of the reaction temperature. Thus, the calculation related to the fluorescence emission was carried out, assessing the analytical specificity of the primers by denaturing the PCR product generated.

The efficiency of the SGlu2 gene was determined by performing a curve with serial dilutions of 1:3; 1:30; 1:300; 1:3000. After the dilutions it was possible to perform the calculation of efficiency through the slope of each curve, following the formula: $E = |10^{(-1/slope)}|-1$ (Zhao and Fernald, 2005), obtaining the value of *slope* (S) of -3.687 and Efficiency (E) of 0.867339.

Quantitative analysis of the gene expression in real-time of the target gene was performed in the equipment LightCycler 480 Instrument II (96) (*Roche Applied Science®*) using *SYBR® Green*.

At the end of the tests of reaction was obtained the C_q (*Quantification Cycle*) of the increase in fluorescence occurring during the reaction cycles. The optical data were subsequently analyzed using the program Light Cycler® 480 *Gene Scanning Software*.

The relative gene expression was calculated based on the amplification efficiency (E) and in the PCR cycle, in which was found the increase in fluorescence above the baseline signal (Pfaffl, 2001). After obtaining the values of relative expression of the SGlu2 gene, this one was normalized to the values observed in control, adopting for such a development stage of 40 days after anthesis (first stage of collection) for the

IAC genotype, because it is a rustic material, which has not undergone any process of artificial selection or improvement, and it may present some features of interest.

The gene expression results were submitted to analysis of variance and then compared by a test of means, applying Scott-Knott, at 5% probability.

Results and Discussion

The analysis of variance indicated an interaction among genotypes with seeds showing different colors of coats and stages of their development, for the assessed gene (Table 1). Thus, the average expression data were compared among genotypes for each stage of development of the coats and among the developmental stages of the coats, for each genotype (Figure 1).

Table 1. Summary of analysis of variance for the relative expression of the SGlu2 gene in contrasting coats of four soybean genotypes.

Sources of variation	DL	Medium square SGlu2
Genotypes (F1)	3	6821.69779*
Coat collection times (F2)	3	567.79720*
Interaction F1 x F2	9	487.13848*
Treatments	15	1770.18208
Residue	16	37.10850

DL – Degrees of liberty. *Significant at the 1% level of probability by F test.

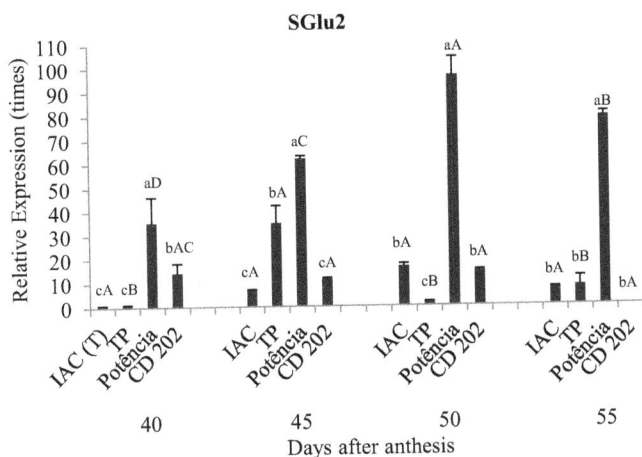

Figure 1. Relative expression of the SGlu2 gene in soybean seed coats collected in four stages of development after anthesis in four contrasting genotypes for the characteristics of coats.
[1]Means followed by different letters, smaller lower case among genotypes, within each collection time, and larger upper case within each genotype and among sampling times, differ by the Scott-Knott test at 5% probability.

After 40 days after anthesis (DAA), the highest transcript accumulation was observed in the coat of cultivar BMX Potência RR, followed by another cultivar with a yellow coat, CD 202. The lowest expression was observed in the genotypes of black coat, IAC and TP, which did not differ (Figure 1).

At 45 DAA, the expression of the gene under study remained higher in cultivar BMX Potência RR. However, cultivar CD 202 decreased in expression, together with the IAC strain, with no differences among them. Intermediate value of expression occurred in the TP strain (Figure 1).

After fifty days of development of the seeds, the lowest SGlu2 gene expression took place in the strain of TP black coat. Intermediate values were observed in the coat of the seeds of cultivar CD 202 and IAC strain, which did not differ. The higher expression of the gene remained seen in cultivar BMX Potência RR, as at 55 DAA. However, in this last stage of development, gene expression did not differ among the other genotypes assessed (Figure 1).

In the individual comparison of genotypes, the expression of the SGlu2 gene in IAC did not vary among the different stages of seed development, showing no significant difference (Figure 1).

For the TP black coat strain, the highest gene expression took place at 45 DAA, differing from the other coat development stages, which did not differ (Figure 1).

Regarding the BMX Potência RR, the highest accumulation of transcripts of the SGlu2 gene took place at 50 DAA, followed by 55 DAA, both differing. As for the lowest expression, it was observed at 40 DAA and with an intermediate value among this one and 55 DAA, the expression that took place at 45 DAA (Figure 1).

A situation similar to the one seen for the IAC genotype took place at the CD 202 genotype, in which there was not a significant difference regarding the gene expression among the different development stages of the coats (Figure 1).

Although the literature does not provide information on the SGlu2 gene expression in soybean seeds coats, it is known, based on studies performed in other plant tissues, in tobacco, that it is related to class III of proteins related to the pathogenesis, PR-Q, showing a higher transcript accumulation in defense systems against attack by pathogens (Payne et al., 1990); therefore, it is of interest to the seed area to identify genotypes that have a higher expression of these genes in the coats, indicating the possibility of them having higher resistance against biotic factors that might affect the seed quality yield.

In a study seeking to analyze and map the gene family that encodes β-1.3-glucanases in soybeans, Jin et al. (1999) found that pathogen invasion is not a prerequisite for the expression of genes of this family, such as for SGlu2, and may be an intrinsic characteristic of the genotype, which is in agreement to what was observed by Memelink et al. (1990). This finding is important since the design of the primers of this gene was based on access from work developed by the authors. They also show that this particular gene showed a higher accumulation of mRNAs in young roots and hypocotyl of soybeans, being probably located in the extracellular spaces of tissues.

Although the results from Jin et al. (1999) and Memelink et al. (1990) indicate a role for preventive or proactive defense (regardless of microbial attack), providing some measure of protection against infection by pathogens during the crucial stages of the life cycle in which the tissues of plants are more susceptible, it is not possible to rule out the role of the β-1.3-glucanases in plant defense post-infection of microorganisms (Knogge et al., 1987).

In order to better assess the role of β-1.3-glucanase in *Arabidopsis thaliana* under infection of plant parasitic nematodes, Hamamouch et al. (2012) found that in mutants that overexpressed the genes encoding these pathogenesis-related enzymes there was a lower susceptibility of the plants to *Heterodera schachtii*, homologous to *Heterodera glycines* nematodes, which attack soybean plants. As for mutants in which there was a silencing of these genes, there was an increased susceptibility of *Arabidopsis* plants to attack by nematodes of the species *Heterodera schachtii*, once again proving the important role played by genes encoding glucanases enzymes to the defense system of plants under attack by pathogen microorganisms.

However, it is noteworthy that the expression of genes encoding glucanase enzymes, giving greater protection of plants to attack by microorganisms, is something very complex and dynamic, which deserves to be analyzed with discretion, since there is a constant evolution in the pathogen/host relationship regarding this system. The phytopathogenic microorganisms of the genus *Phytophthora* are an example, which, to infect plants, have glucanases inhibitory of proteins encoded, thus allowing infection to occur in the host (Damasceno et al., 2008).

Besides being related to the pathogenesis, another attribution has been given to glucanases, in this case to β-1.3-glucanase 1. According to Leubner-Metzger (2005), in seeds of *Nicotiana tabacum*, the genes encoding these enzymes have been playing a key role after the maturity of the seeds, under a minimum threshold of moisture, giving these a break of dormancy imposed by the seed coat, which is an important line of research in the area of seeds with such characteristics, and should be further investigated.

Conclusions

In comparing the genotypes, the SGlu2 gene has a higher expression in the BMX Potência RR coats, at all stages of development studied.

The highest expression of SGlu2 gene occurs 50 days after anthesis in BMX Potência RR genotype, and in the other genotypes, the expression is constant at different stages of development of the coats, except for TP, at 45 days after anthesis.

Acknowledgments

To CNPq, FAPERGS and CAPES for granting the scholarship and other financial aid.

References

BISHOP, J.G.; RIPOLL, D.R.; BASHIR, S.; DAMASCENO, C.M.B.; SEEDS, J.D.; ROSE, J.K.C. Selection on *Glycine* β-1,3-endoglucanase genes differentially inhibited by a *Phytophthora* glucanase inhibitor protein. *Genetics*, v.169, n.2, p.1009-1019, 2005. http://www.ncbi.nlm.nih. gov/pmc/articles/PMC1449112/pdf/5098.pdf.

BUSTIN, S.A.; BENES, V.; GARSON, J.A.; HELLEMANS, J.; HUGGETT, J.; KUBISTA, M.; MUELLER, R.; NOLAN, T.; PFAFFL, M.W.; SHIPLEY, G.L.; VANDESOMPELE, J.; WITTWER, C.T. *The MIQE guidelines*: minimum information for publication of quantitative real time PCR experiments. *Clinical Chemistry*, v.55, n.4, p.611-622, 2009. http://www.clinchem. org/content/55/4/611.full.pdf+html.

DAMASCENO, C.M.B.; BISHOP, J.G.; RIPOLL, D.R.; WIN, J.; KAMOUN, S.; ROSE, J.K.C. Structure of the glucanase inhibitor protein (GIP) family from *Phytophthora* species suggests coevolution with plant endo-β-1,3-Glucanases. *Molecular Plant-Microbe Interactions*, v.21, n.6, p.820-830, 2008. http://kamounlab.dreamhosters.com/pdfs/MPMI_08b.pdf.

HAMAMOUCH, N.; LI, C.; HEWEZI, T.; BAUM, T.J.; MITCHUM, M.G.; HUSSEY, R.S.; VODKIN, L.O.; DAVIS, E.L. The interaction of the novel 30C02 cyst nematode effector protein with a plant β-1,3-glucanase may suppress host defense to promote parasitism. *Journal of Experimental Botany*, v.63, n.10, p.3683-3696, 2012. http://jxb.oxfordjournals.org/content/early/2012/03/21/jxb.ers058.full.pdf+html.

INVITROGEN CORPORATION™. User Manual Vector NTI Advance 11.0, 2008.

JIN, W.; HORNER, H.T.; PALMER, R.G.; SHOEMAKER, R.C. Analysis and mapping of gene families encoding β-1,3-glucanases of soybean. *Genetics*, v.153, n.1, p.445-452, 1999. http:// www.ncbi.nlm.nih.gov/pmc/articles/PMC1460737/pdf/10471725.pdf.

KNOGGE, W.; KOMBRINK, E.; SCHMELZER, E.; HAHLBROCK, K. Occurrence of phytoalexins and other putative defense-related substances in uninfected parsley plants. *Planta*, v.171, n.2, p.279-287, 1987. http://libgen.org/scimag/?s=Occurrence+of+phytoalexins+and+ other+putative+defense-related+substances+in+uninfected+parsley+plants.+&siteid=&v=&i=&p=&redirect=1.

LEUBNER-METZGER, G. β-1,3-Glucanase gene expression in low-hydrated seeds as a mechanism for dormancy release during tobacco after-ripening. *The Plant Journal*, v.41, n.1, p.133-145, 2005. http://www.seedbiology.de/pdf/PlantJ2005.pdf.

LIU, B.; FUJITA, T.; YAN, Z.E.; SAKAMOTO, S.; XU, D.; ABE, J. QTL Mapping of domestication-related traits in soybean (*Glycine max*). *Annals of Botany*, v.100, n.5, p.1027-1038, 2007. http://aob.oxfordjournals.org/content/100/5/1027.full.pdf+html.

MARCOS-FILHO, J. *Fisiologia de sementes de plantas cultivadas*. Piracicaba: FEALQ, 2005. 495p.

MEMELINK, J.; LINTHORST, H.J.M.; SCHILPEROORT, R.A.; HOGE, J.H.C. Tobacco genes encoding acidic and basic isoforms of pathogenesis-related proteins display different expression patterns. *Plant Molecular Biology*, v.14, n.2, p.119-126, 1990. http://libgen.org/scimag/ ?s=Tobacco+genes+encoding+acidic+and+basic+isoforms+of+pathogenesis-related+ proteins + display +different+expression+patterns.&siteid=&v=&i=&p=&redirect=1.

MERTZ, L.M.; HENNING, F.A.; BORSUK, S.; MAIA, L.C.; DELLAGOSTIN, O.A.; PESKE, S.T.; ZIMMER, P.D. cDNA-AFLP analyses between black and yellow soybean seed coats. *Seed Science and Technology*, v.38, p.88-95, 2010. http://www.ingentaconnect.com/content/ista/sst/2010/00000038/00000001/art00009.

MOISE, J.A.; HAN, S.; GUDYNAITE-SAVITCH, L.; JOHNSON, D.A.; MIKI, B.L.A. Seed coats: structure, development, composition, and biotechnology. *In Vitro Cellular and Developmental Biology – Plant*, v. 41, n. 5, p. 620-644, 2005. http://libgen.org/scimag/?s=.+Seed+ coats%3A+-structure%2C+development%2C+composition%2C+and+biotechnolo-gy.+&siteid=&v=&i=&p=&redirect=1.

PAYNE, G.; WARD, E.; GAFFNEY, T.; GOY, P.A.; MOYER, M.; HARPER, A.; MEINS, F.; RYALS, J. Evidence for a third structural class of β-1,3-glucanase in tobacco. *Plant Molecular Biology*, v.15, n.6, p.797-808, 1990. http://libgen.org/scimag/?s=Evidence+for+third+structural+class+of+b-1%2C3-glucanase+in+tobacco&siteid=&v=&i=&p=& redirect = 1.

PFAFFL, M.W. A new mathematical model for relative quantification in real time RT – PRC. *Nucleic Acids Research*, v.29, n.9, p.2003-2007, 2001. http://nar.oxfordjournals.org/content/29/9/ e45.full.pdf+html.

RANATHUNGE, K; SHAO, S.; QUTOB, D.; GIJZEN, M.; PETERSON, C.A.; BERNARDS, M.A. Properties of the soybean seed coat cuticle change during development. *Planta*, v.231, n.5, p.1171-1188, 2010. http://libgen.org/scimag/?s=Properties+of+the+soybean+seed+coat+cuticle+change+during+development.+&siteid=&v=&i=&p=& redirect=1.

SENDA, M.; MASUTA, C.; OHNISHI, S.; GOTO, K.; KASAI, A.; SANO, T.; HONG, J.S.; MACFARLANE, S. Patterning of virus-infected *Glycine max* seed coat is associated with suppression of endogenous silencing of chalcone synthase genes. *The Plant Cell*, v.16, n.4, p. 807–818, 2004. http://www.plantcell.org/content/16/4/807.full.pdf+html.

TUTEJA, J.H.; CLOUGH, S.J.; CHAN, W.C.; VODKIN, L.O. Tissue-specific gene silencing mediated by a naturally occurring chalcone synthase gene cluster in *Glycine max*. *The Plant Cell*, v.16, n.4, p.819–835, 2004. http://www.plantcell.org/content/16/4/819.full.pdf+html.

WEBER, H.; BORISJUK, L.; WOBUS, U. Molecular physiology of legume seed development. *Annual Review of Plant Biology*, v.56, n.1, p.253–279, 2005. http://libgen.org/scimag/index. php?s=&siteid=1040-2519and1545-2123&v=56&i=1.

ZHAO, S.; FERNALD, R.D. Comprehensive algorithm for quantitative real-time polymerase chain reaction. *Journal of Computational Biology*, v.12, n.8, p.1047-1064, 2005. http://www.ncbi.nlm. nih.gov/pmc/articles/PMC2716216/pdf/nihms114913.pdf.

20

Relationship between pod permeability and seed quality in soybean

Carolina Maria Gaspar de Oliveira[1]*, Francisco Carlos Krzyzanowski[2],
Maria Cristina Neves de Oliveira[2,] José de Barros França-Neto[2],
Ademir Assis Henning[2]

ABSTRACT – The aim of this study was to evaluate the influence of pod wall permeability on the physiological quality of soybean seed. The cultivars studied were Sant'Ana, FT-2, FT-10, Bossier, Davis and the breeding line F 84-7-30, with a black seed coat. Pods were collected from plants at the R4, R5, R6, R7 and R8 development stages, which composed the treatments in regard to time of harvest. The parameters of permeability and the lignin content of the pods and the seeds within the pods were evaluated. The seeds were collected just after full maturity (R8), and the following tests were performed: germination, electrical conductivity, and tetrazolium, which determined seed viability and vigor. A randomized complete block design in a split-plot in time arrangement was used, with four replications per treatment. The soybean genotypes (six) composed the plots, and the split-plots consisted of the development stages (R4, R5, R6, R7 and R8). In seed evaluation, the same design was used, reducing the number of treatments to three in the split-plots (R6, R7 and R8). Pod permeability varied with the genotype and stage of development; this affected seed vigor, but not the viability of newly-harvested seeds. The pod lignin content did not show any influence on pod permeability.

Index terms: *Glycine max* (L.) Merrill., imbibition, lignin.

Relação entre a permeabilidade da vagem e a qualidade da semente de soja

RESUMO - O objetivo do trabalho foi avaliar a influência da permeabilidade da parede da vagem na qualidade fisiológica das sementes de soja. As cultivares utilizadas foram Sant'Ana, FT-2, FT-10, Bossier, Davis e a linhagem F 84-7-30 com tegumento na cor preto. Coletaram-se vagens nos estádios de desenvolvimento R4, R5, R6, R7 e R8, sendo que os mesmos compuseram os tratamentos quanto à época de coleta. Avaliaram-se a permeabilidade e o conteúdo de lignina das vagens e das sementes no interior das mesmas. As sementes para os estudos foram obtidas logo após a maturidade plena (R8), sendo avaliadas a germinação, a condutividade elétrica e a viabilidade e o vigor pelo teste de tetrazólio. O delineamento experimental na avaliação das vagens foi em blocos ao acaso com parcelas subdivididas, sendo nas parcelas genótipos de soja (seis) e nas subparcelas estádios de desenvolvimento (R4, R5, R6, R7 e R8) com quatro blocos. Na avaliação das sementes utilizou-se o mesmo delineamento reduzindo o número de tratamentos para três nas subparcelas (R6, R7 e R8). A permeabilidade da vagem variou com o genótipo e com o estádio de desenvolvimento, e influenciou o vigor das sementes, entretanto, sem afetar a viabilidade das sementes recém-colhidas. O conteúdo de lignina nas vagens não teve influência sobre a permeabilidade das mesmas.

Termos para indexação: *Glycine max* (L.) Merrill., embebição, lignina.

Introduction

Production of high quality soybean seeds in tropical and subtropical regions, especially at low latitudes, is hindered by diverse factors, among which is the occurrence of high temperatures, associated with high rainfall during seed maturation. These unfavorable environmental conditions lead to a fluctuation in seed moisture, resulting in reduction of germination and vigor. In addition, they favor the incidence of fungi associated with soybean seeds, such as *Phomopsis* spp., *Colletotrichum truncatum*, *Cercospora kikuchii* and *Fusarium* spp. (Pereira et al., 1985; Henning and França-Neto, 1980; França-Neto et al., 1994, 2007).

Among some of the properties that may contribute to improving the quality of soybean seeds, seed coat characteristics, semi-permeability of the pod walls, seed resistance to fungi, seed size, and the permeability of the cells of the component tissues of the seeds stand out (França-Neto et al., 1994).

[1]IAPAR, Instituto Agronômico do Paraná , 86047-902 - Londrina, PR, Brasil.

[2]Embrapa Soja, Caixa Postal 231, 96001-970 - Londrina, PR, Brasil.
*Corresponding author < carolina@iapar.br>

Thus, waxy pods, or those that through another reason have less permeability to moisture exchange between the environment and the seeds, may provide extra protection to the seeds against deterioration in the field (França-Neto and Krzyzanowski, 2000).

Tully (1982) proposed selection of soybean genotypes with pods impermeable to moisture. If these genotypes are found, and this characteristic exhibits a high degree of heritability, then it may be introduced in already existing cultivars.

Diverse researchers have observed genetic variability for the characteristic of pod wall permeability for soybean genotypes (Yaklich and Cregan, 1981; Tully, 1982; Dassou and Kueneman, 1984; Pereira et al., 1985). Among them, Pereira et al. (1985) observed a relationship between the lower permeability exhibited by the pod walls and the excellent physiological and health quality of the seeds since the low permeability of the pods provided greater protection to the seeds in the field during the final stages of maturation.

Tully (1982) developed a methodology for soybean genotype selection in regard to pod permeability, based on the principle of monitoring electron passage through the pod wall. Yaklich and Cregan (1981) measured the pod permeability of forty-six soybean cultivars through immersion of intact pods in water and subsequent evaluation of the moistening rate of the pods and the seeds; they observed significant differences in permeability among the different cultivars. Dassou and Kueneman (1984) used an incubation chamber to expose the seeds in pods to high temperature and moisture conditions, and the method allowed identification of soybean genotypes with high indices of tolerance to deterioration.

In relation to resistance to mechanical damage, this factor in the soybean seed is directly related to the lignin content present in the seed coat because this component is responsible for maintaining the integrity and structural cohesion of the plant fiber. Genotypes with more than 5% lignin in the seed coat are considered promising in relation to tolerance to mechanical damage (Alvarez et al., 1997).

The study of lignin content in the pod may also be of great worth for seed technology because, hypothetically, pods with greater lignin content may exhibit less permeability to water, resulting in less seed deterioration in the field, especially in regions where there are inclement weather conditions during the maturation and pre-harvest stages. In addition, lignin may make the pod walls more resistant to insect and fungus attack, preserving seed quality. Romkaew et al. (2008) observed variability in lignin content in the pods among soybean genotypes, and that this is one of the factors to be considered in selection of material resistant to dehiscence of the pods.

Therefore, the study of the relationship of the permeability of the pod wall with soybean seed quality may contribute to genetic breeding of the crop, assisting in selection of materials with the ability to produce high quality seeds for low latitude tropical and subtropical regions. Thus, the aim of this study was to evaluate the effect of permeability of the pod wall on the physiological qualities of the soybean seed and verify the relationship between lignin content of the pod and its permeability.

Materials and Methods

The present study was developed in the experimental fields and in the Seed Analysis Laboratories of Embrapa Soja in the city of Londrina, PR, Brazil. The cultivars were selected as a result of previous information available in the literature in regard to their physiological and plant health qualities. It was used the cultivar Sant'Ana, due to the low permeability of its pods (Pereira et al., 1985) and the good quality of its seeds (Gilioli et al., 1978); the cultivars FT-2, FT-10, Bossier and Davis, due to the distinct lignin content in the seed coat and its potential effect on physiological quality (Alvarez et al., 1997); and the breeding line F 84-7-30 of seeds with a black seed coat, due to the information available in regard to the high level of lignin in the seed coat and in regard to the physiological quality of its seeds (Krzyzanowski et al., 1999).

In the field, the experiment was set up in plots of 8 rows of 15 m, with 0.45 m spacing between rows and a population of 20 plants per meter for harvesting of pods and seeds from each cultivar. Seeds were sown on November 5, 2008 by the conventional method. The daily climate data of mean temperature and total rainfall during the harvest period were gathered from the meteorological station of Embrapa Soja and supplied by the Agrometeorological Laboratory of the same institution (Figure 1). Mean daily temperature oscillated from 20 to 28 °C. The occurrence of rainfall was relatively small, being restricted to a few scattered days in the harvest period. The experimental area was irrigated when necessary.

The treatments for study of the pods were the periodical manual collections in the various stages of development, according to Ritchie et al. (1985), as follows: R4 (full pod), collected from February 1 to 9; R5 (beginning seed), collected from February 13 to 26; R6 (full seed), collected from March 6 to 20; R7 (beginning maturity), collected from March 18 to 25; and R8 (full maturity), collected from March 23 to April 6. In the study of seed quality, the pods were collected at R8 (full maturity) and threshed manually.

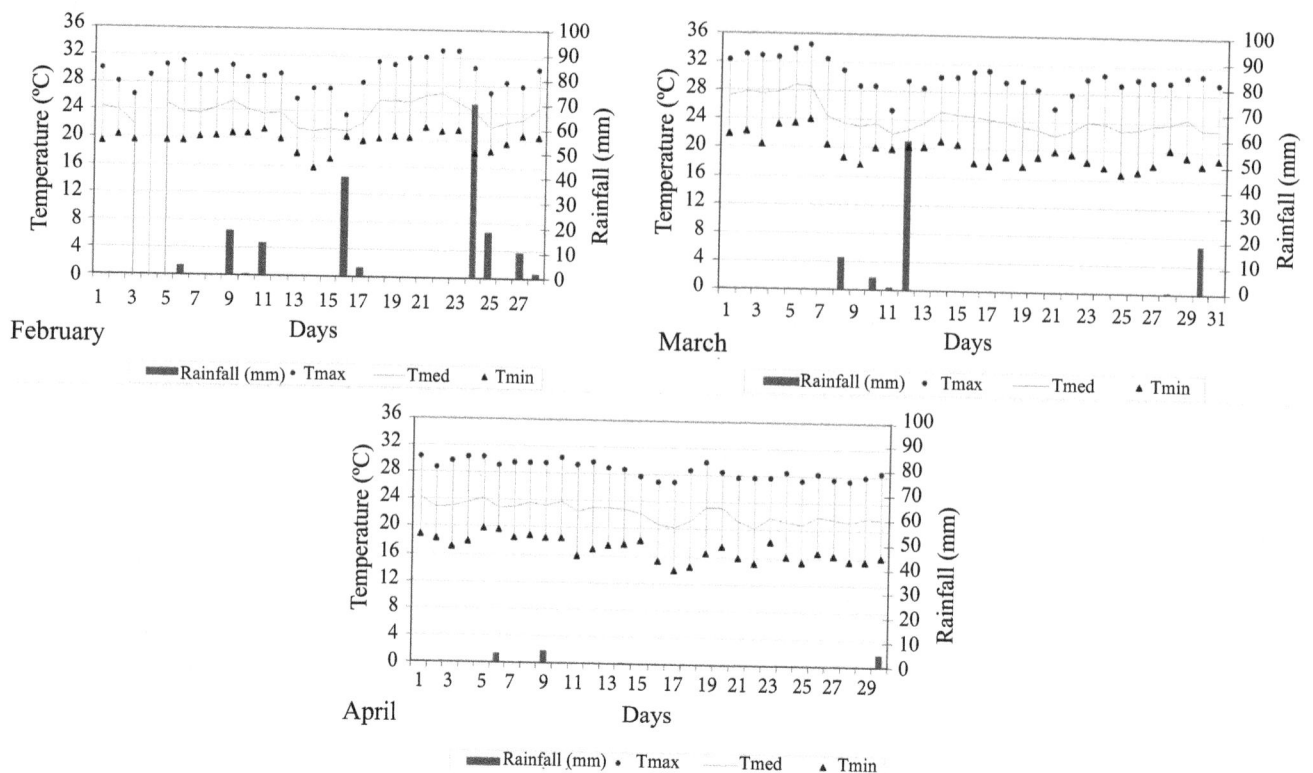

Figure 1. Rainfall and maximum and minimum temperatures on the Experimental Farm of Embrapa Soja, Londrina, PR, Brazil, in February, March and April 2009.

For each cultivar and in each treatment, the following factors were evaluated in the pods:

a) Pods permeability: evaluated by water absorption of the pods and seeds within them, which were immersed in water following the method proposed by Yaklich and Cregan (1981). Four replications of two subsamples of 20 newly-harvested pods were used, selecting those that did not exhibit visible damage. The pods were immersed in distilled water for 1, 6 and 24 h at 25 °C and then placed on paper toweling to dry their surface for fifteen minutes. It was weighed the wet material and then the dry material (dried in an air-circulation laboratory oven at 100 °C for 24 h). Only in the R6, R7 and R8 stages the pods and the seeds were separated after imbibition. Water absorption was determined by the difference between the initial water content of the pods and corresponding seeds, and after 1, 6 and 24 h of imbibition.

b) Pod lignin content: evaluated using four replications of 20 pods divided into two subsamples. After sterilization in NaClO (2%) for three minutes, the pods were rinsed in distilled water and placed for drying in a laboratory oven at 80 °C for 24 h. The resulting dry material was ground and homogenized, and a 0.3 g sample was weighed for extraction of proteins. After that, from the material free of the proteins bound to the cell wall, the quantity of lignin was determined by the thioglycolic acid (TGA) method (Ferrarese et al., 2002).

In the analysis of pod lignin content and in evaluation of seed quality, after harvest, the pods and the seeds were packaged in paper bags and kept in dry cold storage (10 °C and 50% relative humidity) up to the time of evaluation.

In the seeds from each cultivar, the following quality parameters were evaluated:

a) Water content: previously determined by the laboratory oven method at 105 °C ± 3 °C for 24 h (Brasil, 2009), using two replications of 20 seeds.

b) Germination: evaluated by the germination test, undertaken with four replications of 100 seeds, divided into two subsamples, as prescribed by the Rules for Testing Seeds (Brasil, 2009).

c) Level of deterioration: evaluated by the electrical conductivity test conducted with four replications of 200 seeds, divided into four subsamples, according to Vieira and Krzyzanowski (1999).

d) Probable causes of reduction in seed vigor and the level of seed deterioration: evaluated by the tetrazolium test, conducted with four replications of 100 seeds divided into two subsamples, according to the methodology of França-Neto et al. (1998). The seeds were classified to vigor and viability at levels from 1 to 8, and the types of damage were noted. Viability was represented by the sum of the seeds percentage

Relationship between pod permeability and seed quality in soybean

belonging to classes 1 to 5; the vigor level, by classes 1 to 3; and loss of viability by classes 6 to 8. In this last category, the causes of loss of seeds physiological quality were characterized, i.e., through mechanical damage, deterioration from moisture, and damage brought about by stinkbugs.

A randomized block experimental design with split-plots was used in evaluation of the pods; the plots consisted of six genotypes (Sant´Ana, FT-2, FT-10, Bossier, Davis, and the breeding line F 84-7-30), and the split-plots consisted of the five stages of development (R4, R5, R6, R7 and R8), with four replications. In seed evaluation, the same design was used, varying only the number of treatments in the split-plots, which were R6, R7 and R8. All necessary tests were applied for use of analysis of variance to the distribution and independence of errors and homogeneity of variance of the errors of the treatments. Statistical analysis of the data for evaluations of pod permeability for each period of imbibition and lignin content of the pods and seeds was carried out in split-plots in time, with analysis of variance and comparison

of mean values by the Scott-Knott test at 5% probability. Correlation was made of the mean values of pod and seed permeability, physiological quality of the seeds, and lignin content of the pods and seeds by the Student t test at 5% (Banzatto and Kronka, 2006).

Results and Discussion

In the results of permeability of the pods and their seeds (Table 1), among the stages of development, it was observed greater water absorption in R8 and R7. In the initial stages (R4, R5 and R6), the pods and seeds absorbed little water, i.e., less than 5% of their initial weight, even after 24 h of imbibition, because the pods and seeds naturally displayed high moisture content (Table 2). Consequently, the statistical difference seen among the genotypes in these stages should not be considered a factor of importance to characterize them about pod permeability. Thus, these treatments were removed from this and from the other comparisons of this study.

Table 1. Water absorption (%) by the pods collected in the R4, R5, R6, R7 and R8 stages of development, and by the seeds within the pods collected in the R6, R7 and R8 stages of six soybean genotypes for the imbibition periods of 1, 6 and 24 h.

Periods	Genotype	Pods: water absorption (%)					Seeds: water absorption (%)			
		R4	R5	R6	R7	R8	R6	R7	R8	Mean Values
1 h	Sant'Ana	0.6 Ac[1]	0.4 Ac	0.0 Ac	11.6 Ab	34.4 Ba	0.0	0.2	0.3	0.2 B
	Bossier	0.0 Ac	0.2 Ac	0.1 Ac	11.1 Ab	29.7 Ca	0.0	0.2	1.5	0.6 B
	Davis	0.0 Ad	0.5 Ad	1.9 Ac	7.1 Cb	22.1 Ea	1.1	2.5	1.9	1.8 A
	FT 2	0.1 Ab	0.1 Ab	0.7 Ab	4.7 Db	27.3 Da	0.1	0.3	0.8	0.4 B
	FT 10	0.0 Ad	0.0 Ad	1.2 Ac	9.7 Bb	42.8 Aa	0.1	0.0	1.3	0.5 B
	F 84-7-30	0.2 Ac	0.8 Ab	0.7 Ab	1.7 Eb	35.5 Ba	0.5	0.4	1.0	0.6 B
	Mean Values	-	-	-	-	-	0.3 b[1]	0.6 b	1.2 a	
	CV%			9.6				167.1		
6 h	Sant'Ana	0.8 Ac	0.7 Bc	1.0 Bc	14.7 Bb	49.2 Ca	0.1 Bb	0.2 Cb	5.5 Ba	
	Bossier	0.1 Ad	2.3 Ad	1.0 Bc	13.0 Bb	38.6 Da	0.1 Bb	0.3 Cb	8.4 Aa	
	Davis	0.6 Ad	1.0 Bd	1.9 Bc	7.8 Db	52.1 Ba	0.5 Bb	2.9 Ba	3.6 Ca	
	FT 2	0.1 Ad	0.1 Bd	2.8 Ac	23.3 Ab	52.6 Ba	0.3 Bb	4.6 Aa	5.3 Ba	
	FT 10	0.2 Ad	0.0 Bd	2.5 Ac	13.3 Cb	56.2 Aa	0.6 Ab	0.0 Cb	7.5 Aa	
	F 84-7-30	0.6 Ac	2.0 Ab	1.7 Bb	2.6 Eb	38.5 Da	1.9 Ab	2.6 Bb	7.9 Aa	
	CV%			5.6				35.4		
24 h	Sant'Ana	3.0 Ad	1.8 Bd	2.3 Ac	19.2 Bb	57.5 Aa	0.4 Ac	4.0 Bb	25.6 Aa	
	Bossier	1.2 Bd	3.9 Ac	1.5 Ad	16.2 Cb	57.3 Aa	0.1 Ab	1.4 Cb	25.6 Aa	
	Davis	2.2 Ac	1.0 Bd	3.0 Ac	20.4 Bb	55.8 Ba	2.3 Ac	4.6 Bb	14.4 Ba	
	FT 2	0.5 Bd	1.7 Bd	3.8 Ac	34.7 Ab	58.5 Aa	1.1 Ac	5.4 Bb	27.5 Aa	
	FT 10	0.4 Bd	0.4 Bd	2.4 Ac	20.0 Bb	53.9 Ca	0.5 Ab	0.6 Cb	23.1 Aa	
	F 84-7-30	0.4 Bd	1.6 Bc	2.4 Ac	12.9 Db	54.6 Ca	2.0 Ac	10.9 Ab	26.3 Aa	
	CV%			6.3				14.7		

[1]Mean values followed by the same uppercase letter in the column and lowercase letter in the line do not differ among themselves by the Scott-Knott test at 5% probability.

Table 2. Moisture content (%) in the pods and corresponding seeds at the time of harvest of six soybean genotypes collected at different stages of development.

Genotype	Initial pod moisture content (%)					Initial seed moisture content (%)		
	R4	R5	R6	R7	R8	R6	R7	R8
Sant'Ana	87.4 Ba[1]	81.3 Ab	75.3 Cc	50.9 Cd	9.0 Be	64.1 Ba[1]	48.4 Cb	8.2 Ac
Bossier	87.9 Ba	79.5 Cb	78.0 Ac	50.3 Dd	10.0 Ae	66.6 Aa	49.4 Cb	7.4 Ac
Davis	87.5 Ba	80.4 Bb	76.5 Bc	55.4 Bd	9.6 Ae	63.3 Ba	53.4 Ab	7.2 Ac
FT 2	87.6 Ba	80.4 Bb	74.0 Dc	31.2 Ed	8.7 Be	60.3 Da	41.2 Db	7.0 Ac
FT 10	88.7 Aa	80.0 Bb	77.5 Ac	51.2 Cd	8.4 Ae	61.1 Da	51.5 Bb	7.3 Ac
F 84-7-30	88.1 Ba	80.2 Bb	75.2 Cc	56.7 Ad	9.3 Be	62.0 Ca	41.1 Db	7.9 Ac
CV%	0.64					2.06		

[1]Mean values followed by the same uppercase letter in the column and lowercase letter in the line do not differ among themselves by the Scott-Knott test at 5% probability.

In R7, in the pod permeability parameter (Table 1), the line F 84-7-30 showed the least water absorption in the three periods. The cultivars Sant'Ana and Bossier had similar behavior with greater water absorption than the other cultivars in the period of 1h. The cultivar FT-2 showed little absorption at 1 h, but after 6 and 24 h of imbibition, it showed greater water absorption than the others. This may be explained by the fact of the FT-2 cultivar having been collected with the lowest initial moisture content (Table 2), which may be seen as a characteristic inherent to this cultivar since in the period of the pods collection in R7 (March 18 to 25) there was no rain (Figure 1) which could have altered the initial moisture content of the pods collected.

In relation to water absorption by the seeds within the pods in R7, the period of 24 h stood out as most efficient in exhibiting the differences among the genotypes, and the genotype that exhibited greatest absorption was the line F 84-7-30, followed by the cultivars FT-2, Davis and Sant´ana (Table 1).

In R8, in pod permeability, the cultivar FT-10 exhibited the greatest values of water absorption for 1 and 6 h, and FT-2 for 24 h, not differing statistically from Bossier and Sant'Ana.

In water absorption by the seeds within the pods, only 6 h of imbibition presents statistical difference among the genotypes. Bossier, FT-10, and the line F 84-7-30 had greater absorption than Davis, and the others were intermediate. At 24 h, Davis maintained the least absorption among the genotypes, which were not statistically different among themselves.

In general, it may be affirmed that in the R8 stage for the period of 24 h, there was imbibition stabilization of the pods and the seeds. Therefore, the period of imbibition of 6 h may be considered the treatment with most significant results to characterize the permeability of pods in R8 because it showed the greatest variation in imbibition among the genotypes. In this context, the cultivar FT-10 showed the greatest absorption of water by the pods and seeds. The cultivars Bossier and Davis exhibited the lowest imbibition of pods and seeds, respectively.

As described by Krul (1978) and by Yaklich and Cregan (1981), it was seen that the water passed through the pod wall and was absorbed by the seeds and, therefore, the water flux of soybean pods and seeds is selectively controlled by the pods. Nevertheless, imbibition of the pods followed a different pattern from that shown by the corresponding seeds, indicating that, in spite of the moisture content of the seeds being dependent on the quantity of moisture available in the pod, there was no direct relationship between water absorption by the pod wall and by the seeds within them; it is therefore a characteristic inherent to each genotype.

It may be affirmed that pod permeability was variable among the genotypes and also varied in the different stages of development, mainly due, in this case, to the moisture content at the time of harvest. The greatest variation among the genotypes was observed in the R7 stage, and the greatest absorption occurred in the R8 stage, when the pods exhibited the lowest water content at the time of harvest.

Similar to this study, Tully (1982), Dassou and Kueneman (1984), Pereira et al. (1985), and Yaklich and Cregan (1981) observed a difference among the genotypes for the pods studied. These differences were not totally foreseeable because according to the study of Pereira et al. (1985), the Sant'Ana cultivar exhibited the least permeability of the pod walls in R8, Bossier exhibited intermediate permeability, and Davis had one of the greatest values of water absorption by the seeds, indicating greater permeability of the pod walls. These results differ from those observed in this study; however, different methodologies were used to evaluate pod permeability since Pereira et al. (1985) evaluated the permeability of the pod walls through exposing the pods to an environment with high relative air humidity (95%).

In analysis of variance of the physiological quality of the seeds (Table 3), a significant statistical difference was seen among the genotypes. In the viability tests (G, TVS G, and TZ Viab), the genotypes Davis and FT-2 exhibited the lowest values, while in the vigor tests, the cultivar FT-10 had worse performance than the others. Thus, the electrical conductivity test identified the level of deterioration of the seed lots, and the highest quality seeds were those derived from the cultivars Davis and FT-2, followed by the line F 84-7-30 and Sant'Ana, and finally, by Bossier and FT-10, which had greater leaching than the others. Nevertheless, the values exhibited were at most 100 μS $cm^{-1}.g^{-1}$, indicating that all the lots exhibited viable seeds (Vieira et al., 1999).

Table 3. Physiological quality of the seeds of six soybean genotypes by the tests of electrical conductivity (EC) (μS cm^{-1} .g^{-1}), normal seedlings from the germination test (G) (%), total of viable seeds from the germination test (sum of normal seedlings and hard seeds) (TVS G) (%), Tetrazolium test for vigor (TZ Vigor) (%), viability (TZ Viab) (%), mechanical damage (MD) (%), deterioration from moisture (DetM) (%), and stink bug damage (SbD) (%), classes 1 - 8 and 6 - 8.

Genotype	EC	G	TVS G	TZ Vigor	TZ Viab	MD (1-8)	MD (6-8)	DetM (1-8)	DetM (6-8)	SbD (1-8)	SbD (6-8)
	μS $cm^{-1}.g^{-1}$	(%)	(%)	(%)	(%)	(%)	(%)	(%)	(%)	(%)	(%)
Sant'Ana	74.2 C[1]	94.3 A	97.3 A	68.5 A	93.0 A	0 B	0	73.5 A	1.9 B	16.5 B	3.3 A
Bossier	88.5 D	97.3 A	97.3 A	70.8 A	94.5 A	0 B	0	58.0 B	1.3 B	19.8 B	4.0 A
Davis	55.3 A	86.7 C	93.8 B	65.3 A	87.0 B	0 B	0	71.0 A	7.7 A	16.0 B	3.3 A
FT-2	61.4 B	91.8 B	94.3 B	74.5 A	96.0 A	0 B	0	53.8 B	0.3 C	20.0 B	2.5 A
FT-10	100.8 E	96.0 A	96.0 A	54.5 C	92.0 A	0 B	0	51.3 B	1.8 B	36.8 A	6.3 A
F 84-7-30	72.5 C	96.8 A	97.3 A	62.0 B	94.3 A	9.5 A	0	21.0 C	0.3 C	34.3 A	5.0 A
CV%	4.9	1.8	1.6	7.4	3.0	170.3	-	13.2	35.7	29.2	74.1

[1]Mean values followed by the same letter in the column do not differ among themselves by the Scott-Knott test at 5% probability.

There was significant correlation between pod and seed permeability and seed quality (Table 4). The electrical conductivity test showed positive and significant correlation with water absorption by the pod in R8 in the period of 1 h, and by the corresponding seeds for 6 h. The germination test also showed this correlation, indicating that the permeability of the pods does not affect seed viability.

Among the correlations, it was also observed that water absorption by the pod in R8 for 1 h reflected the damages by stink bug seen in the seeds by the tetrazolium test, which resulted in a decline especially in seed vigor. Krul (1978) also observed that the differences in pod moisture content may also have an effect on seed deterioration.

The significant negative correlation between deterioration from moisture, characterized in the tetrazolium test, and germination and viability by the tetrazolium test indicates that this was the main cause in the decline in germination potential of the seeds. The significant positive correlation between stink bug damage, for its part, and electrical conductivity and vigor determined by the tetrazolium test indicate that this was the main cause in the decline in seed vigor.

In the Davis cultivar, the presence of hard seeds in the germination and tetrazolium tests (Table 3) justifies the lower permeability of the pods of this cultivar, observed by lower water absorption of the corresponding seeds of the pods in R8, after the periods of 6 and 24 h of immersion (Table 1).

For all these reasons, it may be affirmed that the presence of hard seeds in the seed lots is an interesting characteristic for, as Potts et al. (1978) affirmed, genotypes with high frequency of hard seeds proved to be resistant to deterioration in the field. Nevertheless, selecting lines that exhibit hard seeds is not a practical solution since these seeds soak up moisture slowly and germinate irregularly, unless they are scarified. Yaklich and Cregan (1981) also observed lower permeability of the pods through lower water content of the corresponding seeds in the cultivars with a tendency toward hard seeds.

Pereira et al. (1985), in their study with fifteen genotypes of soybean, observed that the cultivar Sant'Ana exhibited seeds with the best germination and vigor indices, and the lowest fungal infection rates. The authors attributed the excellent physiological and health qualities of the seeds of this cultivar to the lower permeability exhibited by their pod walls. These results were not observed in this study, though the Sant'Ana cultivar showed good quality seeds; the high percentage of seeds with moisture damage decrease seed quality and indicated that the pods were not totally effective in protecting the seeds against deterioration in the field.

In analysis of lignin content in the pods and in the coat of the seeds within the pod, there was statistically significant interaction between the genotypes and the stages of development (Table 5), therefore indicating that this is a characteristic that varies with the genotype, in agreement with the studies of Baldoni et al. (2013), Romkaew et al. (2008), and Alvarez et al. (1997).

Table 4. Simple correlation between permeability of the pods and corresponding seeds, for different periods of imbibition, and physiological quality of the seeds [tests of electrical conductivity (EC), germination (G), total of viable seeds from the germination test (sum of normal seedlings and hard seeds) (TVS G), test of tetrazolium for vigor (TZ Vigor), viability (TZ Viab.), deterioration from moisture (DetM), and stink bug damage (SbD), classes 1 to 8 and 6 to 8] of six soybean genotypes collected at different stages of development.

Correlation		EC	G	TVS G	TZ Viab.	TZ Vigor	DetM (1-8)	DetM (6-8)	SbD (1-8)	SbD (6-8)
		Physiological Quality of the Seeds								
Germination		0.78*								
TVS G		0.59	0.87*							
TZ Viab.		0.26	0.68	0.46						
TZ Vigor		-0.54	-0.22	-0.16	0.39					
DetM (1-8)		-0.18	-0.53	-0.33	-0.45	0.33				
DetM (6-8)		-0.43	-0.81*	-0.59	-0.96*	-0.13	0.57			
SbD (1-8)		0.60	0.57	0.31	0.20	-0.78*	-0.77	-0.44		
SbD (6-8)		0.79*	0.58	0.41	-0.04	-0.90*	-0.51	-0.21	0.90*	
		Permeability								
Pod R7	1 h	0.49	0.14	0.26	-0.16	0.01	0.74*	0.16	-0.35	0.02
	6 h	-0.01	-0.10	-0.31	0.44	0.54	0.44	-0.29	-0.36	-0.47
	24 h	-0.38	-0.45	-0.69	0.24	0.50	0.29	-0.08	-0.33	-0.57
Pod R8	1 h	0.81*	0.74*	0.60	0.31	-0.69	-0.40	-0.55	0.79*	0.82*
	6 h	-0.06	-0.53	-0.66	-0.38	-0.23	0.49	0.35	-0.04	-0.04
	24 h	-0.42	-0.19	-0.12	0.41	0.96*	0.48	-0.16	-0.04	-0.04
Seed R7	1 h	-0.68	-0.87*	-0.67	-0.85*	0.02	0.34	0.93*	-0.43	-0.34
	6 h	-0.80*	-0.57	-0.68	-0.06	0.43	-0.23	0.08	-0.19	-0.54
	24 h	-0.55	-0.04	0.07	0.17	0.09	-0.64	-0.18	0.17	-0.14
Seed R8	1 h	0.00	-0.33	-0.43	-0.60	-0.26	0.03	0.62	0.04	0.21
	6 h	0.79*	0.95*	0.77	0.58	-0.29	-0.62	-0.71	0.64	0.66
	24 h	0.33	0.75*	0.58	0.98*	0.29	-0.44	-0.98*	0.26	0.03

*significant by the T test at 5%.

Table 5. Lignin (%) in the pods and in the seed coat of seeds of six soybean genotypes collected at different stages of development.

Genotype	Lignin Pods (%)					Lignin Seeds (%)		
	R4	R5	R6	R7	R8	R6	R7	R8
Sant'Ana	5.1 Bc[1]	5.2 Bc	8.9 Bb	10.6 Ba	10.4 Ba	2.9 Ab[1]	2.9 Ab	3.8 Fa
Bossier	7.0 Ad	5.2 Be	9.2 Bc	12.2 Aa	10.4 Bb	2.5 Bc	3.0 Ab	4.9 Da
Davis	5.4 Bc	5.8 Ac	8.8 Bb	9.8 Ba	10.0 Ba	2.4 Bc	3.0 Ab	4.5 Ea
FT-2	5.3 Bd	6.4 Ac	9.2 Bb	11.5 Aa	9.9 Bb	2.8 Ab	3.1 Ab	6.0 Ba
FT-10	7.2 Ac	6.1 Ad	9.8 Bb	11.3 Aa	11.7 Aa	2.7 Ac	3.2 Ab	5.3 Ca
F 84-7-30	5.4 Bb	5.4 Bb	11.1 Aa	10.3 Ba	10.3 Ba	2.9 Ab	3.3 Ab	10.0 Aa
CV%	7.0					6.7		

[1]Mean values followed by the same uppercase letter in the column and lowercase letter in the line do not differ among themselves by the Scott-Knott test at 5% probability.

The percentage of lignin in the pods increased with their development up to R7, and then stabilized or declined in R8. The exception was the line F 84-7-30, with seeds with a black seed coat, which reached maximum lignin concentration in R6. In seeds, the lignin content in the seed coat also increased with maturity, reaching a maximum in R8, for all the genotypes studied. These results show that there was lignin transport from the pod to the seed during maturation.

Among the data of maximum lignin accumulation, it canbe observed that Bossier exhibited the greatest lignin content in the pod and Davis the least, while the other genotypes were intermediate. In the seeds, it was observed that the line F 84-7-30 showed the greatest percentage of lignin in the seed coat, followed by the cultivars FT-2, FT-10, Bossier, Davis and Sant'Ana. Alvarez et al. (1997) observed similar results among the cultivars they studied, with the greatest lignin content in the seed coat for the seeds of the cultivar FT-2, followed by FT-10, Bossier and Davis.

Correlation analysis was carried out between the lignin content in the pod and in the seed coat of the corresponding seeds at the different stages and the parameters studied (Table 6). Among the statistically significant correlations, those that exhibited the most relevant results, which may contribute to research, stand out. There was significant positive correlation of the lignin content in the pod in R6 with the seed in R7 and R8. Therefore, the characteristic of the genotype in relation to the lignin content in the seed coat would already be defined in R6. In agreement with this assertion, Baldoni et al. (2013) also observed expression of the genes C4H and PAL, precursors of lignin, in soybean seeds in the R5 and R6 stages.

Table 6. Simple correlation between lignin content in the pods and in the seed coat of seeds and the permeability of the pods and corresponding seeds at different periods of imbibition, and the physiological quality of the seeds (tests of electrical conductivity, germination, tetrazolium for vigor and viability, deterioration from moisture, and stink bug damage, classes 1 to 8 and 6 to 8), of six soybean genotypes collected at different stages of development.

Lignin Content		Pod					Seeds		
		R4	R5	R6	R7	R8	R6	R7	R8
Lignin Content Seed	R6	-0.34	-0.02	0.52	-0.08	0.09		0.43	0.55
	R7	0.22	0.37	0.85*	-0.02	0.33			0.83*
	R8	-0.16	0.00	0.94*	-0.17	-0.12			
Pod Permeability R7	1 h	0.47	-0.34	-0.75*	0.36	0.41	-0.39	-0.70	-0.85*
	6 h	0.00	0.51	-0.82*	0.58	-0.04	0.03	-0.34	-0.52
	24 h	-0.29	0.79*	-0.50	0.24	-0.27	0.02	-0.12	-0.28
Pod Permeability R8	1 h	0.49	-0.08	0.37	0.19	0.85*	0.58	0.48	0.23
	6 h	-0.02	0.72	-0.41	-0.17	0.33	-0.10	-0.10	-0.52
	24 h	-0.40	0.03	0.72	0.38	-0.63	-0.06	-0.67	0.42
Seed Permeability R7	1 h	-0.37	0.11	0.50	-0.68	-0.45	-0.62	-0.24	-0.17
	6 h	-0.63	0.58	0.22	-0.25	-0.69	0.07	0.24	0.37
	24 h	0.69	-0.11	0.72	-0.53	-0.55	0.55	0.43	0.81*
Seed Permeability R8	1 h	0.44	0.15	0.24	-0.05	0.05	-0.82*	0.13	0.04
	6 h	0.66	-0.35	0.51	0.59	0.49	0.29	0.48	0.44
	24 h	0.03	-0.10	-0.01	0.60	0.05	0.72	0.23	0.34
Physiological Quality of the seeds									
Electrical Conductivity		0.88*	-0.17	-0.01	0.56	0.88*	0.09	0.26	-0.06
Germination		0.52	-0.41	0.47	0.57	0.49	0.48	0.36	0.36
Tetrazolium Viability		0.05	0.00	-0.05	0.68	-0.04	0.61	0.25	0.36
Tetrazolium Vigor		-0.46	-0.06	-0.54	0.31	-0.79*	-0.11	-0.56	-0.23
Det. from Moisture (1-8)		-0.07	-0.04	-0.96*	-0.02	-0.09	-0.55	-0.90*	-0.97*
Det. from Moisture (6-8)		-0.15	0.04	-0.53	-0.60	-0.18	-0.73	-0.40	-0.45
Stink bug Damage (1-8)		0.45	0.18	0.78*	0.05	0.68	0.46	0.90*	0.64
Stink bug Damage (6-8)		0.69	-0.05	0.56	0.06	0.85*	0.18	0.66	0.35

*significant by the T test at 5%.

In correlation of lignin content with permeability of the pods and seeds, negative correlations are considered relevant because they indicate that the greatest lignin content resulted in low permeability.

Thus, significant negative correlation was observed between permeability of the pod in R7 for 1 and 6 h of imbibition and the lignin content of the pods in R6, and between imbibition of seeds in R8 for 1h and the percentage of lignin in the seed in R6. Nevertheless, there were more positive correlations than negative. Therefore, it can not be concluded that the results indicated a clear relationship between pod and seed permeability and lignin content. In a similar manner, Romkaew et al. (2008) observed the relationship between lignin content and pod dehiscence, but did not observe a relationship between moisture retention in the cultivars resistant and susceptible to pod dehiscence.

In the study of correlation between the quantity of lignin and the physiological quality of the seeds, there was significant negative correlation between lignin content in the pod in R6 and in the seed in R7 and R8 for deterioration from moisture, evaluated in the tetrazolium test. This result indicates that greater lignin content in the pod and in the seeds may contribute to a decrease in deterioration from moisture. Nevertheless, a significant positive correlation was observed between lignin content in the seed in R7 and in the pod in R6 and R8 and stink bug damage, indicating that the lignin content in the pod did not make the pod walls more resistant to insect attack, disagreeing with the suppositions of Alvarez et al. (1997).

Thus, it may be inferred that the lignin content confers resistance to the pods and seeds, but does not have an effect on their permeability, i.e., it does not contribute to the oscillation in the moisture content of the seeds within the pods.

Conclusions

The permeability of the pod varies with the genotype and with the stage of development, and it has an effect on seed vigor, though it does not affect the viability of the newly-harvested seeds.

The lignin content in the pods does not have an effect on the permeability of the pods.

References

ALVAREZ, P.J.C.; KRZYZANOWSKI, F.C.; MANDARINO, J.M.G.; FRANÇA-NETO, J.B. Relationship between soybean seed coat lignin content and resistance to mechanical damage. *Seed Science and Technology*, v.25, n.2, p.209-214, 1997. http://cat.inist.fr/?aModele=afficheN&cpsidt=2045873

BALDONI, A; VON PINHO, E.V.R; FERNANDES, J.S; ABREU, V.M.; CARVALHO, M.L.M. Gene expression in the lignin biosynthesis pathway during soybean seed development. *Genetics and Molecular Research*, v.12, n.3, p.2618-2624, 2013. http://www.funpecrp.com.br/gmr/year2013/vol12-3/pdf/gmr2170.pdf

BANZATTO, D.A.; KRONKA, S.N. *Experimentação agrícola*. 4.ed. Jaboticabal: FUNEP, 2006. 237p.

BRASIL - Ministério da Agricultura, Pecuária e Abastecimento. *Regras para análise de sementes*. Ministério da Agricultura, Pecuária e Abastecimento. Secretaria de Defesa Agropecuária. Brasília, DF: MAPA/ACS, 2009. 395p. http://www.agricultura.gov.br/arq_editor/file/2946_regras_analise__sementes.pdf

DASSOU, S.; KUENEMAN, E.A. Screening methodology for resistance to field weathering of soybean seed. *Crop Science*, v.24, n.4, p. 774-779, 1984. https://www.crops.org/publications/cs/abstracts/24/4/CS0240040774

FERRARESE, M.L.L.; ZOTTIS, A.; FERRARESE FILHO, O. Protein-free lignin quantification in soybean (*Glycine max*) roots. *Biologia*, v.4, n.57, p.541-543, 2002. http://agris.fao.org/agris-search/search/display.do?f=2003/SK/SK03002.xml;SK2002000611

FRANÇA-NETO, J.B.; HENNING, A.A.; KRZYZANOWSKI, F.C. Seed production and technology for the tropics. In: *Tropical soybean:* improvement and production. Rome: FAO, p.217-240, 1994. (EMBRAPA-Soja, Londrina-PR.)

FRANÇA-NETO, J.B.; KRZYZANOWSKI, F.C. Tecnologia de sementes e o melhoramento de plantas. In: BORÉM, A.; GIÚDICE, M.P.; DIAS, D.C.F.S. & ALVARENGA, E.M. (eds.) *Biotecnologia e produção de sementes*. BIOWORK III. Viçosa. Universidade Federal de Viçosa. p.75-101. 2000.

FRANÇA-NETO, J.B.; KRZYZANOWSKI, F.C.; COSTA, N.P. *The tetrazolium test for soybean seeds*. Londrina: EMBRAPA-CNPSo, 1998. 71p. (EMBRAPA-CNPo. Documentos, 115).

FRANÇA- NETO, J.B.; KRZYZANOWSKI, F.C.; PÁDUA, G.P.; COSTA, N.P.; HENNING, A.A. *Tecnologia da produção de semente de soja de alta qualidade – Série Sementes*. Londrina: Embrapa Soja, 2007. 12 p. (Embrapa Soja. Circular Técnica, 40).

GILIOLI, J.L.; PALUDZYSZYN FILHO, E.; KIIHL, R.A.S.; GAZZIERO, D.L.P; BORDIN, E. Escolha e recomendação de cultivares. In: IAPAR, Londrina-PR. *Manual Agropecuário para o Paraná*. v.2, 1978.

HENNING, A.A.; FRANÇA-NETO, J.B. Problemas na avaliação da germinação de sementes de soja com altos índices de *Phomopsis* sp. *Revista Brasileira de Sementes*, v.2, n.3, p.9-22. 1980.

KRUL, W.R. Diffusible inhibitor(s) of imbibition from senescent soybean pods. *Hortscience*, v.13, n.1, p.41-42, 1978.

KRZYZANOWSKI, F.C.; FRANÇA-NETO, J.B.; KASTER, M.; MANDARINO, J.M.G. *Metodologia para seleção de genótipos de soja com semente resistente ao dano mecânico - relação com o conteúdo de lignina*: determinação do conteúdo de lignina nos tegumentos de sementes de soja com tegumento preto e amarelo. In: EMBRAPA-Soja, Londrina - PR. Resultados de Pesquisa de Soja 1998. Londrina, PR: Embrapa Soja, p.214-215. 1999. (Série Documentos, 125).

PEREIRA, L.A.G.; FRANÇA-NETO, J.B.; COSTA, N.P.; HENNING, A.A.; MAGALHÃES, C.V. Teste de metodologia para identificação de genótipos de alta qualidade fisiológica de sementes de soja. In: EMBRAPA-Soja, Londrina-PR. *Resultados de pesquisa de soja 1984/1985*. Londrina, PR: Embrapa Soja, p.407-408, 1985. (Série Documentos, 15).

POTTS, H.C.; DUANGPATRA, J.; HAIRSTON, W.G.; DELOUCHE, J.C. Some influences of hardseededness on soybean seed quality. *Crop Science*, v.18, n.2, p.221-224, 1978. https://www.crops.org/publications/cs/abstracts/18/2/CS0180020221

RITCHIE, S.W.; HANWAY, J.J; THOMSON, H.E.; BENSON, G.O. *How a soybean plant develops*. Ames: Iowa State University of Science and Technology, Cooperative Extension Service, 1985. 20p. (Special Report, 53).

ROMKAEW, J.; NAGAYA, Y.; GOTO, M.; SUZUKI, K.; UMEZAKI, T. Pod dehiscence in relation to chemical components of pod shell in soybean. *Plant Production Science*, v.11, n.3, p.278-282, 2008. http://cat.inist.fr/?aModele=afficheN&cpsidt=20588464

TULLY, R.E. A new technique for measuring permeability of dry soybean pods to water. *Crop Science*, v.22, n.2, p.437-440 1982. https://www.crops.org/publications/cs/abstracts/22/2/CS0220020437?search-result=1

VIEIRA, R.D.; KRZYZANOWSKI, F.C. Teste de condutividade elétrica. In: *Vigor de sementes*: conceitos e testes. KRZYZANOWSKI, F.C.; VIEIRA, R.D.; FRANÇA-NETO, J.B. (eds.). Londrina: ABRATES, 1999. Cap.4, p.1-26.

VIEIRA, R.D.; PAIVA-AGUERO, J.A.; PERECIN, D.; BITTENCOURT, S.R.M. Correlation of electrical conductivity and other vigor tests with field emergence of soybean seedlings. *Seed Science and Technology*, v.27, n.1, p.67-75, 1999. http://cat.inist.fr/?aModele=afficheN&cpsidt=1895090

YAKLICH, R.W.; CREGAN, P.B., Moisture migration into soybean pods. *Crop Science*, v.21, n.5, p.791-793, 1981. https://www.crops.org/publications/cs/abstracts/21/5/CS0210050791?search-result=1

21

Production of reactive oxygen species in *Dalbergia nigra* seeds under thermal stress

Antônio César Batista Matos[1*], Eduardo Euclydes de Lima e
Borges[1], Marcelo Coelho Sekita[2]

ABSTRACT – Seed germination is dependent on abiotic factors, temperature being one of the main ones, whose influence causes seed damage under extreme conditions. The aim of the present study was to investigate the effect of different temperatures during germination of *D. nigra* seeds and their physiological and biochemical implications. We assessed germination percentage and production of superoxide anion (O_2^-) and hydrogen peroxide (H_2O_2) in seeds subjected to temperatures of 5, 15, 25, 35 and 45 °C for different periods of time. Hydration is promoted at 45 °C and inhibited at 5°C, without germination in either, whereas it is minimal at 15 °C and at a maximum level at 25 °C. Superoxide production increases at higher temperatures (25 and 35 °C) after 72 hours of hydration, coinciding with the beginning of radicle protrusion. Production of hydrogen peroxide decreases at all temperatures, except for 5 °C, with values near each other at temperatures of 15, 25, and 35 °C, where there was radicle protrusion.

Index terms: forest, physiology, biochemistry, rosewood.

Produção de espécies reativas de oxigênio em sementes de *Dalbergia nigra* sob estresse térmico

RESUMO - A germinação de sementes é dependente de fatores abióticos, sendo a temperatura um dos principais, cuja influência, em condições extremas, causa danos às sementes. O presente trabalho teve por objetivo investigar o efeito das diferentes temperaturas durante a germinação de sementes de *D. nigra* e suas implicações fisiológicas e bioquímicas. Avaliaram-se o porcentual de germinação e a produção de ânion superóxido (O_2^-) e peróxido de hidrogênio (H_2O_2) em sementes submetidas às temperaturas de 5, 15, 25, 35 e 45 °C por diferentes tempos. A hidratação é estimulada a 45 °C e inibida a 5 °C, não havendo germinação em ambas, enquanto é mínima a 15 °C e máxima a 25 °C. A produção de superóxido aumenta nas temperaturas mais altas (25 e 35 °C) após 72 horas de hidratação, coincidindo com o início da protrusão radicular. A produção de peróxido de hidrogênio decresce em todas as temperaturas, à exceção da de 5 °C, com valores próximos entre si nas temperaturas de 15, 25 e 35 °C, onde houve protrusão radicular.

Termos para indexação: floresta, fisiologia, bioquímica, jacarandá-da-bahia.

Introduction

Dalbergia nigra (Vell.) Fr.All. ex Benth, also known as rosewood, is a tree species that occurs in different Brazilian states, especially in areas of the Atlantic Forest Formation. Due to intense exploitation and the lack of reforestation programs, this species has been included as vulnerable on the Red List of the International Union for Conservation of Nature since 1998 (IUCN, 2013), with prohibition of its trade since the 1990s (CITES, 2008), as well as being included on the official list of endangered species of Brazilian flora (IBAMA, 2013). It is propagated through seeds and is indicated for programs of recovery of degraded areas, with high potential for sustainable forest management (Lorenzi, 2002).

Knowledge of seed physiology is fundamental. For each species, specific environmental conditions are necessary to ensure germination, as shown by Araújo Neto et al. (2003) and Rego et al. (2009) for the species *Acacia polyphylla* and *Blepharocalyx salicifolius*, respectively. The range of environmental adaptation is related to the cardinal temperatures (minimum, optimum, and maximum) that each species requires for germination, determining the distribution limits of the species (Orozco-Almanza et al., 2003; Borghetti, 2005; Oliveira and Garcia, 2005; Bewley et al., 2013). In relation to

[1]Departamento de Engenharia Florestal, UFV, 36570-000 – Viçosa, MG, Brasil.

[2]Departamento de Fisiologia Vegetal, UFV, 36570-000 – Viçosa, MG, Brasil.
*Corresponding author <batistamatos@gmail.com>

the seeds of native forest species, the thermal range suitable for germination is frequently from 20 to 30 °C (Mello and Barbedo, 2007; Brancalion et al., 2010; Pimenta et al., 2010; Dousseau et al., 2013).

Studies in respect to physiological and biochemical aspects during seed germination of tropical species under abiotic stress conditions, especially thermal stress, are highly relevant in the face of environmental adversities found in tropical ecosystems, and also through the lack of information in regard to the mechanisms involved in seed tolerance to determined levels of stress. Plants under abiotic stress conditions, such as drought, salinity, and high and low temperatures, produce reactive oxygen species (ROSs), as observed by Luo et al. (2011), in which low temperature continually increased the formation of superoxide anion and hydrogen peroxide in leaves of *Fragaria ananassa* Duch., Zoji and Toyonaka cultivars, up to 48 hours, decreasing after that. Airaki et al. (2012) observed an increase in the level of the non-enzymatic antioxidants ascorbate and glutathione and in the activity of the enzyme NADPH dehydrogenase during acclimatization of *Capsicum annum* L. plants, indicating the action of these substances in the cell antioxidant system. Even in storage under low temperature, there is production of superoxide anion and hydrogen peroxide, as observed by Pukacka and Ratajczak (2005) in *Fagus sylvatica* seeds stored at temperatures of 4, 20 and 30 ºC. The loss of viability that occurred in nine weeks under all conditions was related to the increase in the two compounds.

Studies show that the ROSs may not be as harmful to the seed life cycle as previously portrayed, but may have a key role in signaling in response to the different possible stresses during germination (Gomes and Garcia, 2013). The ROSs may be signalers of various biological processes, including responses to biotic and abiotic stresses and programmed cell death (Bailly, 2004; Mittler et al., 2004; Foyer and Noctor, 2005; Fujita et al., 2006). Seeds pass through an infection-sensitive period during germination, and it is believed that the ROSs play an important protective role against attacks from incompatible pathogens (Schopfer et al., 2001; Oracz et al., 2009). Thus, the ROSs play a key role in the seed germination process.

Accordingly, the study of the effect of different temperatures is important, especially those outside of the optimum range for germination, in the germination process and in the production of substances arising from thermal stress, which may result in loss of quality or death of the seeds. Thus, the aim of this study was to assess the germination of *Dalbergia nigra* seeds under thermal stress conditions in association with the production of reactive oxygen species.

Materials and Methods

The seeds were collected in September 2012 in the region of Viçosa, MG, Brazil, processed, and placed in cold storage (5 °C/60% relative humidity-RH).

Percentage of water gain (%) was calculated in relation to the initial seed weight of each treatment. For that purpose, the seeds were placed in a laboratory oven at 105 ± 3 ºC for 24 hours (Brasil, 2009), using three replications of 20 seeds. Calculation was made on a wet basis, with the degree of moisture expressed in percentage.

The seeds were weighed on a digital balance with 0.0001 g precision and subsequently placed for soaking in a Petri dish on two sheets of filter paper moistened with 4.0 mL of distilled water and kept under constant light at the temperatures of 5, 15, 25, 35 and 45 °C. The seeds were weighed at two-hour intervals in the first 12 hours and subsequently weighed at 12-hour intervals until they reached 50% germination or until the tenth day (240 hours) after the beginning of soaking. Before each weighing, the surface of the seeds was dried with absorbent paper, and then they were once more placed in Petri dishes with paper moistened with distilled water. Five replications of 20 seeds were used.

In the germination test, the seeds were immersed in 0.5% Captan® solution for three minutes and placed in Petri dishes on two sheets of paper moistened with 4.0 mL of distilled water and kept in a BOD type germinator at the temperatures of 5, 15, 25, 35 and 45 °C under constant light for 240 hours. Daily assessments were made, with protrusion of the primary root as the criterion of germination. Each treatment (temperature) consisted of five replications of 20 seeds.

The effect of temperature on production of reactive oxygen species (ROSs) was assessed by comparing production levels during germination. Seeds began primary root protrusion at 132 hours of soaking at the temperature of 25 ºC, used as the standard. That way, analyses were made of the embryonic axes of seeds that were soaked for 0 hours (dry seeds) and 24, 72 and 120 hours, which correspond to the end of phase I, and the middle and end of phase II, respectively. The same sample taking times were used for the temperatures of 5, 15 and 35 °C. For the temperature of 45 ºC, samples were taken at 8 and 24 hours since preliminary tests had indicated death of the seeds after 24 hours at that temperature. The seeds were placed under the same conditions of the germination test.

In quantification of superoxide anion, samples of 20 embryonic axes were weighed on a balance with 0.0001 g precision and cut in half in the transversal direction in two segments and incubated in 2.0 mL of reaction medium consisting of disodium salt of 100 µM Ethylenediamine tetraacetic acid (Na_2EDTA), 20 µM β-Nicotinamide adenine dinucleotide reduced (NADH), and 20 mM sodium phosphate buffer, pH 7.8 (Mohammadi and Karr, 2001) in hermetically sealed tubes. The reaction was started by the introduction of 100 µL of 25.2 mM epinephrine in 0.1N HCl, using a chromatography syringe. Samples were incubated at 28 °C, remaining under shaking for 5 minutes. After that, the segments were removed and, as of the seventh minute, reading of absorbance at 480 nm was begun in a Thermo Scientific EVOLUTION 60S spectrophotometer, which was carried out for five minutes. The blank was performed under the same conditions, but without plant tissue. Production of superoxide anion was assessed by determination of the amount of accumulated adrenochrome (Misra and Fridovich, 1971), using the molar absorption coefficient of $4.0 \times 10^3\ M^{-1}$ (Boveris, 1984).

For quantification of hydrogen peroxide, samples of 20 embryonic axes were weighed on a 0.0001 g-precision balance, ground in liquid nitrogen, and then homogenized in 2.0 mL of 50 mM potassium phosphate buffer, pH 6.5, containing 1 mM hydroxylamine, followed by centrifugation at 10,000 g for 15 minutes at 4 °C, and the supernatant was collected (Kuo and Kao, 2003).

Aliquots of 100 µL of the supernatant were added to the reaction medium consisting of 250 µM ferrous ammonium sulfate, 25 mM sulfuric acid, 250 µM xylenol orange, and 100 mM sorbitol in a final volume of 2 mL (Gay and Gebicki, 2000), homogenized, and kept in the dark for 30 minutes. Absorbance was determined in a Thermo Scientific EVOLUTION 60S spectrophotometer at 560 nm, and quantification of H_2O_2 was made based on a calibration curve, using peroxide concentrations as a standard. Blanks for the reagents and plant extracts were prepared in a parallel manner and taken from the sample.

The data were subjected to analysis of variance (ANOVA). For germination, a single-factor completely randomized design (5 temperatures) was used, and subjected to regression analysis.

For ROS production data, a completely randomized design in a 4 x 4 factorial arrangement (5, 15, 25 and 35 °C x 0, 24, 72, and 120 hours) was used, plus 4 additional treatments (45 °C x 0, 8, 24 and 48 hours). The statistical model $Y_{ij}=m+t_i+e_{ij}$ was used for the 20 treatments, with one factorial with 15 degrees of freedom (5, 15, 25 and 35 °C x 0, 24, 72 and 120 hours), the additional treatments with 3 degrees of freedom (45 °C x 0, 8, 24 and 48 hours), and a contrast between factorial x additional with 1 degree of freedom, exhibiting the ANOVA tables (Tables 1 and 2). The regression models were chosen based on biological logic, at the significance of the regression coefficients, using the t test at 5% probability and in the coefficient of determination. The software *Statistica 8* (STATSOFT Inc., 2009) was used.

Table 1. Summary of analysis of variance (ANOVA) for the data on production of superoxide anion (O_2^-) in embryonic axes of *Dalbergia nigra* seeds under different temperatures.

Sources of Variation	DF	SS	MS	F	p-value
Treatments	19	12846.73	676.14	329.46	0.0000
Factorial	15	10147.25	676.48	329.63	0.0000
Temperature (T)	3	33.63	11.21	5.46	0.0030
Period (P)	3	10028.39	3342.80	1628.83	0.0000
T x P	9	85.24	9.47	4.61	0.0003
Additional	3	2691.70	897.23	437.19	0.0000
Factorial vs Additional	1	7.78	7.78	3.79	0.0585
Error	40	82.0909	2.0523		
Total	59	12928.8242			

DF=degrees of freedom; SS=sum of squares; MS=mean square; F=value of the F test; p-value=probability of significance.

Table 2. Summary of analysis of variance (ANOVA) for the data on production of hydrogen peroxide (H_2O_2) in embryonic axes of *Dalbergia nigra* seeds under different temperatures.

Sources of Variation	DF	SS	MS	F	p-value
Treatments	19	4730.26	248.96	73.92	0.0000
Factorial	15	3600.22	240.01	71.27	0.0000
Temperature (T)	3	44.98	14.99	4.45	0.0086
Period (P)	3	3453.93	1151.31	341.86	0.0000
T x P	9	101.31	11.26	3.34	0.0039
Additional	3	1125.19	375.06	111.37	0.0000
Factorial vs Additional	1	4.85	4.85	1.44	0.2371
Error	40	134.71	3.37		
Total	59	4864.97			

DF=degrees of freedom; SS=sum of squares; MS=mean square; F=value of the F test; p-value=probability of significance.

Results and Discussion

The *Dalbergia nigra* seeds were dispersed with moisture content of 8.86%, which was similar to the values of moisture content found by Ataíde et al. (2013) for two lots of seeds of the same species collected in 2010 and 2011 (7.92 and 8.98%, respectively) in the region of Viçosa, MG, Brazil. According to the imbibition curves, temperature stimulated the rate of water uptake; weight gain of the seeds increased through increase in temperature. At the temperature of 5 °C, the lowest rate of water uptake was observed, requiring 72 hours to reach phase II. At the temperature of 15 °C, 36 hours were necessary to characterize phase II, while at the temperatures of 35 and 45 °C, 24 hours were necessary. However, deterioration of the seeds began over the period at the temperature of 45 °C. The seeds remained at phase II at the temperatures of 5 and 45 °C. For the temperature of 25 °C, the three phase imbibition pattern was observed, reaching phase III in 132 hours (Figure 1).

Significant effects of temperatures on germination of *D. nigra* seeds were observed (Figure 2). At 25 °C, the seeds reached the maximum estimated percentage of germination for the 12 days (94%). For the temperatures of 5 and 45 °C, there was no radicle protrusion during the period of observation. Seed imbibition at low temperatures has a harmful effect because it results in irreparable damage to the membranes and leaching of solutes (Castro et al., 2004). High temperatures may allow seed imbibition; however, they do not ensure embryo expansion and seedling establishment (Bradbeer, 1988). In regard to temperatures of 15 and 35 °C, the *D. nigra* seeds reached the maximum estimated percentages of germination at 12 days, with 5% and 39% of germination, respectively.

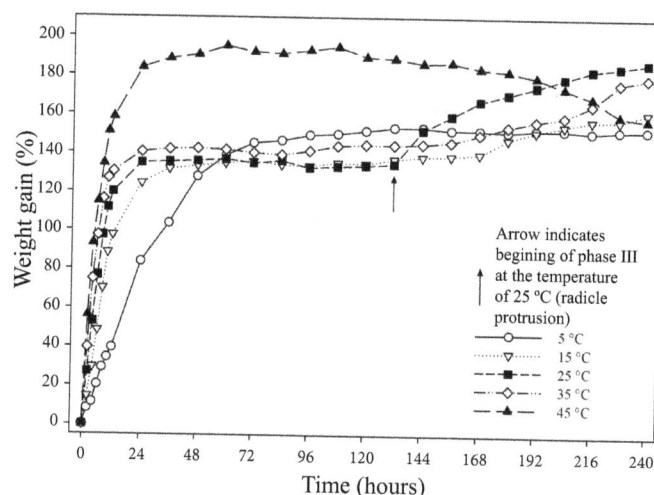

Figure 1. Imbibition curves of *Dalbergia nigra* seeds under different temperatures.

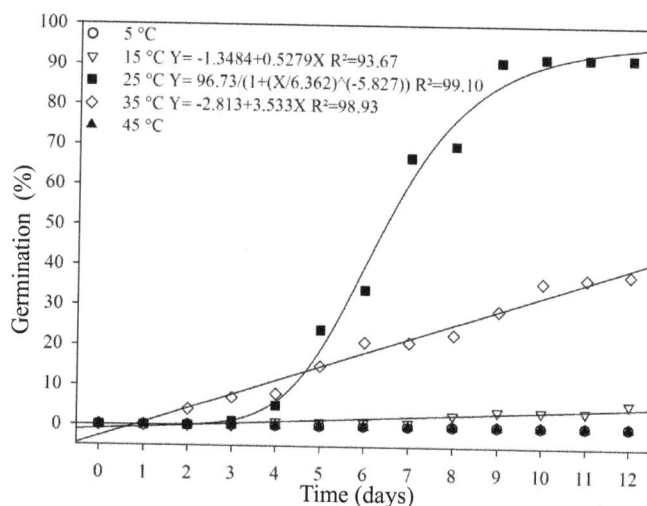

Figure 2. Regression equations for germination of *Dalbergia nigra* seeds under different temperatures.

The balance between ROS production and the capacity of the defense system in their removal indicates the plant response to stress and reflects adaptation and/or tolerance to the different types of environmental conditions (Mittler, 2002; Apel and Hirt, 2004). The radical O_2^- may lead to cell death when there is no specific antioxidant defense system to remove it (Gill and Tuteja, 2010). The conversion of O_2^- to H_2O_2 is the natural route for removal of both substances due to the fact that the latter is less toxic than the former, as well as the possibility of being taken from the production location. Moreover, according to Gill and Tuteja (2010), the accumulation of H_2O_2 in the cells may cause damage to cell metabolism since this compound has the ability of oxidizing the thiol groups (-SH) of enzymes, deactivating them.

The production of O_2^- (34.01 nmol.min^{-1}.g^{-1} FM) and H_2O_2 (34.20 µmol.g^{-1} FM) in embryonic axes of dry *D. nigra* seeds was observed (Figures 3 and 4). In this case, the presence of ROSs may arise from non-enzymatic reactions, such as lipid peroxidation and the Amadori and Maillard reactions since the enzyme activity of the antioxidant system is probably reduced in this condition (Murthy et al., 2003). Pukacka and Ratajczak (2005) also found hydrogen peroxide production in dry *Fagus sylvatica* seeds. With the beginning of imbibition, there was a reduction in the concentration of O_2^- and H_2O_2 for all the temperatures assessed.

For the temperature of 5 °C, reduction in O_2^- production soon after imbibition was seen, when compared with the value of the dry seed, reaching minimum values up to 120 hours (Figure 3). Luo et al. (2011) observed a continual increase in the formation of superoxide anion in strawberry leaves under low temperature up to 48 hours, decreasing after that. During acclimatization of *Capsicum annum* seedlings to low temperature, Airaki et al. (2012) observed an increase in the level of the non-enzymatic antioxidants ascorbate and glutathione and in the activity of the NADPH dehydrogenase.

In relation to production of hydrogen peroxide, at the temperature of 5 °C, minimum production of this free radical was seen up to 70 hours of imbibition (9.6 µmol.g^{-1} FM) and subsequent increase in this production up to 120 hours (20.5 µmol.g^{-1} FM) (Figure 4). The high moisture content of the seeds associated with the increase in the hydrogen peroxide content compromised the germination process. Okane et al. (1996) observed that H_2O_2 production in calluses of *Arabidopsis thaliana* plants was greater at 4 °C than at 23 °C. Similar results were found by Sun et al. (2010), who observed accumulation of H_2O_2 in leaf tissue under low temperature conditions in *Nicotiana tabacum* seedlings. According to Torres and Dangl (2005), there was expression of various genes in different species when the plants were subjected to

low temperature. Thus, there is a clear presence of ROSs as a reaction to temperature stress for *D. nigra* seeds.

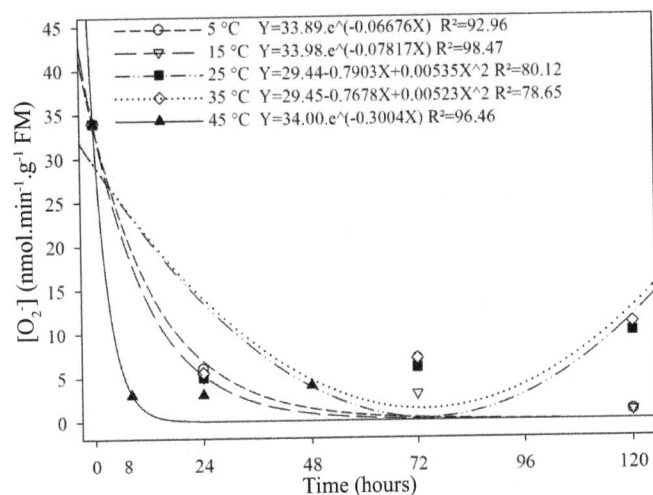

Figure 3. Superoxide anion (O_2^-) production in embryonic axes of *Dalbergia nigra* seeds during germination under different temperatures.

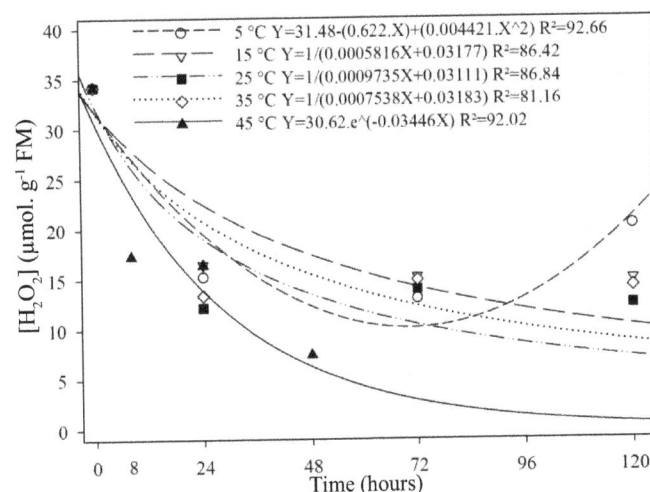

Figure 4. Hydrogen peroxide (H_2O_2) production in embryonic axes of *Dalbergia nigra* seeds during germination under different temperatures.

There was continual reduction in O_2^- production at the temperature of 15 °C up to 120 hours (Figure 3), showing effective action of the enzymes responsible for elimination of the ROSs. The concentration of hydrogen peroxide also decreased up to 120 hours (Figure 4). As it was not under optimum germination conditions (5% radicle protrusion in 120 hours), the production of hydrogen peroxide was still a reaction to thermal stress. It may be seen that the concentration of H_2O_2 at this temperature (9.84 µmol.g^{-1} FM) is greater than at 25 °C (6.75 µmol.g⁻¹ FM) and at 35 °C (8.17 µmol. g⁻¹ FM) up to 120 hours of imbibition. Germination at this temperature indicates that the respiratory system in the mitochondria is occurring, with consequent production of superoxide anion, effectively dismuted by the SOD enzyme.

Minimum production of superoxide anion at the temperature of 25 °C occurred at 74 hours of imbibition (0.27 nmol.min⁻¹.g⁻¹ FM), rising afterwards up to 120 hours (11.70 nmol.min⁻¹.g⁻¹ FM) (Figure 3), when the seeds reached 19% radicle protrusion. Hydrogen peroxide, after a decrease during imbibition in relation to the control, remained relatively high up to 120 hours (Figure 4). The production of ROSs acts against pathogens during radicle protrusion, and also acts as a component of cell growth when it induces the depolymerization of components of the cell well, permitting the cellular expansion (Bailly, 2004; Muller et al., 2009).

At the temperature of 35 °C, minimum production of O_2^- was observed at 73 hours of imbibition (1.27 nmol.min⁻¹.g⁻¹ FM) (Figure 3), increasing continually up to 120 hours (12.62 nmol.min⁻¹.g⁻¹ FM), when 15% protrusion of the primary root was reached. There were small variations in the values of hydrogen peroxide during the period, which were lower than those at the temperature of 15 °C, but greater than those at 25 °C (Figure 4). As the percentage of germination was only 40% after 240 hours (94% germination at the temperature of 25 °C) and the production of superoxide anion was greater among the temperatures assessed, it is clear that at 35 °C there is metabolic damage to the seeds. As the ROSs are produced in the mitochondria, in the glyoxysome, and in the plasmatic membrane, the system for control of the levels of superoxide anion and hydrogen peroxide could be in any one of them. As the increase in temperature corresponds to the increase in respiration, the production of ROSs on a greater scale in the mitochondria would have the consequence of change in their membranes and reduction in production of ATP and death of the seeds. The presence of H_2O_2 could be in the stress signaling system, when at 5 °C or at 35 °C, or in growth and protection at the temperature of 25 °C. This is more certain in the last case since there was no interference in germination.

At the temperature of 45 °C, a 91% reduction was observed in O_2^- production up to 8 hours of imbibition when compared to the value of the dry seeds, until reaching values near zero at 120 hours of imbibition (Figure 3). The H_2O_2 concentration reduced 99% in 8 hours of imbibition when compared with the value of the dry seeds, until reaching 0.49 µmol.g⁻¹ FM after 120 hours of imbibition (Figure 4). Duan et al. (2009) established 40 °C as the optimum temperature for activity of the enzyme NADPH oxidase in rice seeds. Thus, at the temperature of 45 °C, there may still be activity

of this enzyme, which may have an effect on the production of superoxide anion and a consequent harmful effect on seed viability. Maintenance of the H_2O_2 may be by the action of NADPH oxidase since, according to Sagi and Fluhr (2006), the production of hydrogen peroxide between the cell membrane and wall could return in the form of hydroperoxide, formed by the effect of the 5.0 extracellular pH, and, that way, would enable the action of the SOD in catalyzation of the formation of hydrogen peroxide. Thus, it is possible to suppose action of the enzyme already in the initial periods of germination. Kranner et al. (2010) observed an increase in O_2^- production with an increase in temperature during experiments carried out with *Pisum sativum* seeds. Wang et al. (2012) observed an increase in production of ROSs, especially hydrogen peroxide and an increase in lipid peroxidation in soybean seeds under high temperature and moisture conditions. According to Bhattacharjee (2008), high temperature induced a significant increase in the superoxide anion and hydrogen peroxide content in *Amaranthus livindus* seeds and seedlings when compared to the control group. The involvement of other mechanisms in the effects of temperature on the reduction of viability or death of seeds may not be dismissed. According to Larkindale et al. (2005), heat resistance or the acquisition of thermotolerance is a complex multigenic process.

Conclusions

Hydration is stimulated at 45 °C and inhibited at 5 °C, without germination in either, whereas it is minimal at 15 °C and at a maximum level at 25 °C.

Superoxide production increases at higher temperatures (25 and 35 °C) after 72 hours of hydration, coinciding with the beginning of radicle protrusion.

Hydrogen peroxide production decreases at all temperatures, except for 5 °C, with values near each other at the temperatures of 15, 25 and 35 °C, where there was radicle protrusion.

References

AIRAKI, M.; LETERRIER, M.; MATEOS, R.M.; VALDERRAMA, R.; CHAKI, M.; BARROSO, J.B.; DEL RIO, L.A.; PALMA, J.M.; CORPAS, F.J. Metabolism of reactive oxygen species and reactive nitrogen species in pepper (*Capsicum annum* L.) plants under low temperature stress. *Plant Cell Environment*, v.35, n.2, p.281-295, 2012. http://onlinelibrary.wiley.com/doi/10.1111/j.1365-3040.2011.02310.x/full

APEL, K.; HIRT, H. Reactive oxygen species: metabolism, oxidative stress and signal transduction. *Annual Review of Plant Biology*, v.55, p.373-399, 2004. http://www.annualreviews.org/doi/abs/10.1146/annurev.arplant.55.031903.141701?url_ver=Z39.88-2003&rfr_dat=cr_pub%3Dpubmed&rfr_id=ori%3Arid%3Acrossref.org&journalCode=arplant

ARAÚJO NETO, J.C.; AGUIAR, I.B.; FERREIRA, V.M. Efeito da temperatura e da luz na germinação de sementes de *Acacia polyphylla* DC. *Revista Brasileira de Botânica*, v.26, n.2, p.249-256, 2003. http://www.scielo.br/pdf/rbb/v26n2/a13v26n2.pdf

ATAÍDE, G.M.; BORGES, E.E.L.; GONCALVES, J.F.C.; GUIMARÃES, V.M.; FLORES, A. V.; BICALHO, E.M. Alterations in seed reserves of *Dalbergia nigra* ((Vell.) Fr All. ex Benth.) during hydration. *Journal of Seed Science*, v.35, n.1, p.56-63, 2013. http://www.scielo.br/pdf/jss/v35n1/08.pdf

BAILLY, C. Active oxygen species and antioxidants in seed biology. *Seed Science and Research*, v.14, p.93-107, 2004. http://journals.cambridge.org/action/displayAbstract;jsessionid=B80FC3066251AA4A792970AD5C0FA8C9.journals?fromPage=online&aid=709508

BEWLEY, J.D.; BRADFORD, K.J.; HILHORST, H.W.M.; NONOGAKI, H. *Seeds:* physiology of development, germination and dormancy. New York: Springer, 2013, 392p.

BHATTACHARJEE, S. Calcium-dependent signaling pathway in the heat-induced oxidative injury in *Amaranthus lividus*. *Biologia Plantarum*, v.52, p.137-140, 2008. http://link.springer.com/article/10.1007%2Fs10535-008-0028-1

BORGHETTI, F. Temperaturas extremas e a germinação de sementes. In: NOGUEIRA, R.J.M.C.; ARAÚJO, E.L.; WILLADINO, L.G.; CAVALCANTE, U.M.T. (eds.). *Estresses ambientais:* Danos e benefícios em plantas. Recife: UFPE, 2005. p.207-218.

BOVERIS, A. Determination of superoxide radicals and hydrogen peroxide in mithocondria. *Methods in enzymology*, v.105, p.429-435, 1984.

BRADBEER, J.W. *Seed dormancy and germination*. New York: Chapman and Hall, 1988. 146p.

BRANCALION, P.H.S.; NOVEMBRE, A.D.L.C.; RODRIGUES, R.R. Temperatura ótima de germinação de sementes de espécies arbóreas brasileiras. *Revista Brasileira de Sementes*, v.32, n.4, p.15-21, 2010. http://www.scielo.br/pdf/rbs/v32n4/02.pdf

BRASIL. Ministério da Agricultura, Pecuária e Abastecimento. *Regras para análise de sementes*. Ministério da Agricultura, Pecuária e Abastecimento. Secretaria de Defesa Agropecuária. Brasília: MAPA/ACS, 2009. 395p. http://www.agricultura.gov.br/arq_editor/file/2946_regras_analise__sementes.pdf

CASTRO, R.D.; BRADFORD, K.J.; HILHORST, H.W.M. Embebição e reativação do metabolismo. In: FERREIRA, A.G.; BORGHETTI, F. (orgs). *Germinação*: do básico ao aplicado, Porto Alegre: Artmed, 2004. p.149-162.

CITES. Convention on International Trade in Endangered Species of Wild Fauna and Flora. *CITES*: Appendix I, II and III to the Convention on International Trade in Endangered Species of Wild Fauna and Flora. US Fish and Wildlife Service: Washington. 2008. http://www.cites.org/sites/default/files/eng/app/2013/E-Appendices-2013-06-12.pdf Accessed on: 10 Feb. 2013.

DOUSSEAU, S.; ALVARENGA, A.A.; ARANTES, L.O.; CHAVES, I.S.; AVELINO, E.V. Technology of *Qualea grandiflora* Mart. (Vochysiaceae) seeds. *Cerne*, v.19, n.1, p.93-101, 2013. http://www.scielo.br/pdf/cerne/v19n1/12.pdf

DUAN, Z.Q.; BAI, L.; ZHAO, Z.G.; ZHANG, G.P.; CHENG, F.M.; JIANG, L.X.; CHEN, K.M. Drought-stimulated activity of plasma membrane nicotinamide adenine dinucleotide phosphate oxidase and its catalytic properties in rice. *Journal Integrative Plant Biology*, v.51, n.12, p.1104-1115, 2009. http://onlinelibrary.wiley.com/doi/10.1111/j.1744-7909.2009.00879.x/full

FOYER, C.H.; NOCTOR, G. Redox homeostasis and antioxidant signaling: a metabolic interface between stress perception and physiological responses. *Plant Cell*, v.17, p.1866–1875, 2005. http://www.plantcell.org/content/17/7/1866.full.pdf+html

FUJITA, M; FUJITA, Y; NOUTOSHI, Y; TAKAHASHI, F; NARUSAKA, Y; YAMAGUCHI-SHINOZAKI, K; SHINOZAKI, K. Crosstalk between abiotic and biotic stress responses: a current view from the points of convergence in the stress signaling networks. *Currents Opinion Plant Biology*, v.9, p.436–442, 2006. http://www.sciencedirect.com/science/article/pii/S1369526606000884

GAY, C.; GEBICKI, J.M. A critical evaluation of the effect of sorbitol on the ferric-xylenol orange hydroperoxide assay. *Analytical Biochemistry*, v.284, p.217-220, 2000. http://www.sciencedirect.com/science/article/pii/S0003269700946967

GILL, S.S.; TUTEJA, N. Reactive oxygen species and antioxidant machinery in abiotic stress tolerance in crop plants. *Plant Physiology and Biochemistry*, v.48, p.909-930. 2010. http://www.sciencedirect.com/science/article/pii/S0981942810001798

GOMES, M.G.; GARCIA, Q.S. Reactive oxygen species and seed germination. *Biologia*, v.68, p.351-357, 2013. http://link.springer.com/article/10.2478%2Fs11756-013-0161-y

IBAMA. Instituto Brasileiro do Meio Ambiente e dos Recursos Naturais Renováveis. *Lista Oficial de Flora ameaçada de extinção*. http://www.ibama.gov.br/documentos/lista-de-especies-ameacadas-de-extincao Accessed on: 10 Feb. 2013.

IUCN. International Union for Conservation of Nature and Natural Resources. *The IUCN Red List of Threatened Species*. 2013. http://www.iucnredlist.org/ Accessed on: 10 Feb. 2013.

KRANNER, I.; ROACH, T; BECKETT, R.P.; WHITAKER, C; FARIDA, V.M. Extra cellular production of reactive oxygen species during seed germination and early seedling growth in *Pisum sativum*. *Journal of Plant Physiology*, v.167, p.805-811. 2010. http://www.sciencedirect.com/science/article/pii/S0176161710000945#

KUO, M.C.; KAO, C.H. Aluminum effects on lipid peroxidation and antioxidative enzyme activities in rice leaves. *Biologia Plantarum*, v.46, p.149-152, 2003. http://download.springer.com/static/pdf/864/art%253A10.1023%252FA%253A1022356322373.pdf?auth66=1392913488_982040178578e21d9dca009621c469ad&ext=.pdf

LARKINDALE, J.; HALL, J.D.; KNIGHT, M.R.; VIERLING, E. Heat stress phenotypes of *Arabidopsis* mutants implicate multiple signaling pathways in the acquisition of thermotolerance. *Plant Physiology*, v.138, p.882-897, 2005. http://www.plantphysiol.org/content/138/2/882.full.pdf+html

LORENZI, M. *Árvores brasileiras:* manual de identificação e cultivo de plantas arbóreas nativas do Brasil. Nova Odessa: Plantarum, 2002. 352p.

LUO, Y.; TANG, H.; ZHANG, Y. Production of reactive oxygen species and antioxidant metabolism about strawberry leaves to low temperatures. *Journal Agricultural Science*, v.3, n.2, p.89-96, 2011. http://www.ccsenet.org/journal/index.php/jas/article/view/6660/7805

MELLO, J.I.O.; BARBEDO, C.J. Temperatura, luz e substrato para a germinação de sementes de pau-brasil (*Caesalpinia echinata* Lam., Leguminosae-Caesalpiniodeae). *Revista Árvore*, v.31, p.645-655, 2007. http://www.scielo.br/pdf/rarv/v31n4/09.pdf

MISRA, H.P.; FRIDOVICH, I. The generation of superoxide radical during the autoxidation of ferredoxins. *The Journal of Biological Chemistry*, v.246, n.22, p.6886-6890, 1971. http://www.jbc.org/content/246/22/6886.full.pdf

MITTLER, R. Oxidative stress, antioxidants and stress tolerance. *Trends Plant Science*, v.7, p.405–410, 2002. http://www.sciencedirect.com/science/article/pii/S1360138502023129

MITTLER, R; VANDERAUWERA, S; GOLLERY, M; VAN BREUSEGEM, F. Reactive oxygen gene network of plants. *Trends Plant Science*, v.9, p.490–498, 2004. http://www.sciencedirect.com/science/article/pii/S1360138504002043

MOHAMMADI, M.; KARR, A.L. Superoxide anion generation in effective and ineffective soybean root nodules. *Journal of Plant Physiology*, v.158, p.1023-1029, 2001. http://ac.els-cdn.com/S0176161704701261/1-s2.0-S0176161704701261-main.pdf?_tid=b7b81844-98ba-11e3-bfc4-00000aacb35f&acdnat=1392741526_26d7c65a6e5015491654507bf2573b24

MULLER, K.; LINKIES, A.; VREEBURG RAM, F.S.C.; KRIEGER-LISZKAY, A.; LEUBNER-METZGER, G. In vivo cell wall loosening by hydroxyl radicals during cress seed germination and elongation growth. *Plant Physiology*, v.150, p.1855–1865, 2009. http://www.plantphysiol.org/content/150/4/1855.full.pdf+html

MURTHY, U.M.N.; KUMAR, P.P; SUN, W.Q. Mechanisms of seed ageing under different storage conditions for *Vigna radiate* (L.) Wilczek: lipid peroxidation, sugar hydrolysis, Maillard reactions and their relationship to glass state transition. *Journal Experimental of Botany*, v.54, p.1057-1067, 2003. http://jxb.oxfordjournals.org/content/54/384/1057.full.pdf+html

OKANE, D.; GILL, V.; BOYD, P.; BURDON, R. Chilling, oxidative stress and antioxidant responses in *Arabidopsis thaliana* callus. *Planta*, v.198, p.371-377, 1996. http://link.springer.com/article/10.1007%2FBF00620053

OLIVEIRA, P.G.; GARCIA, Q.S. Efeitos da luz e da temperatura na germinação de sementes de *Syngonanthus elegantulus*, *S. elegans* e *S. venustus* (Eriocaulaceae). *Acta Botanica Brasilica*, v.19, n.3, p.627-633, 2005. http://www.scielo.br/pdf/abb/v19n3/27380.pdf

ORACZ, K.; EL-MAAROUF-BOUTEAU, H.; KRANNER, I.; BOGATEK, R.; CORBINEAU, F.; BAILLY, C. The mechanisms involved in seed dormancy alleviation by hydrogen cyanide unravel the role of reactive oxygen species as key factors of cellular signaling during germination. *Plant Physiology*, v.150, p.494–505, 2009. http://www.plantphysiol.org/content/150/1/494

OROZCO-ALMANZA, M.S.; LEON-GARCIA, L.P.; GRETHER, R.; GARCIA-MOYA, E. Germination of four species of the genus *Mimosa* (Leguminosae) in a semi-arid zone of Central Mexico. *Journal of Arid Environments*, v.55, p.75–92, 2003. http://www.sciencedirect.com/science/article/pii/S0140196302002653

PIMENTA, R.S.; LUZ, P.B.; PIVETTA, K.F.L.; CASTRO, A.; PIZETTA, P.U.C. Efeito da maturação e temperatura na germinação de sementes de *Phoenix canariensis* hort. ex Chabaud – ARECACEAE. *Revista Árvore*, v.34, p.31-38, 2010. http://www.scielo.br/pdf/rarv/v34n1/v34n1a04

PUKACKA, S.; RATAJCZAK, E. Production and scavenging of reactive oxygen species in *Fagus sylvatica* seeds during storage at varied temperature and humidity. *Journal Plant Physiology*, v.162, n.8, p.873-885, 2005. http://www.sciencedirect.com/science/article/pii/S0176161705000337

REGO, S.S.; NOGUEIRA, A.C.; KUNIYOSHI, Y.S.; SANTOS, Á.F. Germinação de sementes de *Blepharocalyx salicifolius* (H.B.K.) Berg. em diferentes substratos e condições de temperaturas, luz e umidade. *Revista Brasileira de Sementes*, v.31, n.2, p.212-220, 2009. http://www.scielo.br/pdf/rbs/v31n2/v31n2a25.pdf

SAGI, M.; FLUHR, R. Production of reactive oxygen species by plant NADPH oxidases. *Plant Physiology*, v.141, p.336-340, 2006. http://www.plantphysiol.org/content/141/2/336.full.pdf+html

SCHOPFER, P.; PLACHY, C.; FRAHRY, G. Release of reactive oxygen intermediates (superoxide radicals, hydrogen peroxide, and hydroxyl radicals) and peroxidase in germinating radish seeds controlled by light, gibberellin, and abscisic acid. *Plant Physiology*, v. 125, p.1591-1602, 2001. http://www.plantphysiol.org/content/125/4/1591.full.pdf+html

STATSOFT, Inc. *STATISTICA*. Data analysis software system, version 8, 2009. www.statsoft.com.

SUN, W.H.; DUAN, M.; LI, F.; SHU, D.F.; YANG, S.; WENG, Q.W. Overexpression of tomato tAPX gene in tobacco improves tolerance to high or low temperature stress. *Biologia Plantarum*, v.54, p.614-620. 2010. http://link.springer.com/article/10.1007%2Fs10535-010-0111-2

TORRES, M.A.; DANGL, J.L. Function of the respiratory burst oxidase in biotic interactions, abiotic stress and development. *Current Opinion in Plant Biology*, v.8, p.397-403, 2005. http://www.sciencedirect.com/science/article/pii/S1369526605000750

WANG, L.; MA, H.; SONG, L.; SHU, Y.; GU, W. Comparative proteomics analysis reveals the mechanism of pre-harvest seed deterioration of soybean under high temperature and humidity stress. *Journal of Proteomics*, v.75, p.2109-2127. 2012. http://www.sciencedirect.com/science/article/pii/S1874391912000243

22

Sequential sampling of soybean and beans seeds for *Sclerotinia sclerotiorum* detection (Lib.) DeBary

Alessandra de Lourdes Ballaris[1*], José da Cruz Machado[2],
Maria Laene Moreira de Carvalho[3], Cláudio Cavariani[1]

ABSTRACT- This study aimed to verify the efficiency of a sequential sampling plan for bean and soybean seeds, in the *Sclerotinia sclerotiorum* detection. Firstly the heterogeneity of Pérola and Bolinha Bean cultivars lots samples and IAC-Foscarin soybean was assessed, through H and R testing values, verifying their physiological and sanitary quality. To test the pathogen presence or absence a positive binomial distribution was used, in naturally infected seeds, and recovered by Neon-S and Roll Paper tests. The Neon-S test efficiency was obtained at different levels of infection. Statistical analysis for each test was made in DIC with four replications and the data subjected to variance analysis, and averages compared using the Tukey test ($P \leq 0.05$) probability. The soybean and bean seeds sampling sequence for mycelial *Sclerotinia sclerotiorum* detection was efficient, once at least 800 and 1.000 seeds are evaluated considering 0.01 and 0.005%, incidences respectively. The Neon-S and Towel Paper methods were sensitive to the *Sclerotinia sclerotiorum* mycelial detection in soybean and bean seeds.

Index terms: white mold, mycelial transmission, Neon-S.

Amostragem sequencial de sementes de soja e feijão na detecção de *Sclerotinia sclerotiorum* (Lib.) DeBary

RESUMO - O presente trabalho teve por objetivo verificar a eficiência de um plano de amostragem sequencial para sementes de feijão e da soja, na detecção de *S. sclerotiorum*. Deste modo, primeiramente foi avaliada a heterogeneidade das amostras dos lotes para as cultivares Pérola e Bolinha de feijão e IAC-Foscarin de soja, pelos testes dos valores H e R e, verificadas a sua qualidade fisiológica e sanitária. Para testar a presença ou ausência do patógeno foi utilizada a distribuição binomial positiva, em sementes infectadas naturalmente e recuperadas pelos testes Neon-S e Rolo de Papel. A eficiência do teste Neon-S foi obtida para diferentes níveis de infecção. A análise estatística para cada teste foi feita em DIC, com quatro repetições e, os dados submetidos à análise de variância, sendo as médias comparadas pelo teste de Tukey ($P \leq 0,05$) de probabilidade. O plano de amostragem sequencial de sementes de soja e de feijão para detecção de *Sclerotinia sclerotiorum* na forma micelial foi eficiente, desde que 800 e 1.000 sementes sejam avaliadas, considerando incidências de 0,01 e 0,005%, respectivamente. Os métodos Neon-S e Rolo de Papel foram sensíveis na detecção micelial de *Sclerotinia sclerotiorum* em sementes de soja e de feijão.

Termos para indexação: mofo-branco, transmissão micelial, neon-S.

Introduction

The epidemics' occurrence by the *Sclerotinia sclerotiorum* (Lib.) DeBary in soybean and bean cultures, in regions where favorable climatic conditions to pathogen exist, has lead to a major concern from the producers. Until the 90s the occurrence of white mold, common name of the disease caused by the respective pathogen, was restricted to southern Brazil and sporadically in areas of central Minas Gerais and Goiás states. However, the lack of a proper care in the purchase of seeds, the use of "homemade seeds or from the black market", arising from white mold affected areas, the lack of necessary care with processing areas, as well as the succession of susceptible crops (soy, beans, cotton) made this disease one of the major pest problems nowadays.

Every year massive economic losses are endured by

[1]Departamento de Produção e Melhoramento Vegetal, UNESP, Caixa Postal 237, 18610-307 - Botucatu, SP, Brasil.
[2]Departamento de Fitopatologia, UFLA, Caixa Postal 3037, 37200-000 - Lavras, MG, Brasil.

[3]Departamento de Agricultura, UFLA, Caixa Postal 3037, 37200-000- Lavras, MG, Brasil.
*Corresponding author <alballaris@hotmail.com>

both the farmers and the seed trade, because of the white mold. According to Steadman et al. (1994), an infected seed is capable of forming more than one sclerotium, and this one, forms up to 20 apothecia and it may release 2,000,000 ascospores in the environment for each formed apothecium. The damage from the pathogen/seed association result in reductions in seedling emergence and productivity, as well as damage to the entire agricultural system, through the disease's spread, transforming the regions unsuitable for the vegetative species growing. Therefore, the health issue, demanding the communicable diseases control by seeds is important in the agricultural production, once it refers not only to the damage caused to the seeds, but also to the economic expression of each disease (Menten, 1995).

The seed is a commercial product and to support its negotiation, seeds producers, sellers and buyers demand reliable results from analysis of their evaluation. That explains analyzes relevant procedures uniformity performed in different laboratories, by standardizing the tests to be performed in the seed quality evaluation. Once incorporated into the analysis rules, these, are periodically evaluated regarding performance, fitness, improvement or even development of new methods.

Currently, in the Seed Analysis Rules, three different methods are recommended for the *Sclerotinia sclerotiorum* detection in beans and soy seed lots, which are the paper towel, blotter and Neon tests. Recommended methods used in seeds' analyzes, have precision and accuracy difficulties in getting artificially infected seeds lots, mainly because of the variability of the pathogen distribution. Therefore, studies determining the sample's size submitted to the laboratory analysis, enabling the fungus recovery in the seeds natural infection conditions, become necessary.

In this sense, the objective of this study was to verify sequential sampling plan efficiency for bean and soybean seeds, in the *S. sclerotiorum* detection.

Materials and Methods

This study was conducted at the Plant Pathology Laboratory of Camilo Castelo Branco University - Fernandópolis Campus and by UNESP-Botucatu, together with the plant pathology department of Lavras University. Commercial bean seeds of Certified Category 1 (C1), Pérola cultivar and soybean Certified Category 1 (C1), IAC Foscarin, (2012/2013 harvest), used in this experiment, were provided by the Coordination of Integral Technical Assistance (CITA) - Production Core of Fernandópolis Seeds (PC/FS) and collected by an accredited company sampler. The Pérola and Bolinha "saved seeds" bean varieties were obtained directly from producers and

came from areas with *S. sclerotiorum* historical incidence in the Fernandópolis region-São Paulo state.

The seeds lots sampling were performed as per Brasil (2009), checking out the *Sclerotinia sclerotiorum* homogeneity, using the seeds in containers heterogeneity test (H and R-value tests). The samples from the seed lots had their physiological and sanitary profiles characterized by the Germination Speed Index (GSI), the Germination Percentage and the Filter Paper Test, as per Brasil (2009).

For physiological and sanitary quality assessment of seeds the experimental design was completely randomized with thirty treatments (single samples) and 4 replications. Statistical analyzes of the tests were performed using SISVAR computer program (Ferreira, 2000), and the averages between treatments compared by the Tukey's test at 0.05 level of probability.

Determination of sample size required to detect Sclerotinia sclerotiorum for the seed health test

Sclerotia of *S. sclerotiorum* were submitted to surface disinfection with a 1% sodium hypochlorite solution for 5 minutes, then washed three times in distilled water and transferred to Petri plates containing medium PDA (Potato 200 g, dextrose 20 g and agar / liter 20 g).

The plates were kept in an incubation chamber at 20 ± 2 °C temperature, in the dark for seven days. 9 mm diameter disks cut from growing colonies borders, were transferred to Petri plates containing PDA and incubated at the same conditions previously reported, for a period of five to seven days, obtaining pure cultures (Botelho et al., 2013).

Pure cultures of *S. sclerotiorum* were transferred to Petri plates containing half PDA and incubated in environmental chamber temperature at 20 ± 2 °C for seven days and 12 hours photoperiod. Afterwards, the seeds were placed in monolayer over the fungus culture and kept in the dark for 48 hours, as described by Botelho et al. (2013).

Evaluation of the Neon test sensitivity in the S. sclerotiorum recovery in artificially inoculated seeds

To check the test sensitivity, samples of Pérola cultivars bean and soybean seeds ("saved"), Pérola (commercial) and Bolinha ("saved") were inoculated with infected seeds in the ratio of 1/100, 1/200 1/300......1/1000 respectively, the number of infected seeds varying from 1 to 10 and submitted to the Neon testing as previously described.

The experimental design was completely randomized, with four replications. Data were subjected to variance analysis and the treatment averages were compared using the Tukey test (P ≥ 0,05) probability level.

Determination of the S. sclerotiorum sample size

Sequential sampling was used to determine the number of seeds required to accurately verify the lot infestation by *S. sclerotiorum*. For the pathogen detection the NEON test was used as previously described, on natural and artificially infected seeds, the last one in the proportion of 1/100 to 1/1000 seeds, thereby obtaining the lot infection levels ranging from 0,01 to 0,001%, respectively.

The type I (α) and type II (β) errors, were pre-set at 5%, and percentages of maximum (p_1) and minimum (p_o) infections by *S. sclerotiorum* in bean and soybean seeds, at 1 and 0,5%, respectively, for they were compatible with tolerance standards recommended for this phytopathogenic agent, the *Sclerotinia sclerotiorum*.

Formulas of the Sequential Probability Ratio Test (SPRT) by Wald (1947), adapted by Santana (1994), were used in the experiment for the upper and lower limit, $li = b + an$ e $ls = c + an$, respectively, where:

$$a = \frac{ln(\frac{1-p_0}{1-p_1})}{ln(\frac{p_1(1-p_0)}{p_0(1-p_1)})} \qquad b = \frac{ln(\frac{\beta}{1-\alpha})}{ln(\frac{p_1(1-p_0)}{p_0(1-p_1)})} \qquad c = \frac{ln(\frac{1-\beta}{\alpha})}{ln(\frac{p_1(1-p_0)}{p_0(1-p_1)})}$$

The sample average size may be obtained from the formula $TMA_p = \frac{P_p.(b-c)+c}{p-a}$ where:

$$P_p = \frac{(\frac{\beta}{1-\alpha})^w - 1}{(\frac{1-\beta}{\alpha})^w - (\frac{\beta}{1-\alpha})^w} \quad \text{and} \quad p = \frac{1-(\frac{1-p_0}{1-p_1})^w}{(\frac{p_1}{p_0})^w - (\frac{1-p_0}{1-p_1})^w} .$$

Results and Discussion

Determination of the lot heterogeneity

The results determining the beans and soy lots' heterogeneity concerning the germination percentage are shown in Table 1.

The H lots test values, for simple samples were 1.02, 0.81, 0.65 and 0.19 for bean cultivars, Pérola (C1 seeds), Pérola ("saved seeds") and Bolinha ("saved seeds"), and soybean cultivar IAC-Foscarin (C1 seeds), respectively, while the H tabulated value to check the lots heterogeneity at 1% level of probability is 1.10 (Brasil, 2009). The R test value showed regarding the germination test category at 1% probability and maximum amplitudes tolerated, values of 20, 24, 23 and 18 which were also the respective results for bean seeds, Pérola (C1 seeds), Pérola ("saved seeds") and Bolinha ("saved seeds"), and soybean Foscarin cultivar (C1 seeds).

The analytical results of the heterogeneity verification of species and cultivars lots showed that they should be considered non heterogeneous, based on the normal seedlings percentage in the germination test, due to the lack of significance of H and R tests values. Therefore, as lots have similar characteristic attributes, they can be used for standardized analyzes in laboratory, according to Brasil (2009).

In Table 2 are showed physiological qualitative data of C1 certified soybean and bean seeds and "saved seeds" used in the research. Cv. Pérola C1 bean seeds, had superior germination and vigor (IVG) compared to Pérola and Bolinha "saved seeds", where no statistical difference between them was verified.

Table 1. Normal seedlings percentage in the germination test to check the bean and soybean seed lots heterogeneity.

Simple samples	Bean			Soybean
	cv. Pérola C1 seeds	cv. Bolinha "saved seeds"	cv. Pérola "saved seeds"	cv. Foscarin C1 seeds
1	87.0	81.5	74.0	92.0
2	87.5	82.0	73.0	90.5
3	90.0	77.0	86.5	90.0
4	95.5	76.5	79.0	91.5
5	89.5	85.0	80.0	89.5
6	89.5	80.5	78.0	95.0
7	88.5	80.5	78.0	91.0
8	87.5	82.0	76.5	86.5
9	86.5	77.0	77.0	88.5
10	85.5	78.0	74.0	91.5
11	86.5	76.5	80.0	85.5
12	87.5	86.5	80.0	92.5

continuation....

continuation....

Simple samples	Bean			Soybean
	cv. Pérola C1 seeds	cv. Bolinha "saved seeds"	cv. Pérola "saved seeds"	cv. Foscarin C1 seeds
13	86.0	75.5	83.5	89.0
14	88.5	78.0	83.5	92.5
15	88.0	87.0	87.5	91.0
16	86.0	84.5	77.0	90.0
17	80.5	79.5	84.5	93.0
18	83.5	90.5	79.5	95.0
19	83.0	85.0	84.5	94.5
20	88.0	79.5	81.5	89.5
21	87.0	88.0	84.5	90.0
22	84.5	84.5	79.0	93.5
23	94.0	81.5	79.0	93.5
24	87.5	81.5	78.0	91.0
25	86.5	82.5	80.0	93.0
26	94.5	86.5	76.0	94.5
27	82.0	81.0	86.0	89.0
28	83.0	83.0	83.0	92.5
29	84.5	79.0	83.0	94.0
30	80.25	82.0	88.0	91.5
$\bar{X}$	86.955	81.733	81.217	91.367
H	1.02 [NS(1)]	0.65 [NS]	0.81 [NS]	0.19 [NS]
R	6.75 [NS]	7.50 [NS]	7.50 [NS]	4.75 [NS]
R_t	20[(1)]	23	24	18
H_t.	1.10[(2)]			

[NS]Not significant, [(1)]R tabulated values (Brasil, 2009), [(2)]H tabulated values (Brasil, 2009).

Table 2. Average values obtained in preliminary determinations evaluating bean and soybean seeds physiological quality.

Cultivars	Germination %*	IVG
Bean-cv. Pérola-seeds C1	86.95 a**	8.39 a
Bean-cv. Bolinha-"saved seeds"	81.73 b	7.85 b
Bean-cv. Pérola-"saved seeds"	81.22 b	7.59 b
F	0.15	0.58
CV (%)	2.05	4.03
Soya-cv. IAC Foscarin-seeds C1	91.37	8.77
F	0.20	0.71
CV (%)	1.37	3.48

*Transformed data; **Averages with the same letter in the column do not differ by the Tukey test with 5% probability.

Significant differences between bean seeds lots reveal the importance of using commercial seed for crops development instead of using "saved seeds". Successive plantings of seeds from the same origin ("saved seeds") cause productivity losses by the gradual quality reduction, both genetic and physiological, as well as increased impurities occurrence of microorganisms and varietal mixtures, with damages to producers, consumers as well as to the environment.

The results from the seed quality analysis of IAC-Foscarin soybean cultivar (Chart 2) show high quality, clearly pointing out the fundamental importance of using seeds of known origin, culminating with the plants population establishment required to grow, greater emergence and plant development speed, culminating rapidly in closed canopy, which also results in the weeds efficient control and contributes to high levels of productivity.

The bean crop, according to Menten (1995) and Machado (2000), is subject to a wide pathogens variety, most of which have negative effects on the seeds physiological quality produced, and the inoculums present in them, may result in the increase of field diseases and its introduction in pathogens free areas. The results of the lots' health analysis revealed by the filter-paper incubation method are shown in Table 3. Clearly, the C1 category pathogens incidence was smaller in commercial seeds, compared to those observed in "saved seeds" not affecting the seeds quality.

Fungi recovered from IAC-Foscarin cultivar C1 soybean seeds are coincident to those reported by Braccini et al. (2000),

Machado (2000), Goulart (2005), Henning (2009), Henneberg et al. (2012), Sousa et al. (2011), Barros and Juliatti (2012) as the most common gender in soybean seeds. These species infect the seeds internally or externally, being responsible in most cases for poor seed germination and, as a consequence, the soybean production reduced physiological quality, emphasizing however that the pathogens' incidence transmitted by these seeds are within tolerable levels.

Table 3. Fungi incidence percentage in bean and soybean seeds.

Fungi/Bean	cv. Pérola seeds C1	cv. Pérola "saved seeds"	cv. Bolinha "saved seeds"
Phoma spp.	0	1	0.5
Fusarium spp.	0.25	2	1
Aspergillus spp.	2	13.5	9.75
Penicillium spp.	3	5	8.5
Rhizopus spp.	6	11	7.75
Fungi/Soybean	cv. Foscarin – seeds		
Colletotrichum truncatum	0		
Phomopsis sojae	5		
Fusarium spp.	1.75		
Aspergillus spp.	0.75		
Penicillium spp.	6		

Testing the presence - absence hypothesis of pathogen in the lot

The results obtained in this study concerning natural seed infection by Neon-S test (Table 4) for bean cultivars (0,25% samples -12 cv. Pérola, "saved seeds") and by the Towel paper method (0,25%, samples 19 and 26 cv. Pérola, "saved seeds") is according to those found by Hoffman et al. (1998), Nasser et al. (1999), Parisi et al. (2006), Henning (2009), Botelho et al. (2013) and Henneberg et al. (2012), which means, low levels of the pathogen recovery (0,25 to 2%) by the detection traditional methods.

Differences in pathogen incidence and seeds recovery observed in literature may be due to various intrinsic and extrinsic factors to the seed, as per Yorinori (1982) and Machado (2000), and by the sample's lack of precision, according to Carvalho et al. (2011).

The mycelial form of S. sclerotiorum detection verified in this work is of utmost importance, for it features the contamination in seeds that should be discarded, however, these are reused by the producer in the next crop, leveraging the infected area increase within the property, or other producing areas due to the illegal sale of the "saved seeds".

As a reminder, Steadman et al. (1994), correlating the link action of this pathogen to seeds with potential future damages in production fields, noted that this could yield 2,000,000 ascospores approximately (initial inoculums) through the sclerotium germination attached or mixed to them.

Nasser et al. (1999) also pointed out that one infected seed only is able to give 625 primary outbreaks of the disease in a 250,000 plants/ha field, thus highlighting the problem of the illegal seeds trade and the use of "saved seeds" by the producers.

According to Table 5, the Neon-S test found sensitivity to the presence of the S. sclerotiorum pathogen, detected on 100% of artificially infected seeds, regardless the proportion used, fact that confirms the effectiveness of this semi-selective medium regarding its goal. Similar results were observed by other authors, however without differentiation of the infection level (Napoleão et al., 2006; Henneberg et al., 2012; Botelho et al., 2013). The method ability to acknowledge the pathogen's presence in seeds low percentages of infection, is on its own, relevant. However, the verified data of naturally infected seeds are very small compared to those seen in artificially inoculated seeds; in these ones, the Neon test is always sensitive.

According to ISTA (2013), the lack of due care when sampling and during sample handling in the laboratory is the main factor to determine errors and discrepancies in seed health analyzes results. Therefore, the traditional sampling of seed lots infected by fungi, such as S. sclerotiorum, may not be representative and may indicate a lot as uninfected, when in reality it is infected by the pathogen (Henning et al., 2009).

Table 4. Percentage of seeds infected by *Sclerotinia sclerotiorum* in bean and soybean seeds using two detection tests.

Simple Sample	NEON-S				PAPER ROLL			
	cv. Pérola C1 seeds	cv. Bolinha "saved seeds"	cv. Pérola "saved seeds"	cv. Foscarin C1 seeds	cv. Pérola C1 seeds	cv. Bolinha "saved seeds"	cv. Pérola "saved seeds"	cv. Foscarin C1 seeds
1	0 NS	0 NS	0 NS*	0 NS	0 NS	0 NS	0 NS	0 NS
2	0 NS	0 NS	0 NS	0 NS	0 NS	0 NS	0 NS	0 NS
3	0 NS	0 NS	0 NS	0 NS	0 NS	0 NS	0 NS	0 NS
4	0 NS	0 NS	0 NS	0 NS	0 NS	0 NS	0 NS	0 NS
5	0 NS	0 NS	0 NS	0 NS	0 NS	0 NS	0 NS	0 NS
6	0 NS	0 NS	0 NS	0 NS	0 NS	0 NS	0 NS	0 NS
7	0 NS	0 NS	0 NS	0 NS	0 NS	0 NS	0 NS	0 NS
8	0 NS	0 NS	0 NS	0 NS	0 NS	0 NS	0 NS	0 NS
9	0 NS	0 NS	0 NS	0 NS	0 NS	0 NS	0 NS	0 NS
10	0 NS	0 NS	0 NS	0 NS	0 NS	0 NS	0 NS	0 NS
11	0 NS	0 NS	0 NS	0 NS	0 NS	0 NS	0 NS	0 NS
12	0 NS	0 NS	0.25 NS	0 NS	0 NS	0 NS	0 NS	0 NS
13	0 NS	0 NS	0 NS	0 NS	0 NS	0 NS	0 NS	0 NS
14	0 NS	0 NS	0 NS	0 NS	0 NS	0 NS	0 NS	0 NS
15	0 NS	0 NS	0 NS	0 NS	0 NS	0 NS	0 NS	0 NS
16	0 NS	0 NS	0 NS	0 NS	0 NS	0 NS	0 NS	0 NS
17	0 NS	0 NS	0 NS	0 NS	0 NS	0 NS	0 NS	0 NS
18	0 NS	0 NS	0 NS	0 NS	0 NS	0 NS	0 NS	0 NS
19	0 NS	0 NS	0 NS	0 NS	0 NS	0 NS	0.25 NS	0 NS
20	0 NS	0 NS	0 NS	0 NS	0 NS	0 NS	0 NS	0 NS
21	0 NS	0 NS	0 NS	0 NS	0 NS	0 NS	0 NS	0 NS
22	0 NS	0 NS	0 NS	0 NS	0 NS	0 NS	0 NS	0 NS
23	0 NS	0 NS	0 NS	0 NS	0 NS	0 NS	0 NS	0 NS
24	0 NS	0 NS	0 NS	0 NS	0 NS	0 NS	0 NS	0 NS
25	0 NS	0 NS	0 NS	0 NS	0 NS	0 NS	0 NS	0 NS
26	0 NS	0 NS	0 NS	0 NS	0 NS	0 NS	0.25 NS	0 NS
27	0 NS	0 NS	0 NS	0 NS	0 NS	0 NS	0 NS	0 NS
28	0 NS	0 NS	0 NS	0 NS	0 NS	0 NS	0 NS	0 NS
29	0 NS	0 NS	0 NS	0 NS	0 NS	0 NS	0 NS	0 NS
30	0 NS	0 NS	0 NS	0 NS	0 NS	0 NS	0 NS	0 NS

*NS not significant by the Tukey test with 5% probability.

Table 5. Sensitivity evaluation of the Neon test for *Sclerotinia sclerotiorum* detection at different incidence rates.

Cultivars	Nr. of seeds	Recovery percentage of infected seeds										
		Number of inoculated seeds										
		0	1	2	3	4	5	6	7	8	9	10
	100	0 NS	100 NS	100 NS	100 NS	100 NS	100 NS	100 NS	100 NS	100 NS	100 NS	100 NS
	200	0 NS	100 NS	100 NS	100 NS	100 NS	100 NS	100 NS	100 NS	100 NS	100 NS	100 NS
	300	0 NS	100 NS	100 NS	100 NS	100 NS	100 NS	100 NS	100 NS	100 NS	100 NS	100 NS
	400	0 NS	100 NS	100 NS	100 NS	100 NS	100 NS	100 NS	100 NS	100 NS	100 NS	100 NS
cv. Pérola	500	0 NS	100 NS	100 NS	100 NS	100 NS	100 NS	100 NS	100 NS	100 NS	100 NS	100 NS
C1 seeds	600	0 NS	100 NS	100 NS	100 NS	100 NS	100 NS	100 NS	100 NS	100 NS	100 NS	100 NS
	700	0 NS	100 NS	100 NS	100 NS	100 NS	100 NS	100 NS	100 NS	100 NS	100 NS	100 NS
	800	0 NS	100 NS	100 NS	100 NS	100 NS	100 NS	100 NS	100 NS	100 NS	100 NS	100 NS
	900	0 NS	100 NS	100 NS	100 NS	100 NS	100 NS	100 NS	100 NS	100 NS	100 NS	100 NS
	1000	0 NS	100 NS	100 NS	100 NS	100 NS	100 NS	100 NS	100 NS	100 NS	100 NS	100 NS

continuation....

continuation....

Cultivars	Nr. of seeds	Recovery percentage of infected seeds										
		Number of inoculated seeds										
		0	1	2	3	4	5	6	7	8	9	10
cv. Bolinha "saved seeds"	100	0 NS	100 NS	100 NS	100 NS	100 NS	100 NS	100 NS	100 NS	100 NS	100 NS	100 NS
	200	0 NS	100 NS	100 NS	100 NS	100 NS	100 NS	100 NS	100 NS	100 NS	100 NS	100 NS
	300	0 NS	100 NS	100 NS	100 NS	100 NS	100 NS	100 NS	100 NS	100 NS	100 NS	100 NS
	400	0 NS	100 NS	100 NS	100 NS	100 NS	100 NS	100 NS	100 NS	100 NS	100 NS	100 NS
	500	0 NS	100 NS	100 NS	100 NS	100 NS	100 NS	100 NS	100 NS	100 NS	100 NS	100 NS
	600	0 NS	100 NS	100 NS	100 NS	100 NS	100 NS	100 NS	100 NS	100 NS	100 NS	100 NS
	700	0 NS	100 NS	100 NS	100 NS	100 NS	100 NS	100 NS	100 NS	100 NS	100 NS	100 NS
	800	0 NS	100 NS	100 NS	100 NS	100 NS	100 NS	100 NS	100 NS	100 NS	100 NS	100 NS
	900	0 NS	100 NS	100 NS	100 NS	100 NS	100 NS	100 NS	100 NS	100 NS	100 NS	100 NS
	1000	0 NS	100 NS	100 NS	100 NS	100 NS	100 NS	100 NS	100 NS	100 NS	100 NS	100 NS
cv. Pérola "saved seeds"	100	0 NS*	100 NS	100 NS	100 NS	100 NS	100 NS	100 NS	100 NS	100 NS	100 NS	100 NS
	200	0 NS	100 NS	100 NS	100 NS	100 NS	100 NS	100 NS	100 NS	100 NS	100 NS	100 NS
	300	0 NS	100 NS	100 NS	100 NS	100 NS	100 NS	100 NS	100 NS	100 NS	100 NS	100 NS
	400	0 NS	100 NS	100 NS	100 NS	100 NS	100 NS	100 NS	100 NS	100 NS	100 NS	100 NS
	500	0 NS	100 NS	100 NS	100 NS	100 NS	100 NS	100 NS	100 NS	100 NS	100 NS	100 NS
	600	0 NS	100 NS	100 NS	100 NS	100 NS	100 NS	100 NS	100 NS	100 NS	100 NS	100 NS
	700	0 NS	100 NS	100 NS	100 NS	100 NS	100 NS	100 NS	100 NS	100 NS	100 NS	100 NS
	800	0 NS	100 NS	100 NS	100 NS	100 NS	100 NS	100 NS	100 NS	100 NS	100 NS	100 NS
	900	1 NS	100 NS	100 NS	100 NS	100 NS	100 NS	100 NS	100 NS	100 NS	100 NS	100 NS
	1000	0 NS	100 NS	100 NS	100 NS	100 NS	100 NS	100 NS	100 NS	100 NS	100 NS	100 NS
cv. Foscarin C1 seeds	100	0 NS	100 NS	100 NS	100 NS	100 NS	100 NS	100 NS	100 NS	100 NS	100 NS	100 NS
	200	0 NS	100 NS	100 NS	100 NS	100 NS	100 NS	100 NS	100 NS	100 NS	100 NS	100 NS
	300	0 NS	100 NS	100 NS	100 NS	100 NS	100 NS	100 NS	100 NS	100 NS	100 NS	100 NS
	400	0 NS	100 NS	100 NS	100 NS	100 NS	100 NS	100 NS	100 NS	100 NS	100 NS	100 NS
	500	0 NS	100 NS	100 NS	100 NS	100 NS	100 NS	100 NS	100 NS	100 NS	100 NS	100 NS
	600	0 NS	100 NS	100 NS	100 NS	100 NS	100 NS	100 NS	100 NS	100 NS	100 NS	100 NS
	700	0 NS	100 NS	100 NS	100 NS	100 NS	100 NS	100 NS	100 NS	100 NS	100 NS	100 NS
	800	0 NS	100 NS	100 NS	100 NS	100 NS	100 NS	100 NS	100 NS	100 NS	100 NS	100 NS
	900	0 NS	100 NS	100 NS	100 NS	100 NS	100 NS	100 NS	100 NS	100 NS	100 NS	100 NS
	1000	0 NS	100 NS	100 NS	100 NS	100 NS	100 NS	100 NS	100 NS	100 NS	100 NS	100 NS

* NS= Not significant by the Tukey test with 5% probability.

Determination of the sample size required to detect Sclerotinia sclerotiorum on a seed health test

The lower (l_i) and upper limits (l_s) for each group of 100 sampled seeds (n), are presented in Table 6 and Figure 1, obtained through the formulas $l_i = -4,21728 + 0,007216n$ = ls $4.21728 + 0.007216n$. In it, one can observe that the minimum analyzed number of seeds must be 700, regarding the pathogen presence-absence in seed lots, observing the value ranges, in order to determine the lot's destination.

On Table 7 and Figures 1 and 2, it can be seen that the traditional pathological analysis of 400 seeds is insufficient to determine the lots free of *S. sclerotiorum*, being necessary to analyze 848 seeds for an average of 0,005 infections. By the sequential sampling results (Table 8), was demonstrated that this number is insufficient for the *S. sclerotiorum* recovery, as in the Neon test it took 8 consecutive shots of 100 seeds before the decision of the lot's destination.

The actual sample size for health tests, according to Carvalho et al. (2011), depends on the pathogen occurrence frequency in the seeds. Works that determine the sample actual size for fungi detection in routine analysis are rare and, regarding specifically the *S. sclerotiorum*, they are nonexistent.

Having in mind that this pathogen is monocyclic, and therefore has a low rate of infection and sporulation compared to other seed fungal contaminants, it is relevant to prove that it requires a higher number of seeds than the 400 standardized seeds only, for their detection in routine tests, avoiding consequently false negative results.

Table 6. Calculated values of the sequential sampling plan decision boundaries for p_0=0,01, p_1=0,005 and $\alpha = \beta$=5% for seeds groups size m=100.

Number of groups of tested seeds	Total number of seeds	Lower Limit	Upper limit
1	100	ND[*]	5
2	200	ND	6
3	300	ND	6
4	400	ND	7
5	500	ND	8
6	600	ND	9
7	700	ND	9
8	800	1	10
9	900	2	11
10	1000	3	12

[*]= Not determined.

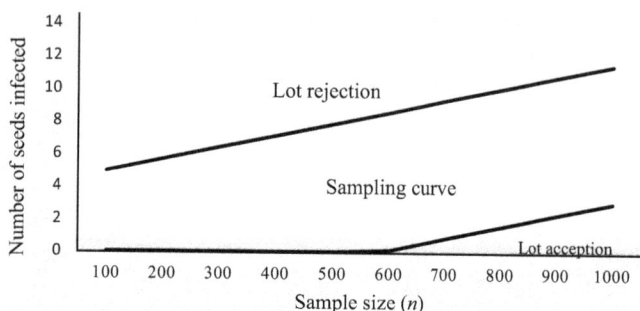

Figure 1. Decision lines of the sequential sampling plan with parameters p_0= 0,005; p_1 = 0.01 and $\alpha=\beta$ = 0.05.

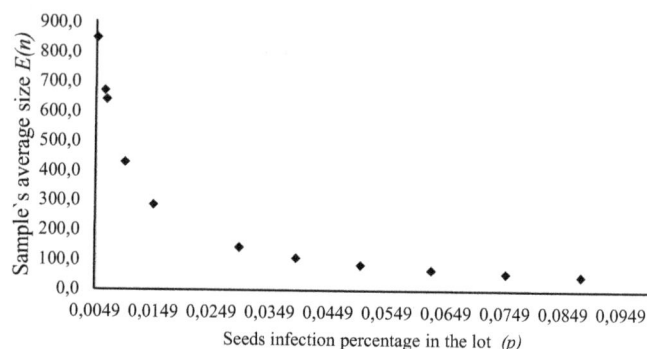

Figure 2. Curve of the average sample size of the sequential sampling plan for parameters p_0 = 0.005; p_1 =0.01 and $\alpha = \beta$ = 5%.

Table 7. Average sample size (ASS) according to values assigned to W in the range of -6 to 6.

W	p	P(p)	ASS (p)
-6	0.0879	0.0000	47.9
-5	0.0750	0.0000	56.3
-4	0.0623	0.0000	67.7
-3	0.0503	0.0001	83.8
-2	0.0393	0.0028	107.3
-1	0.0296	0.0500	142.7
1	0.0148	0.9500	285.4
2	0.0098	0.9972	429.1
3	0.0063	0.9999	670.2
3.5	0.0050	1.0000	848.4
4	0.0039	1.0000	1082.5
5	0.0023	1.0000	1800.4
6	0.0014	1.0000	3069.0

In accordance with Henning et al. (2009), who found eight seeds infected by *Sclerotinia sclerotiorum* only in a lot of 10.400 seeds, in other words, 0,076% pathogen infection, the result obtained for this study's sample size indicates that a minimum of 1.000 seeds, and not more than 10,000, are to be analyzed when one wants to find 1 infected seed in the lot.

The TSRP reliability, concerning routine testing used in seed pathology laboratory, is given by the Operating Characteristic Curve OCC (p) (Figure 3) (Bányai and Barabás, 2002), cited by Santana (1994).

One can see that lots averaging 1% of *S. sclerotiorum* infection have a 95% chance of being rejected; on the other hand, the probability of a seed infected in the lot with fungus at its surface or its inner integument tissues and not detected by the test, is on average 0,5%.

One should have in mind that the infection determination of a lot is done through standardized tests recommended by the Seeds Review Rules (Brasil, 2009), the sequential sampling being responsible for establishing if this infection level for this agent lies within the levels of tolerance suggested. As per Santana's (1994) recommendation, one should use its maximum value for the expected sample average size in order to obtain a reliable decision on the pathogen presence-absence, ensuring the lots health quality to the producer and consumer.

The average sample size determination is the first step in the search for detection adequate methods, targeting efficiency for decision making on the lots' destination, ensuring health quality to the seeds producer and consumer, restricting as well the pathogen spread on free environmental areas and reducing the inoculums potential, where the disease already exists.

Table 8. Number of seeds needed in the sequential sampling plan for the lot acceptation or rejection decision, due to *Sclerotinia sclerotiorum* presence in bean and soybean seed.

	Lots	Total of seeds tested	Total of seeds with *S. sclerotiorum*	Results
Naturally Infected	cv. Pérola C1 seeds	1000	0	Accepted
	cv. Bolinha "saved seeds"	1000	0	Accepted
	cv. Pérola "saved seeds"	800	1	Rejected
	cv. Foscarin C1 seeds	1000	0	Accepted
Artificially Infected	:100	100	1	
	1:200	200	1	
	1:300	200	1	
	1:400	300	1	
	1:500	300	1	
	1:600	500	1	
	1:700	700	1	
	1:800	800	1	
	1:900	900	1	
	1:1000	1000	1	

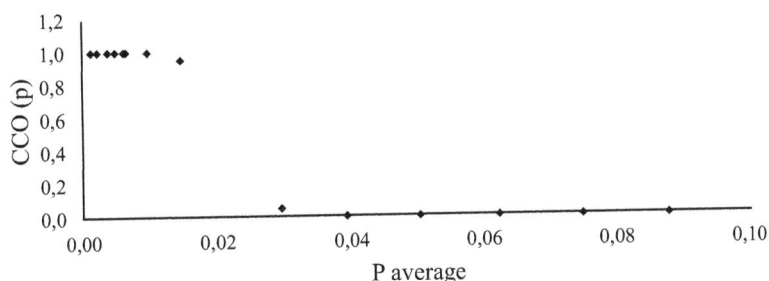

Figure 3. Operation characteristic curve CCO (p) of the likelihood test of *Sclerotinia sclerotiorum* total population.

Conclusions

The sampling sequence of soybean and bean seeds for mycelial *Sclerotinia sclerotiorum* detection was efficient, once 800 and 1,000 seeds were evaluated, with incidences of 0,01 and 0,005%, respectively;

The Neon and paper towel methods were sensitive in the mycelial *Sclerotinia sclerotiorum* detection in soybean and bean seeds.

References

BÁNYAI, J.; BARABÁS, J. *Handbook on statistics in seed testing.* Association International of Seed Testing, 2002. 84p. Available at <http://www.world-seed-project.org/upload/prj/product/stahandbk2002.pdf> Accessed on: Nov. 17th, 2013.

BARROS, F. C.; JULIATTI, F. C. Levantamento de fungos em amostras recebidas no laboratório de micologia e proteção de plantas da Universidade Federal de Uberlândia no período 2001-2008. *Bioscience Journal*, v.28, n.1, p.77-86, 2012. http://www.seer.ufu.br/index.php/biosciencejournal/issue/view/707

BRACCINI, A. L.; REIS, M. S.; BRACCINI, M. C. L.; SCAPIM, C.A.; MOTTA, I. SÁ. Germinação e sanidade de sementes de soja (*Glycine max* (L.) Merrill) colhidas em diferentes épocas. *Acta Scientiarum*, v.22, n.4, p.1017-1022, 2000. http://periodicos.uem.br/ojs/index.php/ActaSciAgron/article/view/2868/2047

BRASIL. Ministério da Agricultura, Pecuária e Abastecimento. *Regras para análise de sementes.* Ministério da Agricultura, Pecuária e Abastecimento. Secretaria de Defesa Agropecuária. Brasília: MAPA/ACS, 2009. 395p. http://www.agricultura.gov.br/arq_editor/file/2946_regras_analise__sementes.pdf

BOTELHO, L. S.; ZANCAN, W. L. A.; MACHADO, J. C.; BARROCAS, E. N. Performance of common bean seeds infected by the fungus *Sclerotinia sclerotiorum. Journal of Seed Science*, v.35, n.2, p.153-160, 2013. http://www.scielo.br/pdf/jss/v35n2/03.pdf

CARVALHO, M.L.M.; VON PINHO, E.V.R.; OLIVEIRA, J.A.; GUIMARÃES, R.M.; MUNIZ, J.A. *Manual do amostrador de sementes.* Lavras-MG, 2011. 143 p.

FERREIRA, D. F. Análises estatísticas por meio do SISVAR for Windows version 4.0. http://www.dex.ufla.br/~danielff/softwares.htm Lavras: UFLA, 2000. 72p.

GOULART, A.C.P. *Fungos em sementes de soja*: detecção, importância e controle. Dourados: EMBRAPA AGROPECUARIA OESTE, 2005. 72p.

HENNEBERG, L.; GRABICOSKI, E. M. G.; JACCOUD-FILHO, D. S.; PANOBIANCO. M. Incidência de *Sclerotinia sclerotiorum* em sementes de soja e sensibilidade dos testes de detecção. *Pesquisa Agropecuária Brasileira*, v.47, n.6, p.763-768, 2012. http://www.scielo.br/pdf/pab/v47n6/47n06a05.pdf

HENNING, A.A. Manejo de doenças da cultura da soja (*Glycine max* L. Merrill*). Informativo ABRATES*, v.19, n.1, p.9-12, 2009.

HENNING, A.A.; PAULA, F.Y.H.; MOMTEMEZZO, C.A.O.; BOSSE, E.J.; BERGONSI, J.S.S. Avaliação de princípios ativos para o controle químico de mofo branco (*Sclerotinia sclerotiorum*) em soja – safra 2008/2009. *Informativo ABRATES*, v.19, n.1, p.29-31, 2009.

HOFFMAN, D. D.; HARTMAN, G. L.; MUELLER, D.M.; LEITZ, R. A.; NICKELL, C. D.; PEDERSEN, W. L. Yield and seed quality of soybean cultivars infected with *Sclerotinia sclerotiorum. Plant Disease*, v.82, p.826-829, 1998. http://apsjournals.apsnet.org/doi/pdf/10.1094/PDIS.1998.82.7.826.

ISTA- INTERNATIONAL SEED TESTING ASSOCIATION. Method Validation for seed testing. Switzerland, 2007. 70p. Available at: <http://www. seed test. org /upload/ cms/ user/ ISTAMethodValidationforSeedTesting - v1.01.pdf > Accessed on May. 30th 2013.

MACHADO, J.C. Patologia de sementes: significado e atribuições. In: CARVALHO, N.M. E NAKAGAWA, J. (Eds.). *Sementes*: ciência, tecnologia e produção. Jaboticabal: FUNEP, 2000. p.522-588.

MENTEN, J. O. M. Prejuízo causado por patógenos associados às sementes. In: MENTEN, J.O.M. *Patógenos em sementes:* detecção, danos e controle químico. São Paulo: Ciba Agro, 1995. p.115-136.

NAPOLEÃO, R.; NASSER, L.C.B.; LOPES, C.A.; C.A.F, FILHO, A.C. Neon-S, a new medium for detection of *Sclerotinia sclerotiorum* on seeds. *Summa Phytopathologica*, v. 32, n.2, p. 180-182, 2006. http://www.scielo.br/pdf/sp/v32n2/v32n2a14.pdf

NASSER, L. C. B.; CAFE FILHO, A. C.; AZEVEDO, J. A.; GOMES, A. C.; VIVALDI, L. J.; ALBRETCH, J. C.; FREITAS, M. A.; KARL, A. C.; FERRAZ, L. L. C.; MEDEIROS, R. G.; CARVAJAL, R. A.; NAPOLEAO, R. L.; JUNQUEIRA, N. T. V. Metodo alternativo para o manejo do cancro-da-haste da soja (*Diaphorte phaseolorum* f. sp. *meridionalis*) e mofo-branco do feijoeiro (*Sclerotinia sclerotiorum*) em sistemas de produção de grãos do Cerrado. Embrapa Cerrados- Comunicado Tecnico, n.11, p.4, 1999.

PARISI, J.J.D.; PATRÖCIO, F.R.A., OLIVEIRA, S.H.F. Modification of the paper towel seed health test for the detection of *Sclerotinia sclerotiorum* in bean seeds (*Phaseolus vulgaris* L.). *Summa Phytopathologica*, v.32, n.3, p.288-290, 2006. ttp://www.scielo.br/pdf/sp/v32n3/a15v32n3.pdf

SANTANA, D. G. Adaptação do teste do pH do exsudato e viabilidade do uso da amostragem sequencial na rápida definição sobre o destino de lotes de sementes de milho (*Zea mays* L.). Lavras:UFLA. 1994. 79p.

SOUSA, T. P.; NASCIMENTO, I. O.; MAIA. C. B.; MORAIS, J.; BEZERRA, G. A.; BEZERRA, J. W. T. Incidência de fungos associados a sementes de soja transgênica variedade BRS Valiosa RR. *Agroecossistemas*, v.3, n.1, p. 52-56, 2011.

STEADMAN, J.R.; MARCINKOWSKA, J.; RUTLEDGE, S. A semi-selective medium for isolation of *Sclerotinia sclerotiorum. Canadian Journal of Plant Pathology*, v.16, n.2, p.68-70, 1994.

YORINORI, J.T. Doenças da soja no Brasil. In: MIYSAKA, S.; MEDINA. J.C. *A soja no Brasil Central*. Campinas: Fundação Cargill, 1982.350p.

WALD, A. *Sequential analysis*. New York: John Wiley and Sons, 1947. 212 p.

23

Dynamics of reserves of soybean seeds during the development of seedlings of different commercial cultivars

Welison Andrade Pereira[1]*, Sara Maria Andrade Pereira[2], Denise Cunha Fernandes dos Santos Dias[3]

ABSTRACT – Physiological quality and vigor of the seeds comprise properties that determine a high level of activity and performance during germination and seedling emergence, having a direct relation with the establishment of the stand of a crop. In this context, the assessment of seedling development, including the analysis of the seed reserves mobilization are a reliable method to investigate the physiological potential of seed lots. In this preliminary study, the aim was to investigate the dynamics of seed reserves mobilization of a sample of soybean commercial cultivars. By means of the seedling length bioassay and weight of dry matter of seeds, cotyledons, hypocotyls and radicles, information on the reserves mobilization during the germination process was obtained. Data were subjected to analysis of variance and Scott and Knott test, and afterwards, phenotypic correlations between traits were obtained. The results have shown that the dry matter of seeds, reserves reduction of seeds and dry matter of seedlings are positively correlated, and thus, seeds containing more reserves may be more effective during the early development of seedlings. In contrast, reserve reduction of seeds and conversion efficiency of the seeds reserves in the dry matter of seedlings are negatively correlated.

Index terms: *Glycine max* (L.) Merrill, germination, seedlings, reserves mobilization.

Dinâmica de reservas das sementes de soja durante o desenvolvimento de plântulas de diferentes cultivares comerciais

RESUMO – Qualidade fisiológica e vigor compreendem propriedades das sementes que determinam alto nível de atividade e desempenho durante a germinação e emergência das plântulas, tendo relação direta com o estabelecimento do estande de uma lavoura. Neste contexto, a avaliação do desenvolvimento das plântulas, incluindo análise da mobilização de reservas das sementes pode se constituir em um método confiável para investigação do potencial fisiológico de lotes. Neste estudo preliminar, objetivou-se investigar a dinâmica de reservas em sementes de cultivares comerciais de soja. Por meio do teste de comprimento de plântulas e peso de massa seca de sementes, cotilédones, hipocótilos e radículas foram obtidas informações sobre a mobilização de reservas durante o processo de germinação. A massa seca de sementes, redução de reservas de sementes e massa seca de plântulas estão positivamente correlacionados entre si, e, logo, sementes contendo mais reservas podem apresentar maior vigor durante o desenvolvimento inicial de plântulas. Em contrapartida, redução de reserva de sementes e eficiência na conversão de reservas das sementes em massa seca de plântulas são negativamente correlacionadas.

Termos para indexação: *Glycine max* (L.) Merrill, germinação, plântulas, mobilização de reservas.

Introduction

Vigor is a set of properties that contribute to indicate the quality of seeds, relating them to the potential for germination, seedling emergence and storage capacity under conditions different from those considered standard (Sun et al., 2007).

Seeds with high vigor give rise to seedlings which emerge quickly, uniformly and completely from the ground (Soltani et al., 2006), which gives them a competitive advantage over weeds (Dias et al., 2011).

The establishment of the stand of a crop is related to seed vigor because seedlings grown from seeds with reduced

[1]Departamento de Biologia, UFLA, Caixa Postal 3037, 37200-000, Lavras, MG, Brasil.
[2]Departamento de Produção Vegetal, UFES, Caixa Postal 16, 29500-000 – Alegre, ES, Brasil.
[3]Departamento de Fitotecnia, UFV, 36570-000 – Viçosa, MG, Brasil.
*Corresponding author < welison.pereira.ufv@gmail.com>

vigor slowly emerge from the soil, are smaller in height and dry matter accumulation (Henning et al., 2010), later as to flowering, and form a lower final population in the field (Vanzolini and Carvalho, 2002). With regard to plant height, lower insertion of the first pod results in lower yield (Câmara et al., 1998). Regarding the faults at the stand, a number of negative implications for crop is triggered, such as the loss in competitiveness with weeds (Dias et al., 2011); lower shading of the soil, thus more evaporation, and thus less available water for the crop; low interception to the light and interruption in the yield potential (Nepomuceno et al., 2007; Valério et al., 2009) and little growth in early seasons (Carvalho et al., 2008).

Given the due importance to seed physiological quality and seedling vigor, the assessment of hypocotyls, radicle and seedling (Vanzolini et al., 2007), as well as the analysis of seed reserves content (Henning et al., 2010), are reliable methods for assessment of the physiological potential of the seeds. Based on the literature, seed lots with high quality and vigor have higher seedling and are able to accumulate more dry matter (Schuab et al., 2002; Henning et al., 2010), and, in general, larger seeds have a better physiological quality (Pereira et al., 2013). These findings may be related to the fact that high vigor seeds have a higher starch content, soluble proteins and sugars, and also greater capacity to mobilize reserves during the germination period (Henning et al., 2010).

The dry matter content of the seedlings can be determined in addition to the test of seedlings length (Pereira et al., 2009), and the reduction of reserves of cotyledons and the conversion efficiency of these reserves into dry matter in the seedlings may also be studied (Soltani et al., 2006).

The dynamics of reserves is noteworthy because it represents a fundamental part of the germination process. First, the embryo starts this process with its own reserves, but maintaining it will depend on a flow of soluble components of its reserves for the regions in full development (Henning et al., 2010). Thus, when the genetic variability for such characters is found, these can be considered in enhancement programs, with the goal of searching for genotypes with greater potential to mobilize and convert reserves contained in the cotyledons in the dry matter of seedlings.

Given the above, the objective was to work with the preliminary information on the dynamics of reserves of soybean seeds commercial cultivars, in order to extend a posteriori such methodology for the assessment of strains obtained in a soybean enhancement program at Universidade Federal de Viçosa. Thus, this study aimed to analyze the mobilization of reserves during the process of germination of seedlings of different soybean cultivars and the existence of correlations between the characteristics studied.

Material and Methods

All assays were conducted in Soybean Genetics Laboratory at Department of Plant Science at Universidade Federal de Viçosa, Viçosa, MG, Brasil). Characteristics were assessed that could show the dynamics of reserves of seeds of 14 soybean cultivars, *Glycine max* (L.) Merrill: CD202, CD206, CD211, CD217, CD219, UFV16, UFV18, UFV19, UFVTN103, UFVTN105, MONARKA, MSOY8605, CONFIANÇA and EMGOPA315.

The seeds used in this study were multiplied in the harvest immediately preceding the experiments, allowing that all tests could be performed with freshly harvested seeds, as performed by Soltani et al. (2006).

To obtain the water content (WC), three replications of 50 seeds were weighed (W1), taken to an oven at a temperature of 105 °C for 24 hours and reweighed (W2). In this procedure, in addition to water content, based on the equation $TA = \frac{W1 - W2}{W1}$, the dry matter of seeds was also obtained (DWS), $W2$ (Soltani et al., 2006), understood, for the purposes of this study, as the total reserve of seeds available to be mobilized to the seedlings.

The germination test (GR) was conducted with four replications of 50 seeds, distributed in paper towel moistened with distilled water at a ratio of 2.5 times the weight of the dry paper (Brasil, 2009), making up rolls which were maintained in a germination chamber (B.O.D. (Biochemical Oxygen Demand)) under controlled temperature (25 ± 1 °C) for 7 days. At the end of that period, the number of normal seedlings was counted.

The seedling length test was conducted with four replications of 20 seeds placed in the upper third of the paper towel (Pereira et al., 2009), with their micropyles directed to the base (Vanzolini et al., 2007). Seven days after seeding, hypocotyls, radicles and seedlings were measured in centimeters. Obtaining the average length of hypocotyl, radicle and seedling of the plot followed two recommendations: ISTA (International Seed Testing Association), the sum of the lengths of the seedlings divided by the number of seeds put to germinate; and AOSA (Association of Official Seed Analysts, 2009), the sum of the lengths divided by the number of seedlings measured, according to Vanzolini et al. (2007). Thus, mean lengths were obtained for radicle (RRL and ARL), hypocotyl (RHL and AHL) and seedling (RLS and ALS), according to recommendations from ISTA and AOSA, respectively.

Once the length measures were taken, the hypocotyl-radicle axes were separated from the cotyledons and taken to the oven at a temperature of 80 ± 1 °C for 24 hours. After this period, the material was weighed, giving rise to the dry matter of the experimental plot (SDMW). This value (SDMW) was divided by the number of seedlings placed to dry, resulting in dry matter weight per individual. Due to the need to sequentially analyze how much dry matter of the 50 seeds was mobilized to the seedlings of each plot, the value found by individual was multiplied by 50, thus estimating an approximate value for 50 seedlings (EDMW - Estimated Dry Matter Weight for 50 seedlings), if the germination had been 100%. For the remaining dry matter of the cotyledons (RDWC), the pairs of cotyledons reserved during the length measurement procedure of the seedlings were placed to dry in an oven at 80 ± 1 °C for 24 hours and then weighed. In order to match all weights, these values were also estimated for 50 pairs of cotyledons, a procedure adapted from the methodologies presented by Soltani et al. (2002) and Soltani et al. (2006).

From the dry matter remaining from the cotyledon (RDWC) and the dry matter weight of the seeds (DWS) it was possible to obtain the seeds reserve reduction value (RRS) by the formula: $RRS = DWS - RDWC$. As soon as the values of the seeds reserve reduction and the seed estimated dry matter weight for 50 seedlings (EDMW) were obtained, the value corresponding to the conversion efficiency of the reserves in seedling dry matter was obtained (CESR) from the equation: $ECR = {EDMW}/{RRS}$. This value allowed for estimating the proportion of the reduced dry matter of the seeds turned into seedling dry matter. As for the relative yield of dry matter (RYDM), it was obtained from the

ratio between the weight of the experimental unit (SDMW) and the number of seeds placed in the test (20). The higher the value the higher the dry matter yield and/or number of developed seedlings. Finally, the seed reserve reduction rate (SRRR) obtained by the equation $SRRR = {RRS}/{DWS}$ has allowed for the identification of the genotypes that most mobilized dry matter (Soltani et al., 2006).

Initially, the data were subjected to analysis of variance and Scott-Knott mean test at 5% probability. Then the phenotypic correlations were obtained between the characteristics assessed.

Results and Discussion

Differences were found among the genotypes by the F test at 5% probability for all the studied traits.

The minimum value for water content was 10.3% and the maximum 11.6%. As for germination, it is observed in Table 1 that the genotypes obtained different germination percentages. Physiological quality and vigor relate to normal seedling development (Dias et al., 2011; Vanzolini et al., 2007), therefore the presence of different lots was important for the interpretation of the results in this regard.

Table 1. Means of 16 characteristics of 14 soybean genotypes and formation of groups by the Scott-Knott criterion.

Cultivars	Assessed characteristics [1]							
	WC[2] %	GR[3] %	RRL[4] cm	ARL[5] cm	RHL[6] cm	AHL[7] Cm	RLS[8] cm	ALS[9] cm
CD217	10.77 d	92.5 a	19.64 a	21.23	11.29 a	12.18 b	30.94 a	33.42 a
CD211	10.18 f	82.5 b	18.33 a	22.22	8.77 b	10.58 c	27.10 b	32.80 b
CD219	10.24 f	83.7 b	17.80 b	21.22	8.98 b	10.74 c	26.79 b	31.97 b
Emgopa315	10.82 d	92.5 a	19.37 a	21.22	11.66 a	12.77 b	31.03 a	34.00 a
UFV19	11.19 b	88.7 a	18.66 a	21.12	10.74 a	12.20 b	29.41 a	33.31 a
CD202	11.13 b	51.2 d	11.14 d	21.75	5.42 c	10.62 c	16.56 d	32.36 b
UFVTN103	11.03 c	83.7 b	17.75 b	21.21	9.77 b	11.66 c	27.51 b	32.87 b
UFV18	10.88 d	80.0 b	16.10 b	20.08	8.94 b	11.16 c	25.04 b	31.24 b
UFV16	10.93 c	70.0 c	14.46 c	20.65	8.37 b	12.01 b	22.83 c	32.66 b
MSOY8605	10.55 e	90.0 a	19.02 a	21.15	11.52 a	12.83 b	30.54 a	33.98 a
Monarka	11.62 a	78.7 b	16.42 b	20.83	10.20 a	12.95 b	26.62 b	33.78 a
Confiança	10.83 d	88.7 a	19.21 a	21.67	11.17 a	12.57 b	30.38 a	34.23 a
UFVTN105	11.15 b	81.2 b	16.67 b	20.51	9.08 b	11.20 c	25.75 b	31.71 b
CD 206	10.30 f	91.2 a	19.52 a	21.37	12.80 a	14.02 a	32.32 a	35.40 a
Mean	10.83	82.5	17.44	21.16	9.90	11.96	27.34	33.12
CV (%)	1.09	4.66	9.82	3.43	10.43	6.35	9.61	3.71
H²	97.85	89.6	86.94	54.41	92.22	86.22	89.82	69.84

[1]Means followed by the same letter in each column belong to the same group according to the grouping criteria of Scott-Knott at 5% probability; [2]WC: Water Content; [3]GR: Germination Rate; [4]RRL: Relative Radicle Length (ISTA); [5]ARL: Average Radicle Length (AOSA); [6]RHL: Relative Hypocotyl Length (ISTA); [7]AHL: Average Hypocotyl Length (AOSA); [8]RLS: Relative Length of Seedlings (ISTA); [9]ALS: Average Length of Seedling (AOSA).

According to Vanzolini and Carvalho (2002), root length and dry matter accumulation are efficient features to classify the vigor of soybean seeds. Schuab et al. (2002) found a positive correlation between the growth of plants and various tests for the estimation of soybean seeds vigor, confirming the analysis efficiency of the length as a means of verification of the physiologic quality. According to Vanzolini et al. (2007), obtaining average lengths of seedlings according to the methodology recommended by ISTA enables the detection of subtle differences between genotypes for seed quality. This is consistent, as the method takes into account the number of normal seedlings in relation to the number of seeds placed in the test.

For such protocol, cultivars CD206, CD211, CD217, UFV19, Msoy8605 and Confiança should be noted, with the highest average length of radicle, hypocotyl and seedling. The CD 202 genotype, with the lowest germination rate (Table 1), was also discriminated from the others on the relative length of radicle (RRL) and seedling (RLS), a relationship of development of seedling – physiological quality already announced by Vanzolini et al. (2007) and Dias et al. (2011).

On the other hand, following the recommendation of AOSA, the identification of the potential of genotypes with a lower rate of germination was favored. Also considering the CD202 genotype as a model, the average lengths of their radicles and seedlings (ARL and ALS) were similar to those of other genotypes that did well in other tests, thus showing good potential (Table 1). This information is useful if a strain in an enhancement program has the potential for initial development, but other factors, such as germination rate, are negatively interfering.

This makes it possible to reflect if the cause of the interruption in germination could be bypassed, and the strain is more carefully assessed. Both calculation recommendations of the average length of an experimental plot (ISTA and AOSA) generate important information and views about the same genotype (Vanzolini et al., 2007).

In terms of biomass, the existence of variability among genotypes for dry matter weight stands out, which was also noticed by Maia et al. (2011) in beans. Genotypes Emgopa315 and UFV19 were the ones that most accumulated dry matter in seedlings in the experimental plot (SDMW) and continued among the top ones after the correction for 50 seedlings (EDMW). The dry matter accumulation in the early stages of the cycle is very important for the crop, especially in seedling emergence, which can be difficult because of the depth of seeding and/or soil crusting. Costa et al. (1999) have stated that seedling emergence depends on the development of the hypocotyl, responsible for the rise of the cotyledons above the surface. Thus, hypocotyl length and diameter are related to the ability of seedling emergence and overcoming any resistance of the ground. Pereira et al. (2009) have simulated this situation in testing seedlings height, and have noticed that when the seedlings find resistance to their growth, they respond with the thickening of the hypocotyl and reduction in size, which in practice means increased strength to overcome the barrier imposed on them.

In terms of mobilization of reserves, two genotypes, CD217 and CD211, were grouped as superior to others for characteristic CESR (Table 2). These genotypes were among the ones that least mobilized reserves (RRS and SRRR); however, proportionally they were the most efficient in converting what they mobilized of reserves into seedlings dry matter. Whereas these were not the genotypes containing more reserves in their remaining cotyledons, their efficiency can be a character to be relieved in the assessment of strains. Soltani et al. (2006) have highlighted that seeds different in size may differ in RRS and CESR. In practice, the dry matter of the seedling originated from a large seed, but with little mobilization of reserves, is equal to that of a seedling from a small seed with high reserve mobilization.

Analyzing the correlations among the characteristics (Table 3), the average length of radicle, hypocotyl, seedling and dry matter yield (RRL, RHL, RLS, RYDM) are positively correlated with the germination rate (GR) of the experimental plot, and are also related to the dry matter accumulation in the experimental plot. These correlations are explained by the fact that the relative values of the lengths and yield of dry matter yield would be greater the more seeds germinated. However, the positive correlation between germination rate (GR) and average length of hypocotyl (AHL) indicated that plots with higher germination rates tend to have higher average length of the hypocotyl. According to this observation, a correlation between the average length of hypocotyl (AHL) and the relative length of seedling, relative dry matter of seedling and relative yield of dry matter (RLS, SDMW, RYDM) was found. According to Pereira et al. (2013), hypocotyl is more sensitive than the radicle under water stress; therefore it is important to observe its development, as it is more sensitive to changes caused by the environment. However, the correlation between the relative lengths of hypocotyl, radicle and seedling points to the plausible fact of maintaining the proportions in the measurements of seedlings in different cultivars.

Table 2. Means of 16 characteristics of 14 soybean genotypes and formation of groups by the Scott-Knott criterion.

Cultivars	Assessed characteristics [1]							
	SDMW [2]	EDMW [3]	DWS [4]	RDWC [5]	RRS [6]	CESR [7]	RYDM [8]	SRRR [9]
	g	g	g	g	g	%	g	%
CD217	.77 b	2.07 c	5.59 l	2.95 b	2.64 e	79.3 a	.038 b	47.3 d
CD211	.61 c	1.87 d	5.22 m	2.72 c	2.49 e	74.8 a	.031 c	47.9 d
CD219	.61 c	1.83 d	5.70 k	2.73 c	2.97 d	62.8 b	.030 c	52.1 c
Emgopa315	.96 a	2.60 a	6.84 e	3.06 b	3.79 b	68.8 b	.048 a	55.3 b
UFV19	.91 a	2.55 a	7.38 a	3.25 a	4.12 a	62.0 b	.045 a	55.9 b
CD202	.45 d	2.22 b	6.40 g	2.91 b	3.48 c	63.7 b	.023 d	54.4 b
UFVTN103	.54 d	1.62 e	4.53 n	1.85 e	2.68 e	61.1 b	.027 d	59.1 a
UFV18	.79 b	2.50 a	6.97 c	3.30 a	3.68 b	68.3 b	.039 b	52.7 c
UFV16	.65 c	2.32 b	6.30 h	2.89 b	3.41 c	68.0 b	.032 c	54.1 b
MSOY8605	.84 b	2.32 b	6.58 f	2.93 b	3.65 b	65.0 b	.042 b	55.5 b
Monarka	.67 c	2.12 c	5.74 i	2.33 d	3.41 c	62.6 b	.033 c	59.3 a
Confiança	.82 b	2.32 b	6.90 d	3.19 a	3.70 b	62.9 b	.041 b	53.7 b
UFVTN105	.79 b	2.44 a	7.05 b	3.34 a	3.71 b	65.6 b	.039 b	52.7 c
CD 206	.78 b	2.13 c	5.72 j	2.32 d	3.40 c	62.8 b	.038 b	59.4 a
Mean	0.73	2.21	6.21	2.84	3.37	66.3	0.04	54.3
CV (%)	10.37	7.03	0.05	5.44	4.59	11.0	9.12	2.43
H²	93.09	92.81	99.99	96.71	97.48	52.80	94.64	87.76

[1] Means followed by the same letter in each column belong to the same group according to the grouping criteria of Scott-Knott at 5% probability;
[2] SDMW: Seedlings Dry Matter Weight (ISTA);
[3] EDMW: Estimated Dry Matter Weight for 50 seedlings;
[4] DWS: Dry Weight of Seeds;
[5] RDWC: Remaining Dry Weight of the Cotyledons;
[6] RRS: Reduction of Reserves of Seeds;
[7] CESR: Conversion Efficiency of Seeds Reserve;
[8] RYDM: Relative Yield of Dry Matter;
[9] SRRR: Seed Reserve Reduction Rate.

The dry matter of seeds (DWS) is positively correlated with the reduction of reserves of seeds (RRS), which is consistent because seeds with more dry matter have more reserves to be mobilized. Soltani et al. (2006) have also witnessed such fact. Importantly, the DWS and RRS characteristics are positively correlated with the dry matter weight of seedlings (SDMW and EDMW), which may indicate that seeds with more reserves result in seedlings with higher dry matter accumulation. In fact, what is confirmed in the analysis of means of genotype UFV19. Pádua et al. (2010) and Pereira et al. (2013) has related the larger size of the seeds with the best seedling development, indicating superiority of large seeds in the formation of the stand (Table 3). These findings make sense in the observation that the reduction in reserves of seeds (RRS) correlates with the relative yields of dry matter (RYDM).

Based on these findings, selecting seeds with more reserves in genetic enhancement can be an interesting strategy, because even if the gains will disappear during the plant cycle

(Camozzato et al., 2009), seeds with more reserves may result in an initial stronger development (Pádua et al., 2010), favoring fast soil coverage (Soltani et al., 2006).

Another important finding of this study was the negative correlation between the seed reserve reduction rate (SRRR) and the conversion efficiency of reserves (CESR). This proves that the reduction of reserves of the seeds does not mean, in direct relation, its conversion into dry matter of seedlings. It is plausible to assume an increased respiratory rate in larger seeds, since practically the whole soybean seed tissue is formed by living cells. Whereas seedlings with strong hypocotyls are more adept at breaking the soil surface (Costa et al., 1999), an interesting situation in the assessment of new strains would be the search for strains able to present a positive correlation between RRS and CESR characteristics, as it would combine a considerable stock of reserves associated with an efficient conversion system in seedling dry matter. Soltani et al. (2006) have shown a priori that these characteristics are under genetic control.

Table 3. Simple (r) phenotypic correlation of coefficients /1 estimated among the different features based on the average of the cultivars.

	WC/2	GR/3	RRL/4	ARL/5	RHL/6	AHL/7	RLS/8	ALS/9	SDMW/10	EDMW/11	DWS/12	RDWC/13	RRS/14	CESR/15	RYDM/16	SRRR/17
WC	-	-.34	-.43	-.44	-.22	.07	-.35	-.15	-.01	.33	.29	.06	.43	-.30	-.01	.42
GR	-	-	.98**	-.04	.91**	.53*	.97**	.46	.72**	.04	-.01	-.01	-.02	.14	.72**	-.03
RRL	-	-	-	.16	.90**	.52	.98**	.55*	.65*	-.04	-.09	-.05	-.10	.16	.65*	-.07
ARL	-	-	-	-	-.02	-.08	.08	.41	-.28	-.42	-.38	-.25	-.41	.12	-.28	-.21
RHL	-	-	-	-	-	.83**	.97**	.75**	.73**	.14	.03	-.10	.14	.01	.73**	.24
AHL	-	-	-	-	-	-	.67**	.88**	.55*	.25	.09	-.21	.33	-.21	.55*	.56
RLS	-	-	-	-	-	-	-	.65*	.71**	.04	-.04	-.08	.01	.09	.71**	.06
ALS	-	-	-	-	-	-	-	-	.37	.03	-.10	-.32	.11	-.13	.37	.42
SDMW	-	-	-	-	-	-	-	-	-	.72**	.64*	.54*	.60*	.10	1.0	.02
SDMAW	-	-	-	-	-	-	-	-	-	-	.94**	.79**	.88**	.01	.72**	.05
DWS	-	-	-	-	-	-	-	-	-	-	-	.87**	.96**	-.13	.64*	-.01
RDWC	-	-	-	-	-	-	-	-	-	-	-	-	.59*	.27	.54*	-.49
RRS	-	-	-	-	-	-	-	-	-	-	-	-	-	-.46	.60*	.41
CESR	-	-	-	-	-	-	-	-	-	-	-	-	-	-	.10	-.81**
RYDM	-	-	-	-	-	-	-	-	-	-	-	-	-	-	-	.021
SRRR	-	-	-	-	-	-	-	-	-	-	-	-	-	-	-	-

1/ *,**: Significant at 1 and 5% probability by the t test.
2/ WC: Water Content;
3/ GR: Germination Rate;
4/ RRL: Relative Radicle Length (ISTA);
5/ ARL: Average Radicle Length (AOSA);
6/ RHL: Relative Hypocotyl Length (ISTA);
7/ AHL: Average Hypocotyl Length (AOSA);
8/ RLS: Relative Length of Seedlings (ISTA);
9/ ALS: Average Length of Seedling (AOSA);
10/ SDMW: Seedlings Dry Matter Weight (ISTA);
11/ EDMW: Estimated Dry Matter Weight for 50 seedlings;
12/ DWS: Dry Weight of Seeds;
13/ RDWC: Remaining Dry Weight of the Cotyledons;
14/ RRS: Reduction of Reserves of Seeds;
15/ CESR: Conversion Efficiency of Seeds Reserve;
16/ RYDM: Relative Yield of Dry Matter;
17/ SRRR: Seed Reserve Reduction Rate.

Conclusions

Dry matter weight of seeds, reduction of seeds reserve and dry matter weight of seedlings are positively correlated, and thus genotypes whose seeds are large have a potential vigor on the initial development of seedlings.

Seeds reserve reduction and conversion efficiency of seeds reserves are negatively correlated.

Acknowledgments

To Conselho Nacional de Desenvolvimento Científico e Tecnológico (CNPq - National Counsel of Technological and Scientific Development) and Fundação de Amparo à Pesquisa e Inovação do Espírito Santo (FAPES - Research Support Foundation of the Brazilian State of Espírito Santo) by granting scholarships to the authors of this research.

References

AOSA. ASSOCIATION OF OFFICIAL SEED ANALYSTS. *Seed vigor testing handbook*, 2009. 341p.

BRASIL. Ministério da Agricultura, Pecuária e Abastecimento. *Regras para análise de sementes*. Ministério da Agricultura, Pecuária e Abastecimento. Secretaria de Defesa Agropecuária. Brasília, DF: MAPA/ACS, 2009. 395p. http://www.agricultura.gov.br/arq_editor/file/2946_regras_analise__sementes.pdf.

CÂMARA, G.M.S.; PIEDADE, S.M.S.; MONTEIRO, J.H.; GUERZONI, R.A. Desempenho vegetativo e produtivo de cultivares e linhagens de soja de ciclo precoce no município de Piracicaba – SP. *Scientia Agrícola*, v.55, n.3, p.403-412, 1998. http://www.scielo.br/scielo.php?pid=S0103-90161998000300008&script=sci_arttext

CAMOZZATO, V.A.; PESKE, S.T.; POSSENTI, J.C.; MENDES, A.S. Desempenho de cultivares de soja em função do tamanho das sementes. *Revista Brasileira de Sementes*, v.31, n.1, p.288-292, 2009. http://www.scielo.br/pdf/rbs/v31n1/a32v31n1.pdf

CARVALHO, J.A.; SOARES, A.A.; REIS, M.S. Efeito de espaçamento e densidade de semeadura sobre a produtividade e os componentes de produção da cultivar de arroz BRSMG Conai. *Ciência e Agrotecnologia*, v.32, n.3, p.785-791, 2008. http://www.scielo.br/pdf/cagro/v32n3/a12v32n3.pdf

COSTA, J.A.; PIRES, J.L.F.; THOMAS, A.L.; ALBERTON, M. Comprimento e índice de expansão radial do hipocótilo de cultivares de soja. *Ciência Rural*, v.29, n.4, p.609-612, 1999. http://www.scielo.br/pdf/cr/v29n4/a06v29n4.pdf

DIAS, M.A.N.; PINTO, T.L.F.; MONDO, V.H.V.; CICERO, S.M.; PEDRINI, L.G. Direct effects of soybean seed vigor on weed competition. Revista Brasileira de Sementes, v.33, n.2, p.346-351, 2011. http://www.scielo.br/pdf/rbs/v33n2/17.pdf

HENNING, F.A.; MERTZ, L.M.; JACOB JUNIOR, E.A.; MACHADO, R.D.; FISS, G.; ZIMMER, P.D. Composição química e mobilização de reservas em sementes de soja de alto e baixo vigor. *Bragantia*, v.69, n.3, p.727-734, 2010. http://www.scielo.br/pdf/brag/v69n3/26.pdf

MAIA, L.G.; SILVA, C.A.; RAMALHO, M.A.P.; ABREU, A.F. Variabilidade genética associada à germinação e vigor de sementes de linhagens de feijoeiro comum. *Ciência e Agrotecnologia*, v.35, n.2, p.361-367, 2011. http://www.scielo.br/pdf/cagro/v35n2/a18v35n2.pdf

NEPOMUCENO, M.; ALVES, P.L.C.A.; DIAS, T.C.S.; PAVANI, M.C.M.D. Períodos de interferência das plantas daninhas na cultura da soja nos sistemas de semeadura direta e convencional. *Planta Daninha*, v.25, n.1, p.43-50, 2007. http://www.scielo.br/pdf/pd/v25n1/a05v25n1.pdf

PÁDUA, G.P.; ZITO, R.K.; ARANTES, N.E.; FRANÇA-NETO, J.B. Influência do tamanho da semente na qualidade fisiológica e na produtividade da cultura da soja. Revista Brasileira de Sementes, v.32, n.3, p.9-16, 2010. http://www.scielo.br/pdf/rbs/v32n3/v32n3a01.pdf

PEREIRA, W.A.; SÁVIO, F.L.; BORÉM, A.; DIAS, D.C.F.S. Influência da disposição, número e tamanho das sementes no teste de comprimento de plântulas de soja. Revista Brasileira de Sementes, v.31, n.1, p.113-121, 2009. http://www.scielo.br/pdf/rbs/v31n1/a13v31n1.pdf

PEREIRA, W.A.; PEREIRA, S.M.A.; DIAS, D.C.F.S. Influence of seed size and water restriction on germination of soybean seeds and on early development of seedlings. *Journal of Seed Science*, v.35, n.3, p.316-322, 2013. http://www.scielo.br/pdf/jss/v35n3/07.pdf

SCHUAB, S.R.P.; BRACCINI, A.L.; FRANÇA-NETO, J.B.; SCAPIM, C.A.; MESCHEDE, D.K. Utilização da taxa de crescimento das plântulas na avaliação do vigor de sementes de soja. *Revista Brasileira de Sementes*, v.24, n.2, p.90-95, 2002. http://www.scielo.br/pdf/rbs/v24n2/v24n2a15.pdf

SOLTANI, A.; GALESHI, S.; ZEINALI, E.; LATIFI, N. Germination, seed reserve utilization and seedling growth of chickpea as affected by salinity and seed size. *Seed Science and Technology*, v.30, p.51-60, 2002.

SOLTANI, A.; GHOLIPOOR, M.; ZEINALI, E. Seed reserve utilization and seedling growth of wheat as affected by drought and salinity. *Environmental and Experimental Botany*, v.55, n.1-2, p.195-200, 2006. http://www.sciencedirect.com/science/article/pii/S0098847204001534#

SUN, Q; WANG, J.H; SUN, B.Q. Advances on seed vigor physiological and genetic mechanisms. *Agricultural Sciences in China*, v.6, p.1060-1066, 2007. http://www.sciencedirect.com/science/article/pii/S1671292707601473#

VALÉRIO, I.P.; CARVALHO, F.I.F.; OLIVEIRA, A.C.; BENIN, G.; SOUZA, V.Q.; MACHADO, A.A.; BERTAN, I.; BUSATO, C.C; SILVEIRA, G.; FONSECA, D.A.R. Seeding density in wheat genotypes as a function of tillering potential. *Scientia Agrícola*, v.66, n.1, p.28-39, 2009. http://www.scielo.br/pdf/sa/v66n1/a04v66n1.pdf

VANZOLINI, S.; ARAKI, C.A.S.; SILVA, A.C.T.M.; NAKAGAWA, J. Teste de comprimento de plântulas na avaliação da qualidade fisiológica de sementes de soja. *Revista Brasileira de Sementes*, v.29, n.2, p.90-96, 2007. http://www.scielo.br/pdf/rbs/v29n2/v29n2a12.pdf

VANZOLINI, S.; CARVALHO, N.M. Efeito do vigor de sementes de soja sobre o seu desempenho em campo. *Revista Brasileira de Sementes*, v.24, n.1, p.33-41, 2002. http://www.scielo.br/pdf/rbs/v24n1/v24n1a06.pdf

24

The germination success of the cut seeds of *Eugenia pyriformis* depends on their size and origin

Juliana Sakagawa Prataviera[1], Edmir Vicente Lamarca[1], Carmen Cinira Teixeira[1], Claudio José Barbedo[1]*

ABSTRACT – Seeds of *Eugenia pyriformis* may produce several seedlings after cutting. Both the type of cutting and the size of the seed can determine the success in obtaining new seedlings. The size of the seeds is dependent on both the number of seeds per fruit and the conditions in which seeds develop, as well as the biometric characteristics of these seeds obtained from different regions and seasons. The seeds from each origin were evaluated in length, width, thickness, water content and dry mass, as well as the average number of seeds per fruit. From one of the regions, seeds were grouped according to the number of seeds per fruit and also according to their size, and then cut into two and four parts, and then analyzed for the fragments germination. The results demonstrated the high capacity of these seed fragments to produce new seedlings, but the capability reduces with the reduction in seed size. This size depends not only on the number of seeds per fruit, but also on the region and the period of the seed production.

Index terms: germination, seed cutting, Myrtaceae, uvaia.

O sucesso da germinação de sementes fracionadas de *Eugenia pyriformis* depende do seu tamanho e da sua origem

RESUMO – Uma semente de uvaieira (*Eugenia pyriformis*) pode produzir várias plântulas quando fracionada. A posição do corte e o tamanho da semente influenciam o sucesso da obtenção de novas plântulas. O número de sementes por fruto e as condições nas quais as sementes se desenvolvem podem promover diferenças nesse tamanho. No presente trabalho foram analisados o grau de influência do número de sementes por fruto sobre a capacidade das sementes de uvaieira em germinar após fracionamento e os aspectos biométricos de sementes de uvaieira obtidas em diferentes regiões e épocas. As sementes de cada origem foram avaliadas quanto ao comprimento, largura, espessura, conteúdo de massa fresca e seca e teor de água. De uma das regiões, sementes foram agrupadas segundo a quantidade de sementes por fruto e também segundo seu tamanho e, a seguir, foram fracionadas em duas e quatro partes e analisadas quanto à capacidade germinativa dos fragmentos obtidos. Os resultados demonstraram a elevada capacidade dos fragmentos dessas sementes em formar novas plântulas, mas tal capacidade reduz concomitante à redução no tamanho da semente. Esse tamanho depende não apenas do número de sementes por fruto, mas também da região e da época de produção das sementes.

Termos para indexação: germinação, fracionamento de semente, Myrtaceae, uvaia.

Introduction

Uvaia (*Eugenia pyriformis* Cambess - Myrtaceae) produces fruits which are of great marketing, gastronomic and pharmacological value and are native of Brazil (Stieven et al., 2009; Ramirez et al., 2012; Lamarca et al., 2013a), with great potential for economic exploitation. However, this exploitation still depends on technology development for a large-scale production, which has limitations starting at the commercial orchards installation. The seedling production of this species is currently low due to, among other factors, the lack of sufficient amount of seeds, for there are few plants producing fruits, few fruits per plant and few seeds per fruit (Silva et al., 2005). The number of seeds per fruit is variable, but generally does not exceed four units (Justo et al., 2007). Therefore, to increase the cuttings production there is the need to deploy orchards producing seeds and/or to maximize the use of seeds produced.

Studies have demonstrated the possibility of having several seedlings from a single uvaia seed because, although

[1]Instituto de Botânica, Núcleo de Pesquisa em Sementes, Caixa Postal, 68041, 04301012 – São Paulo, SP, Brasil.
*Corresponding author< claudio.barbedo@pesquisador.cnpq.br>

monoembryonic, such seed has great potential in producing seedlings after being cut (Delgado et al., 2010; Teixeira and Barbedo, 2012). This potential suggests the presence of meristematic tissues capable of further differentiation, including in new embryos (Delgado et al., 2010). In fact, in general, the embryo has meristematic toti or pluripotent cells, offering all the information needed to grow and give rise to a mature plant (Marcos-Filho, 2005; Batygina and Vinogradova, 2007). In uvaia, this characteristic was described ten years ago and has been studied since then.

Non fractioned uvaia seed produces, as a rule, a single root and single seedlings. It was found, for example, that seeds fractioned into two parts produce two independent root systems (one in each resultant fraction), each one generating a new seedling. This would suggest that there is a potential stimulating effect generated by the cut however, curiously, there is hardly any formation of two root systems in the same fraction, suggesting a method of inhibiting a second germination after the beginning of the first seedling development (Amador and Barbedo, 2011). It was also found that both the cutting positions as well as the size of the seed influence the success of obtaining young seedlings. Therefore, when increasing the number of cuts in the seeds, only the largest ones proved to be able to continue producing young seedlings (Silva et al., 2005; Amador and Barbedo, 2011; Teixeira and Barbedo, 2012). Therefore the resulting mass after fractioning seems to affect the success in the production of seedlings per seed.

One of the factors that influence the size of uvaia seeds is the number of seeds per fruit, for seeds are smaller when more numerous per fruit (Justo et al., 2007). The conditions under which seeds develop can also promote differences in the size of seeds produced. Studies show that water availability and thermal variations, for example, may contribute to the variability of physical aspects such as the size and weight of fruits and seeds, among individuals or populations of the same species (Daws et al., 2004, 2006; Santos et al., 2009; Andrade et al., 2010; Joët et al., 2013). Specifically for uvaia, environmental conditions during seed formation may exert great influence on their various characteristics, but did not assess such influence on the size (Lamarca et al., 2013b). There is no description in literature on uvaia seed size variations according to its origin, nor variations in the potential of these seeds in producing new seedlings after cutting, arising from the size and number of seeds per fruit.

In the present study, the degree of influence of the number of uvaia seeds per fruit on seed germinability after cutting were analyzed, as well as their biometric aspects obtained in different regions and seasons.

Material and Methods

Origin of seeds: uvaia seeds were obtained from freshly dispersed ripe fruits, according to the maturation and dispersal information from Lamarca et al. (2013b). Fruits were collected from 10 different regions, as described in detail in Table 1. Abbreviations have been adopted, according to the area and period or elevation, as follows: Ribeirão Preto (RIB), Lavras (LAV), São Bento do Sapucaí (SBS), Campinas (CAM), Jumirim (JUM), São Paulo (SPA), Ibiúna (IBI), São Bernardo do Campo (SBC), Itaberá (ITA) and Pariquera-Açú (PAR). In four of those regions (CAM, SPA, IBI and SBC) seeds were obtained from the same plants in two years, identified as 1 and 2 and, SBS from three different elevations, indicated by 1, 2 and 3. Each collecting according to the period and region or region and elevation, was seen as a different origin of seeds.

In the period of the species maximum flowering, of each origin, tree inflorescences that had most of the flowers in anthesis were marked. From weather stations, located close to the collection areas, daily rainfall data were obtained (mm), as well as maximum and minimum air temperatures (°C). The data were provided by the Campinas Agronomic Institute, Federal University of Lavras and the Astronomical and Geophysical Institute of São Paulo University. At the period between flowering and seed dispersal the accumulated precipitation (mm) and accumulated degree-days (°C.day), were calculated according to the equations proposed by Villa Nova et al. (1972), taking into consideration the basic temperature of 10 °C (Pedro Junior et al., 1977).

The germinating capacity analysis after cutting according to size and number of seeds per fruit: SPA2 seeds were grouped among those from fruits containing a single seed, fruits containing two seeds and fruits containing three or more seeds. Each group of seed samples was evaluated per size, three orthogonal axes measures being taken, using digital calipers with 0.01 mm accuracy. The largest seed size axis was considered to be its length, perpendicular to the major axis length regarded as width, and the axis perpendicular to the previous two as its thickness.

In a first experiment, seeds of each group were divided into three groups, each receiving different cutting treatments: a) whole seeds (without cutting), b) seeds cut in its middle and c) seeds cut into four parts. Therefore, we obtained a 3 x 3 factorial design (number of seeds per fruit x type of cutting) in a completely randomized design and four replications of 16 seeds in each test.

In a second experiment, the seeds from fruits with one or two seeds were separated according to their size, small and large, keeping track of their origin (number of seeds per fruit). Therefore, four groups of seeds were obtained, according to origin and size: large seeds from fruits with one seed (LGF1), large seeds from fruits with two seeds (LGF2), small seeds from fruits with one seed (SSF1) and small seeds from fruits with two seeds (SSF2). Seeds were also cut as described in the previous experiment, obtaining a 4 x 3 factorial (seed origin x fractionation type), with four replicates of 16 seeds.

Table 1. Uvaia seeds origin. Geographic location, Köppen climate classification, maturation cycle (period between flowering and dispersion) and collection sites meteorological data during the maturation cycle. Min: minimum temperature; Max: maximum temperature; Degree-days; Rain: accumulated rain precipitation; WC: seed water content (% wet basis) at the time of dispersion.

Seed origin	Maturation period (maturation cycle)	Min – Max (°C)	Degree-day (°C day)	Rain (mm)	WC (%)
RIB - Ribeirão Preto, SP (21°10'S, 47°52'O, 593 m; Cwa)	14/08/10 - 17/09/10 (34 days)	14 – 31	440	6.9	62.8
LAV - Lavras, MG (21°13'S, 44°58'O, 949 m; Cwa)	15/08/10 - 25/09/10 (41 days)	12 – 28	417	24.2	53.6
SBS1 - São Bento do Sapucaí, SP (22°41'S, 45°43'O, 884 m; Cfb)	24/08/10 - 07/10/10 (44 days)	12 – 27	426	177.1	65.0
SBS2 - São Bento do Sapucaí, SP (22°41'S, 45°45'O, 1022 m; Cfb)	26/08/10 - 15/10/10 (50 days)	11 – 26	448	177.1	57.3
SBS3 - São Bento do Sapucaí, SP (22°41'S, 45°46'O, 1121 m; Cfb)	26/08/10 - 21/10/10 (56 days)	11 – 26	472	252.4	53.0
CAM1 - Campinas, SP (22°52'S, 47°04'O, 645 m; Cwa)	10/08/10 - 19/09/10 (40 days)	14 – 28	451	6.3	55.4
CAM2 - Campinas, SP (22°52'S, 47°04'O, 645 m; Cwa)	06/08/11 - 14/09/11 (39 days)	14 – 28	442	36.7	58.9
JUM - Jumirim, SP (22°05'S, 47°47'O, 540 m; Cwa)	05/08/10 - 19/09/10 (45 days)	12 – 29	481	12.6	59.8
SPA1 - São Paulo, SP (23°38'S, 46°37'O, 785 m; Cwb)	26/08/10 - 09/10/10 (44 days)	14 – 25	428	128.8	64.5
SPA2 - São Paulo, SP (23°38'S, 46°37'O, 785 m; Cwb)	31/08/11 - 10/10/11 (40 days)	13 – 25	374	99.1	66.6
IBI1 - Ibiúna, SP (23°39'S, 47°09'O, 917 m; Cfb)	12/09/10 - 23/10/10 (41 days)	12 – 27	413	123.6	61.6
IBI2 - Ibiúna, SP (23°39'S, 47°09'O, 917 m; Cfb)	04/09/11 - 12/10/11 (38 days)	10 – 29	380	103.2	62.3
SBC1 - São Bernardo do Campo, SP (23°42'S, 46°33'O, 786 m; Cwb)	16/08/10 - 03/10/10 (48 days)	13 – 25	458	104.3	56.3
SBC2 - São Bernardo do Campo, SP (23°42'S, 46°33'O, 786 m; Cwb)	16/08/11 - 01/10/11 (46 days)	13 – 25	406	68.5	57.9
ITA - Itaberá, SP (23°52'S, 49°06'O, 683 m; Cfa)	17/08/10 - 23/09/10 (37 days)	11 – 26	341	1.8	65.7
PAR - Pariquera-Açú, SP (24°37'S, 47°53'O, 28 m; Af)	23/08/10 - 11/10/10 (49 days)	15 – 25	501	87.4	59.6

Seeds and seed fractions were placed in transparent, colorless plastic boxes (gerbox type) filled with expanded vermiculite of medium particle size, saturated with 70 mL tap water and placed in a growth chamber at 25 °C, with constant light and 100% relative humidity. The number of seeds or seed fractions which produced normal seedlings was registered (Delgado and Barbedo, 2007) to calculate germination and

seeds issuing primary root with at least 2 cm, for germinable seeds evaluation.

The evaluations were performed every 5 days, until there were no more new roots for 30 consecutive days. The final accounting of germination and germinated seeds, for cut seeds, were the number of fractions with normal seedlings or protruded roots, respectively, related to the total number of seeds. Data were subjected to the variance analysis and averages were compared by the Tukey test with 5% probability (Santana and Ranal, 2004).

Biometrics and germination of seeds from different origins: at the end of the period of fruit formation and maturation, freshly dispersed ripe fruits were collected (up to 24 hours from dispersion) for seed biometric evaluations. We selected 40 fruits at random from each origin, opened them, registering the number of seeds per fruit. The seeds were then washed in water, on strainer. Once washed, we removed the excess water on its surface with filter paper. Afterwards the seeds from each origin were evaluated per length, width, thickness, as described above. Dry and fresh mass and water content were also evaluated. The fresh mass content was determined by a scale with 0.001 g accuracy, the results being shown in g .seed^{-1}; the dry mass content and the water content was determined gravimetrically by oven method at 103 °C for 17 hours with the results shown, respectively, in g.seed^{-1} and in % (wet basis), as described in Brasil (2009). The germination test was performed as described previously.

The experimental design was completely randomized with four replications of 10 seeds. The data obtained were subjected to the variance analysis (F test), at 5% level. When relevant, the averages were compared by the Tukey test, also at 5% (Santana and Ranal, 2004). Later, we calculated simple correlation coefficients for all combinations between the meteorological data and biometric data of the seeds and the significance was determined by t test with 5% probability.

Results and Discussion

The seed dimensions according to the number of seeds per fruit (Figure 1A) showed that the greater the number of seeds per fruit, the smaller the final size of each seed, as was also observed by Justo et al. (2007). The germination potential was high for seeds from all sources (Figure 1), reaching values above 200% of germinable seeds, when cut into four parts (Figure 1C). However, it was found that the number of seeds contained in the fruit is related to the germination capacity, specifically when there are three or more seeds per fruit, and also when more than one cutting is performed. One explanation for this difference would be the fact that, as described above, a larger number of seeds per fruit results in smaller seeds and the seed mass are directly related to the number of cutting, capable of ensuring the formation of new seedlings (Silva et al., 2003). Therefore, it would be expected that the greater the number of seeds per fruit, the lower the germination capacity of these seeds. This was observed when comparing seeds from fruits with three or more seeds, to fruits with one or two seeds,

however not when comparing the last two ones. This could be due to the variation in seed size of these two sources, which was high. These are the most frequent categories, specifically the fruits with two seeds, which correspond to 81% of the fruits (Andrade and Ferreira, 2000).

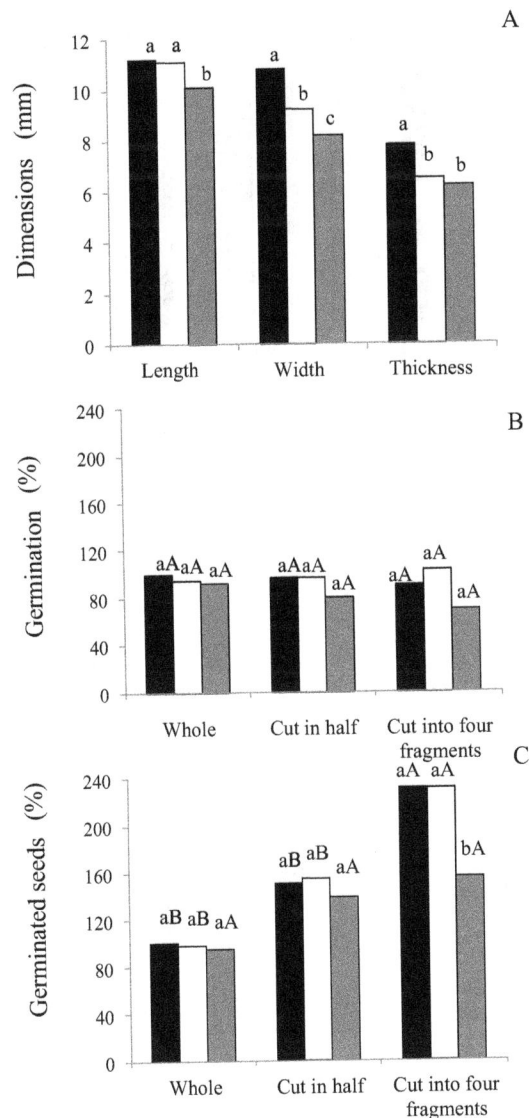

Figure 1. Uvaia seed size (A) germination (B) and germinated seeds (C) according to the number of seeds per fruit and type of cutting. Columns in black: fruit with one seed; in white: with two seeds; gray: three or more seeds. Columns with the same letter (lowercase comparing number of seeds per fruit, upper case comparing cutting) do not differ among them (Tukey, 5%).

The size variation in the seeds of these two groups was evident in the second experiment, in which seeds were visually divided between large and small. Seeds of different sizes within the same sub category were obtained (Figure 2A). The results of germination of these subgroups showed that seed size has greater effect on germination after cutting than the amount of seeds per fruit. Within the same category, large seeds germinated more than the small ones (Figure 2).

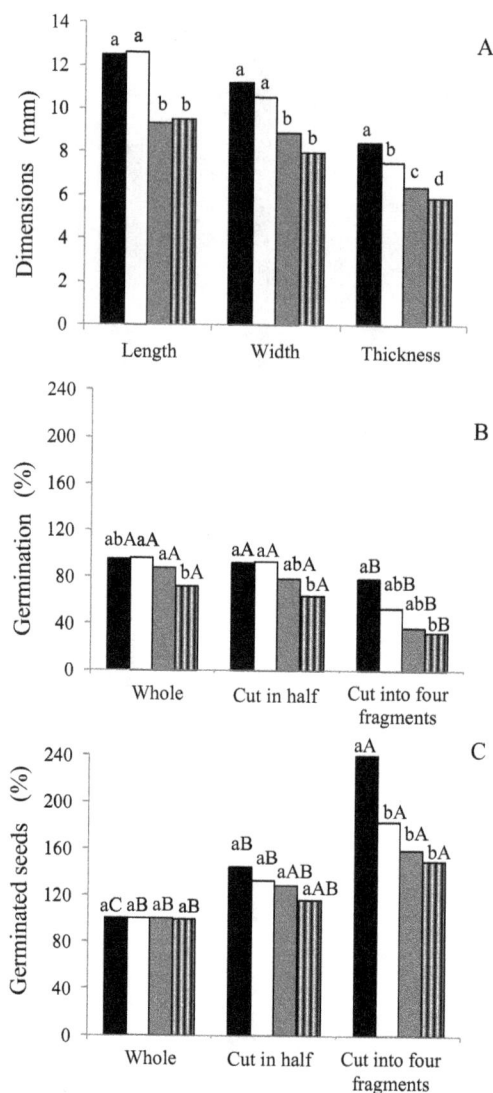

Figure 2. Uvaia seed size (A) germination (B) and germinated seeds (C) according to the number of seeds per fruit and type of cutting. Columns in black: large seeds, fruit with one seed; in white: large seeds, fruits with two seeds; gray: small seeds, fruit with one seed; stripes: small seeds, fruits with two seeds. Columns with the same letter (lowercase comparing size and number of seeds per fruit, upper case comparing cutting type) do not differ among them (Tukey, 5%).

Cutting small seeds SSF1 and SSF2 showed that there is a higher limitation on the plant growth than in the root production. This could be due to: 1) the seed fragmentation reduces the reserves amount that, in the small ones, can become insufficient for the seedling, root and shoot growth, similar to the described in previous studies (Silva et al., 2003; Teixeira and Barbedo, 2012); 2) the distance between the germination region end and its opposite is higher in the larger seeds than on the smaller ones. In this case a possible inhibitory effect, caused by the first germination onset, preventing the totipotent cells differentiation in the opposite area, would take longer to sense in larger fragments, similar to the observed with fissured seeds (partially fractioned) (Amador and Barbedo, 2011).

The results obtained for the cut seeds of different sizes demonstrated the importance of the biometrics knowledge for a better analysis of benefits and harms of this fractionation. However, the variation in seed size not only lies in the number of seeds per fruit, but also in the region and time of seeds production. Seeds of Santa Maria, Rio Grande do Sul state, for example, showed fresh mass between 0.2 and 2.0 g (Andrade and Ferreira, 2000), while those from Lavras, Minas Gerais state, were between 0.5 and 2.5 g (Justo et al., 2007), a variation range similar to the one obtained in the present study, from 0.5 to 2.7 g, considering ten regions and 16 different origins (Table 2).

Seeds obtained from different regions and seasons also demonstrated big changes of other biometric aspects evaluated, such as length (ranging from 8.11 to 19.21 mm), width (7.11 to 16.18 mm), thickness (5.28 to 13.36 mm), beyond the number of seeds per fruit (which ranged from 1.08 to 1.85). The seed size (length, width and thickness) is linked to the fresh and dry mass content and water content, as there were significant correlations between these variables (Table 3). The number of seeds per fruit was always between 1 and 4, similar to the previously observed (Andrade and Ferreira, 2000; Justo et al., 2007). However, the average seeds per fruit little fluctuated, between 1.08 and 1.80 (Table 2). Considering that plants from ten different regions were analyzed, it can be assumed that the uvaia fruit produce mostly, one or two seeds. The number of seeds was negatively correlated to the length and thickness (Table 3). Therefore, when the fruit has two seeds, these are smaller than those with one seed, especially concerning length and thickness and, as seen in the previous experiment, it directly affects the seed germination capacity after fractionation.

The highest values of length, width, thickness and fresh and dry mass were from higher elevation and higher maturation cycle, SBS2 and SBS3 respectively (Tables 1 and 2). On the other side, the source that had length, width and thickness lowest values, had the highest value of seeds number per fruit, as IBI2 (Table 2).

Table 2. Length (mm), width (mm), thickness (mm), Q (quantity of fruits^{-1}), MF (fresh mass, seed.g^{-1}), MS (dried mass, seed.g^{-1}) and G (germination, %) of uvaia seeds from different origins.

Origin	Length	Width	Thickness	Q	MF	MS	G
RIB	14.12 cdef*	12.17 cde	8.79 bcde	1.18 ab	1.29d	0.48 efg	90 abc
LAV	16.06 abc	13.04 bcd	10.08 b	1.33 ab	1.73c	0.81 c	100 a
SBS1	15.63 bcd	13.46 abc	9.89 bc	1.30 ab	1.06def	0.37 g	87 abc
SBS2	17.74 ab	15.56 ab	12.13 a	1.13 b	2.27 b	0.97 b	88 abc
SBS3	19.21 a	16.18 a	13.36 a	1.25 ab	2.70 a	1.27 a	72 abcd
CAM1	13.19 cdefg	11.05 cdef	8.41 cde	1.14 b	1.08def	0.48 efg	92 abc
CAM2	13.45 cdefg	11.29 cdef	8.47 bcde	1.58 ab	1.26d	0.52 ef	85 abcd
JUM	14.85 bcde	12.01 cde	9.02 bcd	1.60 ab	1.06def	0.43 fg	95 ab
SPA1	11.46 fg	9.98 ef	8.02 de	1.48 ab	0.65gh	0.23 h	63 cd
SPA2	10.48 gh	8.99 fg	7.18 e	1.50 ab	0.70gh	0.20 h	87 abc
IBI1	12.49 defg	10.46 def	8.77 bcde	1.08 b	1.00ef	0.39 g	72 abcd
IBI2	8.11 h	7.11 g	5.28 f	1.85 a	0.50h	0.19 h	57 d
SBC1	13.25 cdefg	12.07 cde	9.40 bcd	1.23 ab	1.27d	0.56 e	68 bcd
SBC2	12.13 efg	10.92 cdef	8.51 bcde	1.35 ab	0.92fg	0.39 g	90 abc
ITA	14.65 bcdef	13.17 bcd	9.57 bcd	1.18 ab	1.21de	0.42 fg	78 abcd
PAR	15.60 bcd	13.17 bcd	9.60 bcd	1.10 b	1.67 c	0.68 d	73 abcd
C.V. (%)	9.11	9.12	7.01	20.26	7.82	8.87	11.62

* Averages followed by the same letter do not differ by the Tukey test, at 5%.

Table 3. Simple correlation coefficients (r) between meteorological and biometric data of uvaia seed germination from different origins.

	MIN	MAX	DD	RAIN	LENG	WID	THICK	NUM	FM	DM	WC	G[13]
CYCLE[1]	-0.11	-0.57*	0.59*	0.71*	0.58*	-0.10	0.69*	-0.25	-0.12	-0.53*	0.67*	-0.16
MIN[2]		-0.10	0.42	-0.35	-0.10	-0.11	-0.17	-0.22	-0.04	0.02	-0.19	0.16
MAX[3]			0.05	-0.45	-0.06	0.16	-0.23	0.26	-0.25	-0.02	-0.09	0.31
DD[4]				0.19	0.51*	-0.08	0.45	-0.24	-0.28	-0.56*	0.51*	0.09
RAIN[5]					0.30	-0.23	0.47	-0.11	-0.02	-0.10	0.42	-0.45
LENG[6]						0.22	0.96*	-0.58*	-0.56*	-0.50*	0.87*	0.36
WID[7]							0.14	-0.01	-0.13	-0.41	0.26	0.31
THICK[8]								-0.62*	0.48	-0.52*	0.90*	0.24
NUM[9]									0.09	0.21	-0.45	-0.18
FM[10]										0.26	-0.43	0.18
DM[11]											-0.75*	0.22
WC[12]												0.16

[1]CYCLE = maturation cycle; [2]MIN = minimum temperature; [3]MAX = maximum temperature; [4]DD = degrees-day; [5]RAIN = accumulated rain; [6]LENG = length; [7]WID = width; [8]THICK = thickness; [9]NUM = number of seeds per fruit; [10]FM = fresh mass; [11]DM = dry mass; [12]WC = water content; [13]G = germination. (*) = r significant at 5% probability, without asterisk = non significant correlations.

It was observed also that the origins that were similar in length, width and thickness, for example, RIB, CAM1, CAM2, JUM, SBC1, SBC2 and IBI1 showed as well similar values as the sum of degree-days (Tables 1 and 2). The differences between seed sizes were also observed for samples taken at the same plants, but at different periods (SBS and IBI); however, they were more evident concerning the fresh and dry mass content (SBS, IBI and SBC). Despite the differences in dimensions, seeds from only three regions showed germination values below the highest LAV (100%),

namely, SPA1 (63%), IBI2 (57%) and SBC1 (68%) (Table 2).

In studies of *Tabebuia chrysotricha* seeds, arising from the same population (Santos et al., 2009), but from different plants and with *Eugenia dysenterica* seeds (Silva et al., 2001) and *Eugenia calycina* (Cardoso and Lomônaco, 2003), from different populations, variations in the biometric of seeds were observed, which have also been associated, among other factors, to phenotypic differences in response to environmental conditions. In this study, the biometric variations of uvaia seeds from different regions and periods were correlated with environmental conditions, specifically among accumulated degree-days and length, dry mass content and water content variables (Table 3).

The maturation cycle also exerted influence on the biometric variables (length, thickness, dry mass content and water content), with significant correlations (Table 3). This cycle, however, was positively correlated with accumulated rainfall and degree-days. The influence of environmental conditions on seed maturity and physical characteristics has also been demonstrated for *Aesculus hippocastanum* and *Acer pseudoplatanus* (Daws et al., 2004, 2006), coffee (Petek et al., 2009), *Euterpe edulis* (Martins et al., 2009), *Quercus ilex* (Joët et al., 2013) and *Inga vera* (Lamarca et al., 2013c), besides the uvaia itself (Lamarca et al., 2013a).

The results of germination of cut seeds of different sizes and origins (seed number per fruit) as well as the seed biometric variations from different regions and seasons, allow its division into three groups. In the first one, with greater chances of success in obtaining seedlings from cut seeds, they would include the ones coming from SBS, LAV, PAR, JUM, ITA and RIB; in the second one, with higher probability of failure, SPA and IBI; in the third one, CAM and SBC seeds, in borderline position between success and failure, being more dependent on the number of seeds per fruit and environmental conditions, according to the year of seed production. Therefore, although one can get technological gain by cutting uvaia seeds for seedling production, reaching a 50% increase as shown in previous works (Silva et al., 2003), it became clear that this gain depends on a previous analysis of the lot biometric characteristics, which may vary according to the region and the season.

Conclusions

The ability to produce new seedlings in uvaia seeds after cutting depends on the seed size, which also depends on the number of seeds per fruit, the region and the period of the seed production.

Acknowledgments

The authors wish to thank Dr. Domingos S. Rodrigues (IBt), Dr. João J. D. Parisi (IAC), the Regional Pole of Vale do Ribeira - Pariquera-Açú/São Paulo, CATI from Itaberá/SP, CATI from São Bento do Sapucaí/São Paulo, the Agriculture House in Jumirim/São Paulo, USP-University of Ribeirão Preto/ São Paulo, the Botany Institute of São Paulo/SP, the Agronomic Institute of Campinas/São Paulo, the Lavras Federal University/ Minas Gerais, Mrs. Kathya I. B. Costa and Mr. Milton R. Costa, for permission and support in collections; Dr. Lucia Rossi (IBt), for the species identification; the Lavras Federal University, Lavras, Minas Gerais, the Agronomic Institute of Campinas, Campinas/São Paulo and the Astronomical and Geophysical Institute of the University of São Paulo, São Paulo/SP, for the provision of meteorological data; the CNPq and CAPES for scholarships granted to J.S. Prataviera (Scientific Initiation), E.V. Lamarca and C.C. Teixeira (Doctorate) and C.J. Barbedo (Research Productivity).

References

AMADOR, T.S.; BARBEDO, C.J. Potencial de inibição da regeneração de raízes e plântulas em sementes germinantes de *Eugenia pyriformis*. *Pesquisa Agropecuária Brasileira*, v.46, p. 814-821, 2011. http://www.scielo.br/pdf/pab/v46n8/05.pdf

ANDRADE, R.N.B.; FERREIRA, A.G. Germinação e armazenamento de sementes de uvaia (*Eugenia pyriformis* Camb.) - Myrtaceae. *Revista Brasileira de Sementes*, v.22, n.2, p.118-125, 2000. http://hdl.handle.net/10183/23264

ANDRADE, L.A.; BRUNO, R.L.A.; SOARES, L.; OLIVEIRA, B.; SILVA, H.T.F. Aspectos biométricos de frutos e sementes, grau de umidade e superação de dormência de jatobá. *Acta Scientiarum Agronomy*, v.32, p.293-299, 2010. http://www.scielo.br/pdf/asagr/v32n2/a16v32n2.pdf

BATYGINA, T.B.; VINOGRADOVA, G.Y. Phenomenon of polyembryony. Genetic heterogenity of seeds. *Russian Journal of Developmental Biology*, v.38, p.126-151, 2007.http://download.springer.com/static/pdf/882/art%253A10.1134%252FS1062360407030022.pdf?auth66=1419268241_9f33377ab9ce09346fd271c5311779c0&ext=.pdf

BRASIL. Ministério da Agricultura, Pecuária e Abastecimento. *Regras para análise de sementes*. Ministério da Agricultura, Pecuária e Abastecimento. Secretaria de Defesa Agropecuária. Brasília: MAPA/ACS, 2009. 395p. http://www.agricultura.gov.br/arq_editor/file/2946_regras_analise__sementes.pdf.

CARDOSO, G.L.; LOMÔNACO, C. Variações fenotípicas e potencial plástico de *Eugenia calycina* Cambess. (Myrtaceae) em uma área de transição cerrado-vereda. *Revista Brasileira de Botânica*, v.26, p.131-140, 2003. http://www.scielo.br/pdf/rbb/v26n1/v26n1a14.pdf

DAWS, M.I.; LYDALL, E.; CHMIELARZ, P.; LEPRINCE, O.; MATTHEWS, S.; THANOS, C.A.; PRITCHARD, H.W. Developmental heat sum influences recalcitrant seed traits in *Aesculus hippocastanum* across Europe. *New Phytologist*, v.162, p.157-166, 2004. http://onlinelibrary.wiley.com/doi/10.1111/j.1469-8137.2004.01012.x/pdf

DAWS, M.I.; CLELAND, H.; CHMIELARZ, P.; GORIN, F.; LEPRINCE, O.; MATTHEWS, S.; MULLINS, C.E.; THANOS, C.A.; VANDVIK, V.; PRITCHARD, H.W. Variable dessication tolerance in *Acer pseudoplatanus* seeds in relation to developmental conditions: a case of phenotypic recalcitrance? *Functional Plant Biology*, v.33, p.59-66, 2006. http://dx.doi.org/10.1071/FP04206

DELGADO, L.F., BARBEDO, C.J. Tolerância à dessecação de sementes de espécies de *Eugenia*. *Pesquisa Agropecuária Brasileira*, v.42, p.265-272, 2007. http://www.scielo.br/pdf/pab/v42n2/16.pdf

DELGADO, L.F.; MELLO, J.I.O.; BARBEDO, C.J. Potential for regeneration and propagation from cut seeds of *Eugenia* (Myrtaceae) tropical tree species. *Seed Science and Technology*, v.38, p.624-34, 2010.http://dx.doi.org/10.15258/sst.2010.38.3.10

JOËT, T.; OURCIVAL, J. M.; DUSSERT, S. Ecological significance of seed desiccation sensitivity in *Quercus ilex*. *Annals of Botany*, v.111, p.693-701, 2013. http://aob.oxfordjournals.org/content/early/2013/02/05/aob.mct025.full.pdf

JUSTO, C.F.; ALVARENGA, A.A.; ALVES, E.; GUIMARÃES, R.M.; STRASSBURG, R.S. Efeito da secagem, do armazenamento e da germinação sobre a micromorfologia de sementes de *Eugenia pyriformis* Camb. *Acta Botanica Brasilica*, v.21, p.539-551, 2007. http://www.scielo.br/pdf/abb/v21n3/a04v21n3.pdf

LAMARCA, E.V.; BAPTISTA, W.; RODRIGUES, D.S.; OLIVEIRA JUNIOR, C.J.F. Contribuições do conhecimento local sobre o uso de *Eugenia* spp. em sistemas de policultivos e agroflorestais. *Revista Brasileira de Agroecologia*, v.8, p.119-130, 2013a. http://www.aba-agroecologia.org.br/revistas/index.php/rbagroecologia/article/view/13256/9905

LAMARCA, E.V.; PRATAVIERA, J.S.; BORGES, I.F.; DELGADO, L.F.; TEIXEIRA, C.C.; CAMARGO, M.B.P.; FARIA, J.M.R.; BARBEDO, C.J. Maturation of *Eugenia pyriformis* seeds under different hydric and thermal conditions. *Anais da Academia Brasileira de Ciências*, v.85, p.223-233, 2013b. http://www.scielo.br/pdf/aabc/v85n1/0001-3765-aabc-85-01-223.pdf

LAMARCA, E.V.; BONJOVANI, M.R.; FARIA, J.M.R.; BARBEDO, C.J. Germinação em temperatura sub-ótima de embriões de *Inga vera* subsp. *Affinis* obtidos sob diferentes condições ambientais. *Rodriguésia*, v.64, p.877-885, 2013c. http://www.scielo.br/pdf/rod/v64n4/v64n4a15.pdf

MARCOS-FILHO, J. *Fisiologia de sementes de plantas cultivadas.* Piracicaba: FEALQ, 2005. 495p.

MARTINS, C.C.; BOVI, M.L.A.; NAKAGAWA, J.; MACHADO, C.G. Secagem e armazenamento de sementes de juçara. *Revista Árvore*, v.33, p.635-642, 2009. http://www.scielo.br/pdf/rarv/v33n4/v33n4a06.pdf

PEDRO JUNIOR, M.J.; BRUNINI, O.; ALFONSI, R.R.; ANGELOCCI, L.R. Estimativa de graus-dia em função de altitude e latitude para o estado de São Paulo. *Bragantia*, v.36, p.89-92, 1977. http://www.scielo.br/pdf/brag/v36n1/05.pdf

PETEK, M.R.; SERA, T.; FONSECA, I.C.B. Exigências climáticas para o desenvolvimento e maturação dos frutos de cultivares de *Coffea arabica*. *Bragantia*, v.68, p.169-181, 2009. http://www.scielo.br/pdf/brag/v68n1/a18v68n1.pdf

RAMIREZ, M.R.; SCHNORR, C.E.; FEISTAUER, L.B.; APEL, M.; HENRIQUES, A.T.; MOREIRA, J.C.F.; ZUANAZZI, J.A.S. Evaluation of the polyphenolic content, anti-inflammatory and antioxidant activities of total extract from *Eugenia pyriformes* Cambess (uvaia) fruits. *Journal of Food Biochemistry*, v.36, p.405-412, 2012. http://onlinelibrary.wiley.com/doi/10.1111/j.1745-4514.2011.00558.x/pdf

SANTANA, D.G.; RANAL, M.A. *Análise da germinação:* um enfoque estatístico. Brasília: Universidade de Brasília, 2004. 248p.

SANTOS, F.S.; PAULA, R.C.; SABONARO, D.Z.; VALADARES, J. Biometria e qualidade fisiológica de sementes de diferentes matrizes de *Tabebuia chrysotricha* (Mart. ex A. DC.) StandI. *Scientia Forestalis*, v.37, p.163-173, 2009. http://www.ipef.br/publicacoes/scientia/nr82/cap06.pdf

SILVA, R.S.M.; CHAVES, L.J.; NAVES, R.V. Caracterização de frutos e árvores de cagaita (*Eugenia dysenterica* DC.) no sudeste do estado de Goiás, Brasil. *Revista Brasileira de Fruticultura*, v.23, p.330-334, 2001. http://www.scielo.br/pdf/rbf/v23n2/7976.pdf

SILVA, C.V.; BILIA, D.A.C.; MALUF, A.M.; BARBEDO, C.J. Fracionamento e germinação de sementes de uvaia (*Eugenia pyriformis* Cambess.- Myrtaceae). *Revista Brasileira de Botânica*, v.26, p.213-221, 2003.http://www.scielo.br/pdf/rbb/v26n2/a09v26n2.pdf

SILVA, C.V.; BILIA, A.C.; BARBEDO, C.J. Fracionamento e germinação de sementes de *Eugenia*. *Revista Brasileira de Sementes*, v.27, p.86-92, 2005. http://www.scielo.br/pdf/rbs/v27n1/25185.pdf

STIEVEN, A.C.; MOREIRA, J.J.; SILVA, C.F. Óleos essenciais de uvaia (*Eugenia pyriformis* Cambess): avaliação das atividades microbiana e antioxidante. *Eclética Química*, v.34, p.7-13, 2009. http://www.scielo.br/pdf/eq/v34n3/01.pdf

TEIXEIRA, C.C., BARBEDO, C.J. The development of seedlings from fragments of monoembryonic seeds as an important survival strategy for *Eugenia* (Myrtaceae) tree species. *Trees*, v.26, p.1069-1077, 2012. http://download.springer.com/static/pdf/86/art%253A10.1007%252Fs00468-011-0648-5.pdf?auth66=1419267966_7e6742fd8463ea9746b548039f7dc051&ext=.pdf

VILLA NOVA, N.A.; PEDRO JÚNIOR, M.J.; PEREIRA, A.R.; OMETTO, J.C. Estimativa de graus-dia acumulados acima de qualquer temperatura base em função das temperaturas máxima e mínima. *Caderno Ciência da Terra*, v.30, p.1-8, 1972.

25

Storage and germination of seeds of *Handroanthus heptaphyllus* (Mart.) Mattos

Thaíse da Silva Tonetto[1*], Maristela Machado Araujo[1], Marlove Fátima Brião Muniz[2], Clair Walker[2], Álvaro Luís Pasquetti Berghetti[1]

ABSTRACT – The aim of this study was to determine the substrate and the most suitable sowing method for germination, as well as the environment for storage of *Handroanthus heptaphyllus* seeds (ipê-roxo), and infer the health quality provided by different packaging. Experiment 1 has assessed the treatments (substrates – blotting paper, filter paper, vermiculite, sand, besides paper roll; and sowing methods – among and on the substrates). Experiment 2 has assessed storage in three environments (air conditioned room – 18 °C and 49% of relative humidity (RH); cold and wet chamber – 8 °C and 80% RH; and dry and cold chamber – 7.5 °C and 55% RH) , for 300 days. *Handroanthus heptaphyllus* seed germination test can be performed using seeding among blotting paper, vermiculite, on sand, between sand, on vermiculite and between filter paper. The storage of the seeds in plastic bags kept in an air conditioned room and/or in a dry and cold chamber is suitable for the preservation of *Handroanthus heptaphyllus* seeds for a period of 300 days. The packaging in a dry and cold chamber environment has provided a lower incidence of fungi associated with the *Handroanthus heptaphyllus* seeds.

Index terms: ipê-roxo, germination potential, storage.

Qualidade de sementes de *Handroanthus heptaphyllus* (Mart.) Mattos armazenadas em diferentes ambientes

RESUMO – O objetivo desse estudo foi determinar o substrato e o método de semeadura mais adequado para a germinação, assim como, o ambiente para o armazenamento de sementes de ipê-roxo (*Handroanthus heptaphyllus*), além de inferir sobre a qualidade sanitária proporcionada pelos distintos acondicionamentos. No experimento 1 foram avaliados os tratamentos (substratos – papel mata-borrão, papel filtro, vermiculita, areia, além do rolo de papel; e métodos de semeadura – entre e sobre os substratos). No experimento 2 foi avaliado o armazenamento em três ambientes (sala climatizada – 18 °C e 49% umidade relativa (UR); câmara fria e úmida – 8 °C e 80% UR; e câmara seca e fria – 7,5 °C e 55% UR), por 300 dias. O teste de germinação de sementes de *Handroanthus heptaphyllus* pode ser realizado utilizando-se a semeadura entre papel mata-borrão, entre vermiculita, sobre areia, entre areia, sobre vermiculita e entre papel filtro. O armazenamento das sementes em sacos de plástico mantidos em ambiente de sala climatizada e/ou em câmara seca e fria é adequado à conservação das sementes de *Handroanthus heptaphyllus* por um período de 300 dias. O acondicionamento em ambiente de câmara seca e fria proporcionou menor incidência de fungos associados às sementes de *Handroanthus heptaphyllus*.

Termos para indexação: ipê-roxo, potencial germinativo, armazenamento.

Introduction

The need for maintenance and conservation of native forest species has become evident given the intense social and scientific concern for recovering degraded areas. Thus arises the need for further studies that subsidize information to qualify the yield of those species. According to Sarmento and Villela (2010), society's concern strengthens the environmental policies to promote greater demand for seeds and seedlings of native forest species, which are a basic input for conservation programs, recovery of ecosystems, plant improvement and biotechnology.

The characteristics of the seeds should be assessed by physical, physiological and health tests in order to prove the quality of the lot, as well as provide information on which to base the seeds sowing and preservation. The procedures

[1]Departamento de Ciências Florestais, Centro de Ciências Rurais, Universidade Federal de Santa Maria, 97105-900 – Santa Maria, RS, Brasil.

[2]Departamento de Defesa Fitossanitária, Centro de Ciências Rurais, Universidade Federal de Santa Maria, 97105-900 – Santa Maria, RS, Brasil.
*Corresponding author<thaisetonetto@hotmail.com>

Storage and germination of seeds of Handroanthus heptaphyllus (Mart.) Mattos

185

for such tests are described in the Regras para Análise de Sementes (Rules for Seed Analysis) (Brasil, 2009a; 2009b) and Instruções para Análises de Sementes de Espécies Florestais (Instructions for Analyses of Forest Species Seeds) (Brasil, 2013). Accordingly, knowing which substrate provides a suitable seed germination is needed, since it varies depending on the species, compromising the analysis result on the quality of the seeds. Brasil (2013) suggests the paper roll substrate for *Handroanthus heptaphyllus*.

The importance of studies related to *Handroanthus heptaphyllus* seeds is enhanced by the fact that their spread is mainly by seeds, requiring its preservation in periods in which there is no fruit. Thus, the seed storage is shown as an important tool to remedy possible deficiencies in the production of seeds and seedlings, and, according to Marcos-Filho (2005), the preservation period of the physiological potential depends in large part on the water content of seeds and storage conditions.

It is noteworthy that among the information in the literature, there is divergence with respect to tolerance to storage and maintenance of quality of seeds of the species *Handroanthus heptaphyllus* (Lorenzi, 2002; Carvalho et al., 2006; Grings and Brack, 2011), which becomes a barrier to ensure the seed preservation. Thus, opposition is observed in studies on the species and genus. In addition, it is necessary to ensure that the supply of seeds be free of pathogens, since their presence can reduce the germination ability of a lot. Among the studies on seed health of the genus *Handroanthus*, the one by Botelho et al. (2008) stands out, who have observed the presence of *Alternaria* sp., *Aspergillus* sp., *Fusarium* sp., *Cladosporium* sp., *Penincilium* sp., among others.

Handroanthus heptaphyllus (ipê-roxo), belonging to the Bignoniaceae family, is widely distributed in Brazil from southern Bahia to Rio Grande do Sul, and that, according to Lorenzi (2002), has a large timber and ecological value. Thus, ipê-roxo emerges as a priority for studies because it is adapted to the soil and climate conditions of much of Brazil.

The germination of ipê-roxo seeds is fast, because it does not have dormancy (Lorenzi, 2002), has a high germination potential (71.8%), and the seeds are classified as orthodox (Wielewicki et al., 2006). On the other hand, Grings and Brack (2011) point out that this species produces lots of seeds with high viability; however, they should be sown right after harvest, since they lose the germination ability in a few weeks if they are not preserved in cold or dry storage.

Borba Filho and Perez (2009), assessing the storage of seeds of *Handroanthus roseo-alba* (ipê-branco; tabebuia roseo-alba, known as white ipê, ipê-branco or lapacho blanco) and *Handroanthus impetiginosus* (ipê-rosa) in different packaging,

have found that packaging in cans and preservation in refrigerators (4 to 6 °C and 38 to 43% RH (relative humidity)) were favorable to the preservation of the seeds of both species, with germination around 80% at 300 days of storage.

The objective of this study was to determine the substrate and the most suitable sowing method for germination, as well as the environment for storage of ipê-roxo (*Handroanthus heptaphyllus*) seeds, and infer the health quality provided by different packaging.

Material and Methods

Ripe fruits of *Handroanthus heptaphyllus* were collected in eight trees in December 2011, in the region of Santa Maria, RS, Brazil (29°47'37"S and 53°40'01"O). The seeds analyses were carried out at the Laboratório de Silvicultura e Viveiro Florestal and at Laboratório de Fitopatologia Elocy Minussi at Universidade Federal de Santa Maria (UFSM), Santa Maria, RS, Brasil.

For processing the fruits and seeds, they were stored in a shaded and airy shed to complete their opening and release the seeds. Thereafter, the processing was performed in order to separate impurities and damaged seeds. Then the lot to be studied was made, which was divided into two sub-lots, one for experiment 1 and the other for experiment 2.

Experiment 1 – Lot characterization and study of a suitable substrate for determination of seed physiological quality

The weight of one thousand seeds was performed according to Regras para Análise de Sementes (Rules for Seeds Analyses) (RAS), using eight replicates of 100 seeds (Brasil, 2009a). The water content of the seeds by the oven method (105 ± 3 °C for 24 h) with four replications of 25 seeds (Brasil, 2009a).

For analysis of the electrical conductivity, the mass method was used. The four samples, each consisting of 25 seeds, were first weighed on a digital scale accurate to 0.001 g, and then placed into 200 mL plastic cups, which were immersed in 75 mL of distilled water for 24 hours and then placed in a germination chamber at 25 °C, with the cups covered with foil (Vieira, 1994).

After 24 hours, homogenization of the solution was done, and then electrical conductivity (EC) was measured for the deionized water in order to calibrate and adjust and for the seed samples. The results were expressed in $\mu S/cm^{-1}/g^{-1}$, obtained by subtracting the EC of the seed sample from the EC of the distilled water and dividing the result of this subtraction by the weight of the seed sample.

Substrate analysis suitable for germination and initial physiological quality of the seeds

The analysis of the initial physiological quality (time zero) of the seed lot was accomplished by means of the germination test in gerbox-type transparent plastic boxes with four replications of 25 seeds, testing the following treatments: T1 – on blotting paper, T2 – between blotting paper; T3 – on filter paper, T4 – between filter paper; T5 – on vermiculite; T6 – between vermiculite; T7 – on sand; T8 – between sand and T9 – paper roll, which has totaled 36 sampling units.

For the amount of water to be added in the "sand" and "vermiculite" treatment, the methodology proposed by Brasil (2009a) and Brasil (2013) was respectively considered, adding a water volume for 60% of field capacity. The other treatments with paper substrate were moistened with 2.5 times the weight of the paper. All substrates and the distilled water were autoclaved (120 °C for 2 h), and the gerbox with alcohol 70 °GL (degrees Gay-Lussac (after the French chemist Joseph Louis Gay-Lussac).

The seed disinfection was performed with five drops of neutral detergent diluted in 100 mL of distilled water for 5 minutes, being subsequently rinsed three times in distilled water (Brasil, 2011). Then the seeds were placed in the treatments, with samples being taken to a Mangelsdorf-type germination chamber (25 ± 3 °C and constant light), which was kept in an air conditioned room (25 °C).

The counts were taken every three days until the end of the test. The technological criteria of normal seedling formation was used (Brasil, 2009a). From these data, the percentage of normal seedlings, abnormal seedlings, firm seeds (hard) and dead seedlings was calculated, obtaining information on the first and last count, and the germination speed index (Maguire, 1962).

Experiment 2 – Quality analysis of the stored seeds in different environments

For the analysis of the seeds before storage, the data obtained in experiment 1 were used. Storage of *Handroanthus heptaphyllus* was held in December 2011 in sealed polyethylene bags of 10 microns, containing 135 g of seeds (≅ 3000 seeds). They were placed in Kraft paper drums, staying in three rooms (treatments) during 300 days (until October 2012). Thus, the treatments consisted of: T1 – air conditioned room (18 ± 3 °C and 49 ± 10% relative humidity – RH); T2 – wet and cold chamber (8 ± 2 °C and 80 ± 10% RH); T3 – dry and cold chamber (7.5 ± 1.5 °C and 55 ± 10% RH), which have been compared to T4, which has constituted the time zero (without storage).

For the tests, initially the seeds of each treatment were homogenized (storage). Afterwards, the working sample was obtained for determining the weight of a thousand seeds, the water content of the seeds and the electrical conductivity, using the same methods used in experiment 1. Germination tests were installed adopting the "on sand" treatment, with four replications of 25 seeds for each storage, following the same procedure as in experiment 1.

Along with that, after storage, at 300 days the presence of fungi associated to the *Handroanthus heptaphyllus* seeds by the health test was assessed. First, the gerboxes were disinfected with a sodium hypochlorite 1% and alcohol 70% solution, were covered with two sheets of sterile filter paper dampened with sterile distilled water and then the seeds were distributed on the paper (Brasil, 2009b). Four replicates of 25 seeds for each treatment were used. Incubation was performed in a chamber with a controlled temperature (25 ± 2 °C and a photoperiod of 12 hours of light using six fluorescent lamps) for seven days. After this period, the identification and quantification of the fungi under stereoscopic and optical microscope were performed, according to Barnett and Hunter (1999).

For the experiments, a completely randomized experimental design was used. The data then went through normality tests of waste and homogeneity of variances by means of the Shapiro-Wilk and Bartlett tests, respectively. In case any of these assumptions had not been met, the processing of the data was performed in an arcsine $(X/100)^{0.5}$ for data in percent and for the others $(x + 0.5)^{0.5}$, where X = variable (Santana and Ranal, 2000).

Statistical analysis was performed with the aid of software SISVAR (Ferreira, 2011), subjecting the data to analysis of variance, and when a discrepancy between the treatments by the F test ($p < 0.05$) was noticed, the comparison of means by Student's t and Scott-Knott tests at 5% of error probability was performed.

Results and Discussions

The seeds of *Handroanthus heptaphyllus* have presented initial figures (time zero) for the weight of 1000 seeds of 45.8 ± 0.10 g (coefficient of variation – CV = 2.24%), which represents, on average, 21,834.06 seeds/kg^{-1}. Similarly, the water content of the seeds (WC) was of 8.79 ± 0.32% (CV = 3.66%).

In the germination test (G%) the highest values were obtained in the treatment between blotting paper (BBP), between vermiculite (BV), on sand (OS), between sand (BS), on vermiculite (OV) and between filter paper (BFP) (Table 1).

Storage and germination of seeds of Handroanthus heptaphyllus (Mart.) Mattos

187

Table 1. Normal seedlings (G), germination speed index (GSI), abnormal seedlings (AS), dead seeds (DS) and hard seeds (HS) of *Handroanthus heptaphyllus* (Mart.) Mattos, on different substrates at time zero.

Treatment	G (%)	GSI	AS (%)	DS (%)	HS (%)
OBP	48 c*	0.64 c	31 a	19^{ns}	2^{ns}
BBP	84 a	1.51 a	0 b	16	0
OPF	64 b	0.82 b	17 a	19	0
BFP	77 a	1.32 a	1 b	22	0
OV	79 a	1.39 a	2 b	19	0
BV	83 a	1.50 a	1 b	16	0
OS	83 a	1.52 a	1 b	16	0
BS	81 a	1.43 a	3 b	16	0
RP	64 b	1.01 b	6 b	29	1
Average	74	1.24	6.89	19.11	0.33
CV (%)	13.04	15.45	86.92	42.78	447.21

*Means followed by the same letter do not differ statistically at 5% of error probability by Scott-Knott test; ns = non-significant; CV = coefficient of variation; OBP (on blotting paper); BBP (between blotting paper); OPF (on filter paper); BFP (between filter paper); OV (on vermiculite); BV (between vermiculite); OS (on sand); BS (between sand) and RP (paper roll).

The treatments mentioned before have differed from the treatment on filter paper (OFP), paper roll (PR) and on blotting paper (OBP), and, in the latter, 48% of the seeds have germinated, and therefore were considered unsuitable for germination analysis of ipê-roxo (Table 1) because they do not reflect the actual quality of the seeds because in the other treatments there was a germination with higher percentages.

The assessment of the germination speed index (GSI) has corroborated the results observed in germination, which have indicated as superior the treatments: BBP, BV, OS, BS, OV and BFP (Table 1), which also allows us to infer that the vigor can be expressed by these treatments.

Analysis of the percentage of abnormal seedlings (AS) in the experiment course make it possible to infer that the OBP (31%) and OPF (17%) treatments account for the higher values of abnormalities, intensifying the understanding that these are not suitable for germination of ipê-roxo seeds. The percentage of dead and hard seeds was not significant at the 5% error probability for the different treatments assessed (Table 1). The high mortality rates of seeds in all treatments may be related to the presence of fungi, a fact that suggests the need to identify the pathogens associated with the seeds.

Accordingly, the use of the treatment on sand (Table 1) is indicated, which, besides a high germination and GSI, can be more easily worked on. Therefore, it was selected for performing experiment 2.

Quality analysis of the seeds stored in different environments

After storage (300 days) in the different environments (air condition room – ACR, wet and cold chamber – WCC and dry and cold chamber – DCC), the figures for the weight of 1000 seeds (WTS) were 32.20 ± 0.07 g (CV = 1.92%), 36.46 ± 0.21 g (CV = 5.71%) and 36.69 ± 0.18 g (CV = 4.93%), in each storage environment, representing 31,055.90; 27,427.32 and 27,255.38 seeds/kg^{-1}, respectively. It is shown that the water content (WC) of the seeds in the different environments (air conditioned room, wet and cold chamber, dry and cold chamber) was, respectively, $7,67 \pm 0,47\%$ (CV = 6.16%); $10.41 \pm 0.94\%$ (CV = 9.07%) and $9.29 \pm 0.37\%$ (CV = 3.96%). The moisture content (MC) has declined relative to time zero only for air conditioned room, because for wet and cold chamber and dry and cold chamber there was an increase in the moisture content of the seeds. In the case of the wet and cold chamber, the result can be related to the high relative humidity (80%) occurring to the moisture absorption by the seeds, because the packaging used for storage is semipermeable (polyethylene plastic). A similar situation may have occurred in the dry and cold chamber.

The highest germination has occurred for non-stored seeds (NS = 83%) compared to any storage (Table 2). However, when analyzing the seed storage, there has been a superiority of the air conditioned room environment as well as in a dry and cold chamber, despite the reduction in germination to 54 and 57% (p < 0.05) at 300 days of storage, representing an average loss of about 28% of the germination potential in relation to non-stored seeds (time zero) (Table 2). The germination of the seeds stored in the wet and cold chamber was low (38%), matching the 45% reduction of the germination potential compared to non-stored seeds. Thus, seeds of *Handroanthus heptaphyllus* that are not used soon after harvesting, must be stored for later use; however, the wet and cold chamber environment is not suitable for this species.

In assessing the germination speed index (GSI), the means of the non-stored seeds treatments, kept in an air conditioned room and in a dry and cold chamber have not differed (Table 2), confirming the maintenance of seed vigor after storage, despite the tendency to reduce the GSI after storage. The result of the vigor also supports the storage superiority of ACR and DCC, regarding WCC. On the other hand, this rate was not enough to detect the difference between seeds non-stored and stored in ACR and DCC. It is suggested that there might be GSI inefficiency to identify differences between lots of *Handroanthus heptaphyllus*, having in this study the percentage of germination as one of the most robust variables to express the physiological ability of the seed.

The percentage of dead seeds was higher in the wet and cold chamber, representing 61% of mortality, differing from the other treatments (Table 2), suggesting that the temperature and high humidity may have damaged the seeds and consequently there were more fungi associated to the seeds in this storage condition (Table 3).

Table 2. Mean values of normal seedlings (G), germination speed index (GSI), abnormal seeds (AS) and dead seeds (DS) and mass electrical conductivity (EC) of *Handroanthus heptaphyllus* (Mart.) Mattos, of the non-stored seeds and after 300 days of storage.

Treatment	G^1 (%)	GSI^1	AS^1 (%)	DS^1 (%)	EC^{2**} (μS.cm^{-1}.g^{-1})
ACR	57 b*	1.39 a	4^{ns}	39 b	14.86 a
WCC	38 c	0.84 b	1	61 a	13.40 a
DCC	54 b	1.28 a	3	43 b	16.19 a
NS	83 a	1.52 a	1	16 c	5.55 b
Average	58	1.25	2.25	39.75	12.50
CV (%)	13.72	13.76	147	23.64	17.09

*Means followed by the same letter do not differ statistically at 5% of error probability by ^{1}Scott-Knott test and ^{2}Student's t test; ** 25 seeds and 50 mL of distilled water containing EC of 0.15 μS/cm^{-1}/g^{-1}; ns = non-significant; CV = coefficient of variation; ACR (air conditioned room, 18 °C and 49% RH); WCC (wet and cold chamber, 8 °C and 80% RH); DCC (dry and cold chamber, 7.5 °C and 55% RH) and NS – non-stored seeds (control).

Table 3. Fungi associated to seeds of *Handroanthus heptaphyllus* (Mart.) Mattos, stored in different environments after 300 days of storage.

Storage environment	*Aspergillus* sp.	*Cladosporium* sp.	*Fusarium* sp.
		%	
1 – ACR	23 a*	85 b	82 b
2 – WCC	74 b	55 a	23 a
3 – DCC	15 a	47 a	9 a
Average	37.33	62.33	38.00
CV (%)	23.82	17.67	26.37

*Means followed by the same letter do not differ significantly from each other by the Scott-Knott test at 5% of error probability. CV = coefficient of variation; 1 – ACR: air conditioned room, 18 °C and 49% RH, 2 – WCC: wet and cold chamber, 8 °C and 80% RH and 3 – DCC: dry and cold chamber, 7.5 °C and 55% RH.

The electrical conductivity (EC) after storage compared to time zero confirms the reduction of seed vigor of ipê-roxo seeds (Table 2). However, it does not distinguish the different storage conditions because a mass test has been applied, and thus a deteriorated seed present in each sample preparation may have shrouded the result. Thus, it is suggested that EC be performed individually for each seed. Marcos-Filho (2005) infers that the EC test has as its principle the increase of the permeability of the membrane as the seeds decay. Thus, the higher the measured value, the higher is the amount of leached electrolytes (sugars, amino acids, proteins, among others) of the seed tissues for the water.

Thus, the results for electrical conductivity were not able to detect changes in vigor of seeds of *Handroanthus heptaphyllus* in different storage environments, and this result was found by Borba Filho and Perez (2009).

In the assessment of seed health held in seeds of *Handroanthus heptaphyllus* at 300 days after storage in different environments (ACR, WCC and DCC), ten different genera of fungi were found: *Cladosporium* sp., *Alternaria* sp., *Penicillium* sp., *Fusarium* sp., *Aspergillus* sp., *Epicoccum* sp., *Bipolaris* sp., *Phomopsis* sp., *Rhizopus* sp. and *Pestalotia* sp.

The highest incidence was of *Cladosporium* sp. (85 and 55%) and *Fusarium* sp. (82 and 23%) for seeds stored in an air conditioned room and in a wet and cold chamber, respectively. In the latter storage environment was found the highest percentage of *Aspergillus* sp. (74%) (Table 3). For other fungal genera there was no significant difference between air conditioned room, wet and cold chamber and dry and cold chamber.

The occurrence of *Aspergillus* sp. was higher when associated with wet and cold chamber storage and may have occurred because of the high humidity of the place, which has favored the emergence of this pathogen. Botelho et al. (2008) have identified and quantified 16 genera of fungi associated with seeds of *Handroanthus serratifolius* (ipê-amarelo, commonly known as yellow lapacho, pau d'arco, yellow poui, yellow ipe, pau d'arco amarelo, or ipê-amarelo), among these:

Cladosporium sp., *Fusarium* sp., *Penicillium* sp., *Aspergillus* sp. Fungi of the genera *Aspergillus*, *Penicillium* and *Rhizopus* are considered to cause damage to storage conditions. Cherobini et al. (2008) explain that the *Penicillium* and *Aspergillus* genera have the ability to reduce seed germination and cause the death of the embryo, but when they have low moisture contents, near the lower limit for the growth of the fungi, the attack is slow; however, as the humidity of the seed increases, the loss of germination becomes more rapid because of the rapid growth of the fungus. Analyzing the results together, it was observed that the species soon after seed harvest has a high initial germination (≈80%), which can be seen in germination tests using substrates such as on and between sand and vermiculite, on and between blotting paper and filter. However, storage in a low humidity controlled

environment (≤ 55%) and temperature (≤ 18 °C), although with a proportional loss of about 28 percentage points of the initial germination potential, maintain 55% of the viable seeds, ensuring availability of seedlings between annual collections, but with greater cost to the producer. In addition, the seed water content (WC) obtained at time zero and after storage, associated with the germination result, suggests that the seeds of *Handroanthus heptaphyllus* do not show a typical behavior of recalcitrant seeds, nor orthodox ones, and this study has found that the species mentioned has an intermediate trend between both classifications in relation to desiccation.

Storage analysis by the Pearson (r) correlation matrix was performed according to Callegari-Jacques (2003) and these can be seen in Table 4.

Table 4.   Simple correlation (r) between variables of physiological and sanitary quality of seeds of *Handroanthus heptaphyllus* (Mart.) Mattos after 300 days of storage.

Variable	Sto	G%	GSI	AS%	DS%	Cla	Fus	Asp
Sto	1.00							
G%	-0.11[ns]	1.00						
GSI	-0.15[ns]	0.97*	1.00					
AS%	-0.12[ns]	0.44[ns]	0.46[ns]	1.00				
DS%	0.12[ns]	-0.97*	-0.96*	-0.63*	1.00			
Cla	-0.82*	0.29[ns]	0.31[ns]	0.21[ns]	-0.30[ns]	1.00		
Fus	-0.91*	0.32[ns]	0.39[ns]	0.20[ns]	-0.33[ns]	0.90*	1.00	
Asp	-0.12[ns]	-0.80*	-0.84*	-0.47[ns]	0.81*	-0.19[ns]	-0.20[ns]	1.00

*Significant at 5%; ns = non-significant; Sto = storage; G% = percentage of germination; GSI = germination speed index; AS% = percentage of abnormal seedlings; DS% = percentage of dead seeds; Cla = *Cladosporium* sp.; Fus = *Fusarium* sp.; Asp = *Aspergillus* sp.

The storage has a correlation that is considered strong and strongly, respectively, with pathogens *Cladosporium* sp. (-0.82) and *Fusarium* sp. (-0.91); these values occur because of the storage time, which suggests an increase in the percentage of pathogens associated with the seeds. The germination percentage correlates strongly and positively with GSI (0.97), and this confirmation is expected, since these parameters infer about the proper quality of the seeds. For the percentage of dead seeds (DS) and *Aspergillus* sp. with the percentage of germination, the correlation is negative (-0.97 and -0.80, respectively), because as it decreases the germination percentage, the incidence of mortality of seeds and of said pathogen increases (Table 4). The values that associate the germination percentage with other variables follow the same trend as in a study conducted by Cherobini et al. (2008), in which the authors infer a higher germination linked to higher GSI and a lower incidence of dead seeds and of *Aspergillus* sp. in cedar seeds.

The correlation infers that the GSI is strongly and

inversely correlated with the dead seeds percentage (-0.96) and with *Aspergillus* sp. (-0.84) (Table 4).

The percentage of abnormal seedlings has a strong negative correlation with the percentage of dead seeds (-0.63), indicating that when one of the variables increases the other one decreases. Dead seeds positively correlate with the pathogen *Aspergillus* sp. (0.81), which indicates that the higher the incidence of this fungus, the higher the mortality of the seeds. Finally, *Cladosporium* sp. and *Fusarium* sp. have a strongly positive correlation (0.90), which implies the association of both pathogens with the seeds of *Handroanthus heptaphyllus* (Table 4).

Conclusions

The seed germination test of *Handroanthus heptaphyllus* can be performed using seeding between blotting paper, between vermiculite, on sand, between sand, on vermiculite and between filter paper.

The storage of the seeds in plastic bags maintained in an

air conditioned room environment and/or in a dry and cold chamber is suitable for the preservation of *Handroanthus heptaphyllus* seeds for a period of 300 days.

The packaging in a dry and cold chamber environment has provided a lower incidence of fungi associated with the seeds of *Handroanthus heptaphyllus*.

Acknowledgement

To CNPq for the scholarship granted to the second author.

References

BARNETT, H.L.; HUNTER, B.B. *Illustred genera of imperfect fungi.* 3ed. Minnesota: Burgess Publishing Company, 1999. 241 p.

BORBA FILHO, A.B; PEREZ, S.C.J.G.A. Armazenamento de sementes de ipê-branco e ipê-roxo em diferentes embalagens e ambientes. *Revista Brasileira de sementes*, v.31, n.1, p.259-269, 2009. http://dx.doi.org/10.1590/S0101-31222009000100029

BOTELHO, L.S.; MORAES, M.H.D.; MENTEN, J.O.N. Fungos associados às sementes de ipê-amarelo (*Handroanthus serratifolius*) e ipê-roxo (*Handroanthus impetiginosus*): incidência, efeito na germinação e transmissão para as plântulas. *Revista Summa Phytopathology*, v.34, n.4, p.343-348, 2008. http://www.scielo.br/pdf/sp/v34n4/v34n4a08.pdf

BRASIL. Ministério da Agricultura, Pecuária e Abastecimento. *Regras para análise de sementes.* Ministério da Agricultura, Pecuária e Abastecimento. Secretaria de Defesa Agropecuária. Brasília: MAPA/ACS, 2009a. 395 p. http://www.agricultura.gov.br/arq_editor/file/2946_regras_analise__sementes.pdf

BRASIL. Ministério da Agricultura, Pecuária e Abastecimento. *Manual de análise sanitária de sementes.* Ministério da Agricultura, Pecuária e Abastecimento. Secretaria de Defesa Agropecuária, Brasília: MAPA/ACS, 2009b. 202 p. http://www.agricultura.gov.br/arq_editor/file/12261_sementes_-web.pdf

BRASIL. Ministério da Agricultura, Pecuária e Abastecimento. *Instrução normativa n° 35*, de 14 de julho de 2011. http://sistemasweb.agricultura.gov.br/sislegis/action/detalhaAto.do?method=consultarLegislacaoFederal. Accessed on Jan. 03rd. 2014.

BRASIL. Ministério da Agricultura, Pecuária e Abastecimento. *Instruções para análise de espécies florestais.* Secretaria de Defesa Agropecuária. Brasília: MAPA/ACS, 2013. 98 p. http://www.agricultura.gov.br/arq_editor/file/Laborat%C3%B3rio/Sementes/FLORESTAL_documento_pdf.pdf

CALLEGARI-JACQUES, S.M. *Bioestatística*: princípios e aplicações. Porto Alegre: ARTMED, 2003. 255 p.

CARVALHO, L.R.; SILVA, E.A.A.; DAVIDE, A.C. Classificação de sementes florestais quanto ao comportamento no armazenamento. *Revista Brasileira de Sementes*, v.28, n.2, p.15-25, 2006. http://dx.doi.org/10.1590/S0101-31222006000200003

CHEROBINI, E.A.I.; MUNIZ, M.F. B.; BLUME, E. Avaliação da qualidade de sementes e mudas de cedro. *Revista Ciência Florestal*, v.18, n.1, p.65-73, 2008. http://coral.ufsm.br/cienciaflorestal/artigos/v18n1/A6V18N1.pdf

FERREIRA, D. F. SISVAR: a computer statical analysis system. *Revista Ciência e Agrotecnologia*, v.35, n.6, p.1039-1042, 2011. http://dx.doi.org/10.1590/S1413-70542011000600001

GRINGS, M; BRACK, P. *Handroanthus heptaphyllus* (ipê-roxo) In: BRASIL, Ministério do Meio Ambiente. *Espécies nativas da flora brasileira de valor econômico atual ou potencial*: plantas para o futuro – Região sul. Orgs: CORADIN, L; SIMINSKI, A.; REIS, A. Brasília: MMA, 2011. 934 p.

LORENZI, H. *Árvores Brasileiras.* São Paulo: Plantarum, 2002. v.1, 378 p.

MAGUIRE, J.B. Speed of germination-aid in selection and evaluation for seedling emergence vigor. *Crop Science*, v.2, n.2, p.176-177, 1962. https://www.crops.org/publications/cs/abstracts/2/2/CS0020020176

MARCOS-FILHO, J. *Fisiologia de sementes de plantas cultivadas.* Piracicaba: FEALQ, 2005. 495 p.

SANTANA, D.G.; RANAL, M.A. Análise estatística na germinação. *Revista Brasileira Fisiologia Vegetal*, v.12, p.205-237, 2000. http://www.cnpdia.embrapa.br/rbfv/pdfs/v12Especialp206.pdf

SARMENTO, M.B.; VILLELA, F.A. Sementes de espécies florestais nativas do Sul do Brasil. *Informativo Abrates*, v.20, n.1,2, p.39-44, 2010. http://www.abrates.org.br/portal/images/stories/informativos/v20n12/artigo05.pdf

VIEIRA, R.D. Teste de condutividade elétrica. In: VIEIRA, RD; CARVALHO, N.M. *Teste de vigor em sementes.* Jaboticabal: FUNEP, 1994. p.103-132.

WIELEWICKI, A.P.; LEONHARDT, C.; SCHLINDWEIN, G.; MEDEIROS, A.C.S. Proposta de padrões de germinação e teor de água para sementes de algumas espécies florestais presentes na região Sul do Brasil. *Revista Brasileira de Sementes*, v.28, n.3, p.191-197, 2006. http://dx.doi.org/10.1590/S0101-31222006000300027

PERMISSIONS

All chapters in this book were first published in JSS, by Associação Brasileira de Tecnologia de Sementes; hereby published with permission under the Creative Commons Attribution License or equivalent. Every chapter published in this book has been scrutinized by our experts. Their significance has been extensively debated. The topics covered herein carry significant findings which will fuel the growth of the discipline. They may even be implemented as practical applications or may be referred to as a beginning point for another development.

The contributors of this book come from diverse backgrounds, making this book a truly international effort. This book will bring forth new frontiers with its revolutionizing research information and detailed analysis of the nascent developments around the world.

We would like to thank all the contributing authors for lending their expertise to make the book truly unique. They have played a crucial role in the development of this book. Without their invaluable contributions this book wouldn't have been possible. They have made vital efforts to compile up to date information on the varied aspects of this subject to make this book a valuable addition to the collection of many professionals and students.

This book was conceptualized with the vision of imparting up-to-date information and advanced data in this field. To ensure the same, a matchless editorial board was set up. Every individual on the board went through rigorous rounds of assessment to prove their worth. After which they invested a large part of their time researching and compiling the most relevant data for our readers.

The editorial board has been involved in producing this book since its inception. They have spent rigorous hours researching and exploring the diverse topics which have resulted in the successful publishing of this book. They have passed on their knowledge of decades through this book. To expedite this challenging task, the publisher supported the team at every step. A small team of assistant editors was also appointed to further simplify the editing procedure and attain best results for the readers.

Apart from the editorial board, the designing team has also invested a significant amount of their time in understanding the subject and creating the most relevant covers. They scrutinized every image to scout for the most suitable representation of the subject and create an appropriate cover for the book.

The publishing team has been an ardent support to the editorial, designing and production team. Their endless efforts to recruit the best for this project, has resulted in the accomplishment of this book. They are a veteran in the field of academics and their pool of knowledge is as vast as their experience in printing. Their expertise and guidance has proved useful at every step. Their uncompromising quality standards have made this book an exceptional effort. Their encouragement from time to time has been an inspiration for everyone.

The publisher and the editorial board hope that this book will prove to be a valuable piece of knowledge for researchers, students, practitioners and scholars across the globe.

LIST OF CONTRIBUTORS

Liana Hilda Golin Mengarda, José Carlos Lopes, Rafael Fonsêca Zanotti and Pedro Ramon Manhone
Departamento de Produção Vegetal, Universidade Federal do Espírito Santo, 29500-000, Alegre, ES, Brasil

Rodrigo Sobreira Alexandre
Departamento de Ciências Florestais, Universidade Federal do Espírito Santo, 29550-00, Jerônimo Monteiro, ES, Brasil

Amanda Ávila Cardoso
Departamento de Biologia Vegetal, Universidade Federal de Viçosa, 36570-000 - Viçosa, MG, Brasil

Amana de Magalhães Matos Obolari and Eduardo Euclydes de Lima e Borges
Departamento de Engenharia Florestal, Universidade Federal de Viçosa, 36570-000 - Viçosa, MG, Brasil

Cristiane Jovelina da Silva
Departamento de Botânica, Universidade Federal de Minas Gerais, 31270-901 - Belo Horizonte, MG, Brasil

Haroldo Silva Rodrigues
Departamento de Fitotecnia, Universidade Federal de Viçosa, 36570-000 - Viçosa, MG, Brasil

Sérgio Macedo Silva, Roberta Camargos de Oliveira, Adílio de Sá Júnior and Carlos Machado dos Santos
Universidade Federal de Uberlândia, Instituto de Ciências Agrárias, Caixa Postal 593, 38400-902 – Uberlândia, MG, Brasil

Risely Ferraz de Almeida
Universidade Estadual Paulista Júlio de Mesquita Filho, Caixa Postal 237, 18610-307 – Jaboticabal, SP, Brasil

João Paulo Naldi Silva and Claudio José Barbedo
Instituto de Botânica, Núcleo de Pesquisa em Sementes, Caixa Postal, 68041, 04301012 – São Paulo, SP, Brasil

Danilo da Cruz Centeno
Universidade Federal do ABC, Centro de Ciências Naturais e Humanas, 09606-070 - São Bernardo do Campo, SP, Brasil

Rita de Cássia Leone Figueiredo-Ribeiro
Instituto de Botânica, Centro de Pesquisa em Ecologia e Fisiologia, Caixa-Postal, 68041, 04301902 - São Paulo, SP, Brasil

Thaís Francielle Ferreira, Valquíria de Fátima Ferreira and João Almir Oliveira
Marcos Vinícios de Carvalho and Leonardo de Souza Miguel
Departamento de Agricultura, UFLA, Caixa Postal 3037, 37200-000 – Lavras, MG, Brasil

Cristian Rafael Brzezinski, Julia Abati and Claudemir Zucareli
Departamento de Agronomia, UEL, Caixa Postal 6001, 86051-990 -Londrina, PR, Brasil

Ademir Assis Henning, Fernando Augusto Henning, José de Barros França-Neto and Francisco Carlos Krzyzanowski
Embrapa Soja, Caixa Postal 231, 86001-970 – Londrina, PR, Brasil

Eduardo Euclydes de Lima e Borges and Antônio César Batista Matos
Departamento de Engenharia Florestal, UFV, 36570-000 – Viçosa, MG, Brasil

Glauciana da Mata Ataíde
Departamento de Silvicultura, UFRRJ, 23895-000 – Seropédica, RJ, Brasil

Carla Regina Baptista Gordin, Silvana de Paula Quintão Scalon and Tathiana Elisa Masetto
Universidade Federal da Grande Dourados, Caixa Postal 533, 79804-970 – Dourados, MS, Brasil

Talita Silveira Amador and Claudio José Barbedo
Instituto de Botânica, Núcleo de Pesquisa em Sementes, Caixa Postal, 68041, 04301-012 – São Paulo, SP, Brasil

Fabrícia Nascimento de Oliveira
Departamento de Ciências Ambientais e Tecnológicas, Universidade Federal
Rural do Semi-Árido, 59515-000 – Mossoró, RN, Brasil

Salvador Barros Torres, Narjara Walessa Nogueira and Rômulo Magno Oliveira de Freitas
Departamento de Ciências Vegetais, Universidade Federal Rural do SemiÁrido, Caixa Postal 137, 59625-900 – Mossoró, RN, Brasil

Elisa de Melo Castro, João Almir Oliveira, Amador Eduardo de Lima, Heloísa Oliveira dos Santos and José Igor Lopes Barbosa
Departamento de Agricultura, UFLA, Caixa Postal 3037, 37200-000 – Lavras, MG, Brasil

Maria Cecília Dias Costa, Wilco Ligterink and Henk W.M. Hilhorst
Laboratory of Plant Physiology, Wageningen University, Droevendaalsesteeg 1, 6708 PB Wageningen, Netherlands

José Marcio Rocha Faria and Anderson Cleiton José
Departamento de Ciências Florestais, Universidade Federal de Lavras, Caixa Postal 3037, 37200-000- Lavras, MG, Brasil

Thaís Francielle Ferreira, João Almir Oliveira, Rafaela Aparecida de Carvalho,
Laís Sousa Resende, Cassiano Gabriel Moreira Lopes and Valquíria de Fátima Ferreira
Departamento de Agricultura, Universidade Federal de Lavras, Caixa Postal 3037, 37200-000 - Lavras, MG, Brasil

Julia Abati, Claudemir Zucareli, José Salvador Simoneti Foloni and Cristian Rafael Brzezinski
Departamento de Agronomia, UEL, Caixa Postal 6001, 86051-990 - Londrina, PR, Brasil

Fernando Augusto Henning and Ademir Assis Henning
Embrapa Soja, Caixa Postal 231, 86001-970 - Londrina, PR, Brasil

Andrea Bittencourt Moura, Aline Garske Santos, Jaqueline Tavares Schafer and Bianca Obes Corrêa
Departamento de Fitossanidade, Universidade Federal de Pelotas, Caixa Postal, 54, 96010970 –Pelotas, RS, Brasil

Juliane Ludwig
Universidade Federal da Fronteira Sul, 97900000 - Cerro Largo, RS, Brasil

Vanessa Nogueira Soares
Departamento de Fitotecnia, Universidade Federal de Pelotas, Caixa Postal, 354, 96010970- Pelotas-RS, Brasil

Andrea Bittencourt Moura, Aline Garske Santos, Jaqueline Tavares Schafer and Bianca Obes Corrêa
Departamento de Fitossanidade, Universidade Federal de Pelotas, Caixa Postal, 54, 96010970 –Pelotas, RS, Brasil

Juliane Ludwig
Universidade Federal da Fronteira Sul, 97900000 - Cerro Largo, RS, Brasil

Vanessa Nogueira Soares
Departamento de Fitotecnia, Universidade Federal de Pelotas, Caixa Postal, 354, 96010970- Pelotas-RS, Brasil

Elizabeth Rosemeire Marques
Departamento de Ciências Florestais, Universidade Federal de Lavras, Caixa Postal 3037, 37200-000 – Lavras, MG, Brasil

Roberto Fontes Araújo and Plínio César Soares
Empresa de Pesquisa Agropecuária de Minas Gerais, Caixa Postal 216, 36571-000 – Viçosa, MG, Brasil

Eduardo Fontes Araújo
Departamento de Fitotecnia, Universidade Federal de Viçosa, 36570-000 – Viçosa, MG, Brasil

Sebastião Martins Filho
Departamento de Estatística, Universidade Federal de Viçosa, 36570-000 – Viçosa, MG, Brasil

Eduardo Gomes Mendonça
Departamento de Bioquímica Agrícola, Universidade Federal de Viçosa, 36570-000 – Viçosa, MG, Brasil

Cheila Cristina Sbalcheiro and Solange Carvalho Barrios Roveri José
Embrapa Recursos Genéticos e Biotecnologia, Caixa Postal 023725, 70770-917 - Brasília, DF, Brasil

Jennifer Carine Rodrigues da Costa Molina Barbosa
Faculdade Anhanguera de Brasília, 72950-000 - Taguatinga, DF, Brasil

Wanderlei Antônio Alves Lima, Ricardo Lopes, Márcia Green,Raimundo Nonato Vieira Cunha, Samuel Campos Abreu and Alex Queiroz Cysne
Embrapa Amazônia Ocidental, Caixa Postal 319, 69010-970 – Manaus, AM, Brasil

Carlos André Bahry
Departamento de Ciências Agrárias, UTFPR, 85.503-390 – Pato Branco, PR, Brasil

Paulo Dejalma Zimmer
Departamento de Fitotecnia, UFPel, Caixa Postal 354, 96010-900 – Capão do Leão, RS, Brasil

Carolina Maria Gaspar de Oliveira
IAPAR, Instituto Agronômico do Paraná , 86047-902 - Londrina, PR, Brasil

Francisco Carlos Krzyzanowski, Maria Cristina Neves de Oliveira, José de Barros França-Neto and Ademir Assis Henning
Embrapa Soja, Caixa Postal 231, 96001-970 - Londrina, PR, Brasil

Antônio César Batista Matos and Eduardo Euclydes de Lima e Borges
Departamento de Engenharia Florestal, UFV, 36570-000 – Viçosa, MG, Brasil

Marcelo Coelho Sekita
Departamento de Fisiologia Vegetal, UFV, 36570-000
– Viçosa, MG, Brasil

Alessandra de Lourdes Ballaris and Cláudio Cavariani
Departamento de Produção e Melhoramento Vegetal,
UNESP, Caixa Postal 237, 18610-307 - Botucatu, SP,
Brasil

José da Cruz Machado
Departamento de Fitopatologia, UFLA, Caixa Postal
3037, 37200-000 - Lavras, MG, Brasil

Maria Laene Moreira de Carvalho
Departamento de Agricultura, UFLA, Caixa Postal
3037, 37200-000- Lavras, MG, Brasil

Welison Andrade Pereira
Departamento de Biologia, UFLA, Caixa Postal 3037,
37200-000, Lavras, MG, Brasil

Sara Maria Andrade Pereira
Departamento de Produção Vegetal, UFES, Caixa
Postal 16, 29500-000 – Alegre, ES, Brasil

Denise Cunha Fernandes dos Santos Dias
Departamento de Fitotecnia, UFV, 36570-000 – Viçosa,
MG, Brasil

**Juliana Sakagawa Prataviera, Edmir Vicente Lamarca,
Carmen Cinira Teixeira and Claudio José Barbedo**
Instituto de Botânica, Núcleo de Pesquisa em Sementes,
Caixa Postal, 68041, 04301012 – São Paulo, SP, Brasil

**Thaíse da Silva Tonetto, Maristela Machado Araujo
and Álvaro Luís Pasquetti Berghetti**
Departamento de Ciências Florestais, Centro de
Ciências Rurais, Universidade Federal de Santa Maria,
97105-900 – Santa Maria, RS, Brasil

Marlove Fátima Brião Muniz and Clair Walker
Departamento de Defesa Fitossanitária, Centro de
Ciências Rurais, Universidade Federal de Santa Maria,
97105-900 – Santa Maria, RS, Brasil

Index

www.ingramcontent.com/pod-product-compliance
Lightning Source LLC
Chambersburg PA
CBHW050448200326
41458CB00014B/5103